FUTURE AUTOMOTIVE FUELS

- **Prospects**

- **Performance**

- **Perspective**

PUBLISHED SYMPOSIA

Held at the
General Motors Research Laboratories
Warren, Michigan

FUTURE AUTOMOTIVE FUELS

- Prospects

- Performance

- Perspective

Edited by
JOSEPH M. COLUCCI and NICHOLAS E. GALLOPOULOS

General Motors Research Laboratories

SPRINGER SCIENCE+BUSINESS MEDIA, LLC
1977

Library of Congress Cataloging in Publication Data

Symposium on Future Automotive Fuels—Prospects, Performance, and Perspective,
General Motors Research Laboratories, 1975.
Future automotive fuels.

Includes bibliographical references and index.
1. Motor fuel—Congresses. I. Colucci, Joseph M. II. Gallopoulos, Nicholas E. III.
General Motors Corporation. Research Laboratories. IV. Title.
TP343.S88 1976   662'.6   76-30757

ISBN 978-1-4684-2336-5   ISBN 978-1-4684-2334-1 (eBook)
DOI 10.1007/978-1-4684-2334-1

Proceedings of the Symposium on Future Automotive Fuels: Prospects,
Performance, and Perspective, held at the General Motors Research
Laboratories, Warren, Michigan, October 6-7, 1975

© 1977 Springer Science+Business Media New York
Originally published by Plenum Press, New York in 1977
Softcover reprint of the hardcover 1st edition 1977

A Division of Plenum Publishing Corporation
227 West 17th Street, New York, N.Y. 10011

PREFACE

In October 1975, while the United States was still acutely feeling the aftermath of the 1973 Arab Oil Embargo, the General Motors Research Laboratories held its nineteenth annual symposium. The proceedings of this timely symposium on "Future Automotive Fuels — Prospects, Performance, and Perspective" are reported in this book. We hope that it will serve not only as a permanent record of the papers and discussions, but also as a stimulus and inspiration for ideas, research, and development in the vital field of automotive fuels.

The economy of the United States and the lifestyle of her people are woven together with energy into a unique fabric. Reducing the energy content of this fabric weakens it and can even destroy it. The Oil Embargo stunningly demonstrated how easy it is to attack this fabric, and exposed for all to see its greatest weakness — reliance on imported petroleum. Since petroleum is the only current source of automotive fuels, and cars and trucks consume about 43 percent of the petroleum used in the United States, the Oil Embargo had its most profound and dramatic impact on automotive transportation: First there were long lines at service stations, and then idle lines in car assembly plants and long lines at unemployment offices. Against this grim setting, we planned the symposium on automotive fuels for the future.

Industrial, governmental, and academic experts with varied backgrounds were asked to present papers on a broad range of subjects related to future fuels. The symposium began with papers on such aspects of fuels as demand, supply, and optimum utilization, and ended with a paper about the potential impact of a new fuels industry on the social and economic life of the United States. In between these economically and socially oriented papers, the symposium dealt with several technological issues: coal, shale, and nuclear energy as sources of future fuels; evaluations in internal combustion engines of hydrocarbons derived from coal and shale, methanol, hydrogen, and hydrazine.

Some of the papers reflect the art of forecasting. Consequently, their conclusions undoubtedly will be modified because either unpredictable events will occur, or new data will become available. However, we hope that the contents of this book will help produce improved forecasts in the future. Even the evaluations of fuels presented herein are subject to future modifications. As new experimental techniques, fuel technologies, and engines are developed, these fuels may have to be reevaluated. Therefore, we view this book as a springboard for future studies, and not as a definitive treastise on the subject of future fuels.

We thank Dave Havelock for skillfully shepherding the conversion of the edited manuscripts into this volume. For cheerfully contributing to the preparation of this book, we thank our colleagues Joe Wentworth, Norm Brinkman, and Ashok Sapre. We also thank Kurt Antonius who expertly arranged the social and physical aspects of the symposium, Jeanne Christensen who handled the typing and administrative details, and Jerry Steadman who handled much of the initial artwork. To our session chairmen Phil Myers, Martin Elliott, John Heywood, and Sol Penner; to Jon Pangborn and Jay Gillis of the Institute of Gas Technology; to John Appeldoorn and Fred Kant of Exxon Research and Engineering; to Graham Hagey of ERDA we express our indebtedness for helping us select topics and speakers for the symposium.

Joseph M. Colucci
Nicholas E. Gallopoulos

CONTENTS

SESSION I
THE NEED, RESOURCES, AND POSSIBILITIES

Session Chairman
P. S. MYERS

University of Wisconsin
Madison, Wisconsin

FUTURE DEMAND FOR AUTOMOTIVE FUELS

R. F. HEMPHILL, JR., and C. DIFIGLIO

Federal Energy Administration, Washington, D.C.

ABSTRACT

A baseline forecast of automotive fuels through 1990 is provided. No distinction is made regarding the nature of this fuel (gasoline, diesel fuel, no lead, octane rating, etc.) since the analysis does not specify or evaluate engine technologies. The baseline forecast is developed from a multi-equation model which explicitly considers the fuel efficiency of new cars, the sales and market shares of new cars, used car scrappage and vehicle miles traveled. The factors which affect the demand for automobiles, the characteristics of automobiles demanded and the demand for travel are discussed. A brief model documentation is provided.

INTRODUCTION

With substantially higher energy costs now and in the projected future, our economy is still undergoing a fundamental change. Resources are being transferred from energy intensive goods and services to those which are relatively more efficient or less energy intensive.

This process is nowhere more evident than in the transportation sector which accounts for 51% of our Nation's petroleum consumption. The private automobile, which accounts for 74% of the petroleum consumed in transportation, has been affected by higher petroleum costs. There has been a general slump in the auto market, and there has also been a substantial increase in the proportion of fuel-efficient cars manufactured and purchased by the public. These current and expected improvements in new car fuel economy will eventually neutralize the impact

References p. 29.

of high fuel costs. This, at least, is the expectation implied by FEA (Federal Energy Administration) econometric research.

This research has been undertaken to better analyze the impacts of alternative government policy in the gasoline and automobile markets. But this policy analysis must rest on an expected base case, which would occur in the expected regulatory environment (emissions and safety) and with no further gasoline taxes, auto excise taxes, or imposition of mandatory fuel economy standards. The purpose of this paper is to provide that base case, despite the wide range of uncertainties associated with interdependent econometric forecasts. Also, there will be an examination of the various factors which influence our expectations for fuel demand.

The approach adopted for these baseline forecasts involves several variables which jointly determine fuel demand. Attempts to forecast fuel demand directly, without explicitly considering auto demand, new car characteristics, scrappage and travel demand as an interdependent system are useful, but do not account for the impact of higher energy prices on vehicle operating efficiency. Since this adjustment will largely mitigate the impact of higher fuel costs in the longer term (3-6 years from a fuel price change) it is desirable to estimate the fuel economy of new cars (sales weighted average), account for retirement in a fleet model which provides estimates of fleet operating characteristics, and use these fleet characteristics to estimate vehicle miles traveled in conjunction with other explanatory variables. In addition, if specific policies are designed to influence consumer choice of autos (fuel economy based excise taxes, fleet standards, etc.) it is important to account for their impact on new car sales (by market class) and their impact on the fuel efficiency of cars that manufacturers offer. These impacts can be similarly traced to fleet characteristics and travel demand resulting in policy sensitive estimates of fuel demand. A description of the resultant model equations* and a brief overview of the model structure are provided.

FACTORS WHICH INFLUENCE AUTO OWNERSHIP AND AUTO CHARACTERISTICS

The motives for auto ownership are a bit more complicated than for most other manufactured products. For some, the automobile is desired as a consumption good with little concern for the transportation services provided. Collectors of antique or classic cars are clear examples of such an outlook, but some degree of this view is often dominant in the ownership of specialty and luxury vehicles. Since auto demand is, to a great extent, a derived demand, the characteristics of travel requirements and the total transportation environment should influence the characteristics of the autos demanded. Autos are clearly an attractive means of conveyance when compared to alternative modes, offering a good combination of comfort, flexibility, reliability,

---

* Developed by Jack Faucett Associates under contract with FEA.

time, cost and privacy. Consequently, most Americans regard the auto as a basic need in order to get around, at least for some part of their trip-making activity. However, the characteristics of autos demanded are not likely to be explained entirely by the transportation need fulfilled by them. For example, the desire to purchase a new car instead of a used car is probably only partly justified by a transportation need. Wykoff (1) notes two reasons that new cars are purchased by consumers:

- the esthetics, freshness and prestige of new car ownership, and

- the reliability and relative certainty of new cars.

In terms of forecasting fuel demand, the distinction between new car and used car ownership is critically important because the impact of new car fuel economy improvements on gasoline consumption depends greatly on the ratio of new car purchases to used cars maintained in the fleet in an operating condition. Consequently, the adopted forecasting procedures produce both an estimate of new car sales and auto scrappage which are sensitive to the price and fuel economy characteristics of new cars (by individual market class). In this way, the speed at which new car fuel economy characteristics become fleet fuel economy characteristics can be estimated without assuming constant survivor rates. Also, since certain policies designed to increase the cost of (large) fuel inefficient cars will have different impacts on the scrappage of large vs. small cars, it is important to estimate separately the scrappage of cars by each market class.

Total auto ownership (new and used) is primarily determined by disposable income available to individuals and families. This was clearly evident from data collected by the Federal Highway Administration (FHWA) in the *Nationwide Personal Transportation Study* (2) and other surveys including the *Survey of Consumer Finances* (3) and *Consumer Buying Indicators* (4). The data from the Nationwide Personal Transportation Study is shown in Fig. 1. The relationship shown between auto ownership (per household) and household income is quite non-linear. The

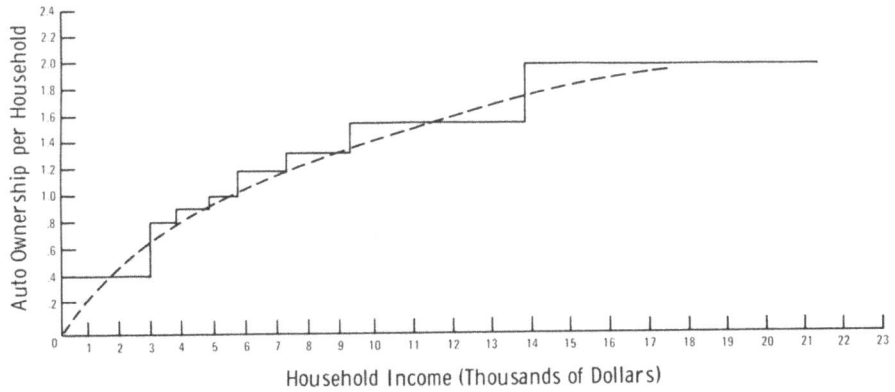

Fig. 1. Auto ownership per household by household income (1969-70). *Source: Reference 2.*

References p. 29.

average number of cars per household increases much faster as income increases from low to moderate levels than when income increases from moderate to high levels. Cars have a higher income elasticity of demand (or ownership) at low incomes than they do at high incomes which suggests a saturation effect. This is further apparent when the percentage of households owning zero, one, two and three or more cars is graphed vs. household income (Fig. 2). As households moves from $2,000 to $5,000 income, most families move from being owners of no cars to being owners of one car. Between $5,000 and $10,000 of household income, families move toward two car ownership. After $12,000 household income, three car ownership becomes significant, but one and two car ownership still dominates.

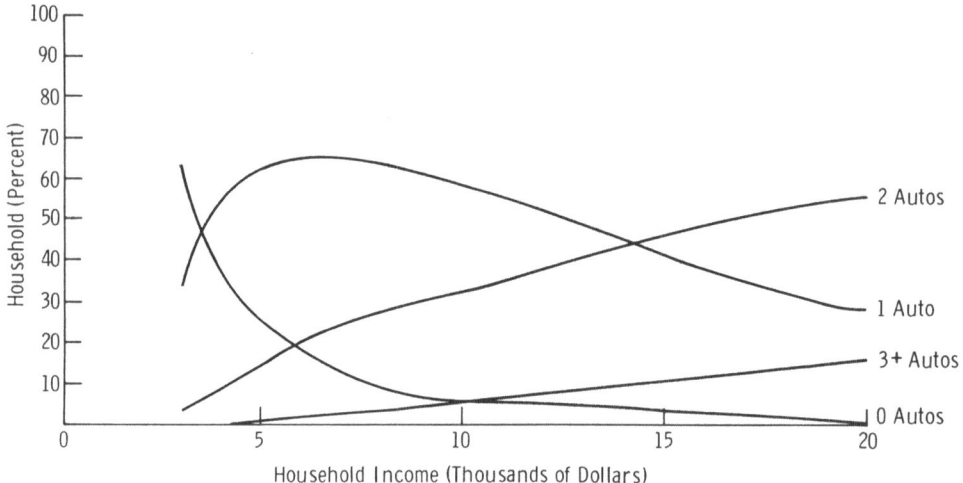

Fig. 2. Car ownership vs. household income. *Source: Reference 2.*

Geographic factors influence auto ownership, but with the exception of the central city vs. suburban geographic classifications, the geographic factor is not strong. Data for 1970 from both the Census and Nationwide Personal Transportation Study indicate less than a 1% variation of household auto ownership between the Standard Metropolitan Statistical Areas (SMSA) and non-SMSA areas as defined by the Census. Rural vs. urban auto ownership shows a greater variation (12%) on a household basis but essentially no variation on a population basis (Census data). The most important geographical distinction regarding the auto is central city vs. suburbs where suburban auto ownership is approximately 40% higher than central city ownership (household basis - Census data). Also, there are variations of auto ownership between the nine Census regions, with the Mountain and Pacific regions having the higherst ownership, and the Middle Atlantic region having the lowest (33% difference). The other six regions have little variation with respect to auto ownership.

Transit availability also has a surprisingly small impact on auto ownership. If the availability of rapid transit is used as a measure, and if New York City is excluded from the comparison, cities without rapid transit average 14% higher auto ownership on a household basis than the five remaining cities with rapid transit (Boston, Chicago, Cleveland, Newark, Philadelphia). On a population basis, this difference drops to 4%. However, when New York City is included, transit cities all appear to have substantially lower auto ownership than non-transit cities since New York's auto ownership is only 60% of the national average. The statistics of auto ownership by region and area type are summarized in Table 1 (page 8).

Because of the factors outlined above, it is permissible to estimate future auto ownership in an aggregate national model which does not account for expected geographical distributional changes or the likely changes in public transit supply. Also, there is little to be gained by using explanatory variables that are, in themselves, very difficult to forecast. Even if transit availability, for example, were believed to have a significant impact on auto ownership, it would not help to include it as an explanatory (exogenous) variable when future levels of public transit supply are uncertain.

The explanatory variables which were used to forecast auto ownership included:

- the real price of automobiles by class,
- the fuel efficiency of automobiles by class,
- the real price of gasoline,
- total real disposable income,
- total number of households in each income group, and
- the unemployment rate.

The variables referring to auto characteristics by market class are not used directly to determine auto sales and total auto ownership. Rather, these are used in the context of a market shares model to determine the proportion of cars sold in each market class. Therefore the sales-weighted characteristics of new cars can be estimated and then used to (with other variables) determine new car sales. Further, the sales-weighted fuel economy of new cars (derived from market shares estimation) is an important characteristic to estimate total fuel consumption. By making shares estimation sensitive to the prices and fuel economy of individual classes of cars, it is possible to analyze the impacts of policies designed to change the fuel economy vs. price relationship (for example, excise taxes on cars which are dependent on the rated fuel economy). These policies will affect differentially the prices and fuel economies of individual classes of cars. The adopted modeling procedures allow the market shifts (to smaller cars) and the improvements in fuel efficiency (higher mpg in all market classes) to be reflected in both the travel and gasoline consumption estimates.

References p. 29.

TABLE 1

Automobile Ownership by Region and Area Type

Ownership Categories[a]	Year and Ownership Basis	Source	Variation in Ownership (percent)
Metropolitan/Non-metropolitan			
SMSA vs. Non-SMSA	1970-H[b]	1970 Census of Housing	0.2
SMSA vs. Non-SMSA	1960-H	1960 Census of Housing	4.0
SMSA vs. Non-SMSA	1969-H	NPTS[b]	0.3
SMSA vs. Non-SMSA	1970-H	Census CPR[b], Series P-65 and P-23	0.7
SMSA vs. Non-SMSA	1970-P[b]	1970 Census of Housing	3.1
SMSA vs. Non-SMSA	1960-P	1960 Census of Housing	2.3
Central City vs. Suburbs	1970-H	1970 Census of Housing	40.2
Central City vs. Suburbs	1960-H	1960 Census of Housing	41.0
Central City vs. Suburbs	1970-H	Census CPR, Series P-65 and P-23	45.7
Central City vs. Suburbs	1970-P	1970 Census of Housing	23.2
Central City vs. Suburbs	1960-P	1960 Census of Housing	24.6
Urban/Rural			
Urban vs. Rural	1970-H	1970 Census of Housing	9.3
Urban vs. Rural	1960-H	1960 Census of Housing	12.5
Urban vs. Rural	1970-P	1970 Census of Housing	0.8
Urban vs. Rural	1960-P	1960 Census of Housing	0.0
Geographic Regions (Census)			
Highest (Mtn) vs. Lowest (Mid. At.)	1970-H	1970 Census of Housing	33.1
Highest (Mtn) vs. Lowest (Mid. At.)	1960-H	1960 Census of Housing	39.4
Highest (Pac.) vs. Lowest (Mid. At.)	1970-P	1970 Census of Housing	34.9
Highest (Pac.) vs. Lowest (Ea. So. Cen.)	1960-P	1960 Census of Housing	48.4
Availability of Rapid Transit			
Rapid Transit vs. No Rapid Transit	1970-H	1970 Census of Housing	14.4
Rapid Transit vs. No Rapid Transit	1960-H	1960 Census of Housing	9.7
Rapid Transit vs. No Rapid Transit	1970-H	1970 Census of Housing	3.6
Rapid Transit vs. No Rapid Transit	1960-H	1960 Census of Housing	6.5

[a] *Underlined item represents category with the higher ownership.*

[b] *H = Household basis; P = Population basis; CPR = Current Population Reports;*
NPTS = Nationwide Personal Transportation Study.

The market shares model used for these forecasts is driven by the relative price of each market class, the fuel economy of each market class and any other operating cost differentials between market classes. The latter two factors were discounted and reflected in the initial purchase price. Thus, the shares model chosen is driven only by cost variables and not by other characteristic differences between market classes. Therefore, the average or typical car in each market class must be assumed to remain relatively homogeneous regarding non-price auto characteristics (roominess, acceleration performance, road holding, styling, etc.) over time. FEA is sponsoring research in models which account for some of these factors in an explicit way (in "Hedonic" models) but these models are not currently available. Also, for the purpose of a base case forecast, such non-price characteristics would not be assumed to change significantly.

FACTORS WHICH INFLUENCE TRAVEL DEMAND

Unlike auto demand, we have better data regarding why people travel. The most comprehensive nationwide survey regarding auto travel is FHWA's *Nationwide Personal Transportation Study* (2). Trip purpose data from this survey are shown in Table 2.

TABLE 2

Automobile Travel by Trip Purpose

(Source — Reference 2)

Trip Purpose	Percent of Total Automobile Trips	Percent of Total Automobile Vehicle Miles	Average Trip Length (Miles)
Earning a Living			
Home-to-work	31.9	33.7	9.4
Related business	4.3	7.9	16.1
Subtotal	36.2	41.6	10.2
Family Business			
Shopping	15.2	7.5	4.4
Medical and dental	1.8	1.6	8.4
Other	14.0	10.2	6.5
Subtotal	31.0	19.3	5.6
Civic, Educational and Religious	9.3	4.9	4.7
Social and Recreational			
Visiting friends and relatives	8.9	12.1	12.0
Pleasure driving	1.4	3.1	20.0
Vacations	0.1	2.5	160.0
Other	12.0	15.3	11.4
Subtotal	22.4	33.0	13.1
Other and Unknown	1.1	1.2	9.4
Total	100.0	100.0	8.9

References p. 29.

The most important use of the automobile is for trips related to earning a living. This purpose accounts for 42 percent of all vehicle miles and 36 percent of all trips. The next largest category, family business, accounts for 31 percent of all trips but only 19 percent of vehicle miles. Social and recreation travel accounts for 33 percent of all vehicle miles but only 22 percent of all trips. Pleasure driving, a subset of social and recreational travel, is explicitly a final demand for auto transportation. It accounts for 3.1 percent of all vehicle miles. Vacation driving accounts for 2.5 percent of all vehicle miles, but only 0.1 percent of all trips. Civic, educational and religious driving accounts for 5 percent of all vehicle miles and 9 percent of all trips. The residual cagegory, other and unknown accounts for slightly more than 1 percent of trips and vehicle miles.

When the trip purpose data are separated by place of residence (Table 3), there is surprisingly little variance among people who live in and out of SMSA's or incorporated areas. In particular, the variation of the proportion of trips by trip purpose is essentially uniform among each geographical type. There is somewhat more variation when the distribution of vehicle miles is considered, since the average trip length varies somewhat with the degree of urbanization. Metropolitan areas have somewhat higher relative trip lengths for "earning a living" and "social and recreational" trips and somewhat lower relative trip lengths for "civic, educational and religious" and "family business" trips.

A much more significant variation of driving among individuals is related to income. Fig. 3 shows the relationship between household income and the number of miles driven per car available to the household. There is a steady increase of miles

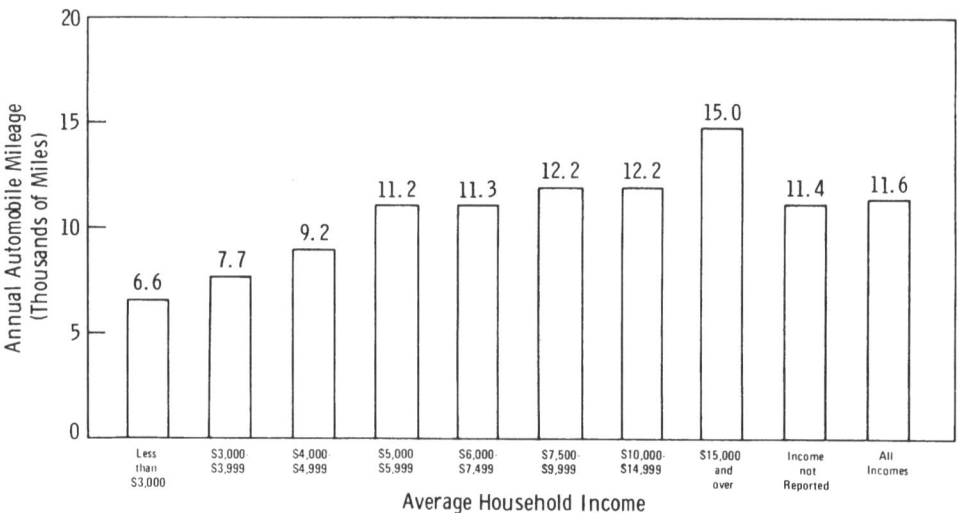

Fig. 3. Average annual miles per automobile by annual household income. *Source: U. S. Department of Transportation, Federal Highway Administration, Nationwide Personal Transportation Study, Report No. 2, April 1972.*

TABLE 3

Automobile Travel by Trip Purpose and Place of Residence
(Source — Reference 2)

Trip Purpose	All Areas	SMSAs	Non-SMSAs[a]	Incorporated Places	Unincorporated Areas
Percent of Automobile Trips					
Earning a Living	36.2	36.3	35.3	36.5	35.8
Family Business	31.0	30.7	31.7	30.8	31.5
Civic, Educational, and Religious	9.3	9.2	9.5	8.9	10.0
Social and Recreational	22.4	22.3	22.6	22.8	21.4
Other	1.1	1.2	0.9	1.0	1.3
Total, All Purposes	100.0	100.0	100.0	100.0	100.0
Percent of Vehicle-Miles of Travel					
Earning a Living	41.6	42.5	39.7	41.5	41.8
Family Business	19.3	18.3	21.4	18.0	21.5
Civic, Educational, and Religious	4.9	4.8	5.1	4.3	6.0
Social and Recreational	33.0	33.4	32.2	35.3	29.0
Other	1.2	1.0	1.6	0.9	1.7
Total, All Purposes	100.0	100.0	100.0	100.0	100.0

[a] *Estimated from data source; assumed to be equal to the difference between data presented for "All Areas" and "SMSAs."*

driven per car as each successively higher income group is considered. Annual mileage for the highest income group is approximately double that of the lower two income groups. When the distribution of car ownership vs. income is considered (Fig. 1), the miles driven vs. household income relationship is even more striking. The combined relationships indicate that the highest income group accumulates approximately six times as many vehicle miles as the lower two income groups.

References p. 29.

Income also affects the distribution of trip purposes among households, but not nearly to the extent that income affects the number of trips or miles traveled. The over $4,000 income and under $4,000 income category show different trip purpose distributions (Table 4). The low-income group represents relatively fewer trips made for "earning a living" and relatively more trips for "family business" and "social and recreational". It must be remembered, however, that the absolute number of these latter two trip purposes is higher for higher income groups than low income groups. On a mileage basis, the same relative distrubitions of trip purposes hold except that they are more pronounced and continuous. Each successively higher income group consumes relatively more miles for "earning a living" indicating that home to work distances are higher and more work related auto travel is required for higher income workers.

TABLE 4

Automobile Travel by Trip Purpose and Household Income

Trip Purpose	Annual Household Income			
	Under $4000	$4000-$9999	$10000-$14000	$15000 and over
	Percent of Automobile Trips			
Earning a Living	29.0	37.0	37.9	36.2
Family Business	34.9	30.8	30.2	30.4
Education, Civic and Religious	7.7	8.6	10.5	10.2
Social and Recreational	27.0	22.8	20.2	21.6
Other	1.4	0.9	1.2	1.6
Total, All Purposes	100.0	100.0	100.0	100.0
	Percent of Vehicle-Miles Traveled			
Earning a Living	31.0	39.6	44.0	47.4
Family Business	27.1	20.0	18.0	17.0
Education, Civic and Religious	4.4	4.9	5.3	5.5
Social and Recreational	35.7	34.2	31.5	28.9
Other	1.8	1.3	1.2	1.2
Total, All Purposes	100.0	100.0	100.0	100.0

Source: U. S. Department of Transportation, Federal Highway Administration, Nationwide Personal Transportation Study, Report No. 7, December 1972.

Age is a factor in average miles traveled per driver (Table 5). The 30-34 year age group maintains the highest average (by 20 percent over the all age average). The 25-59 year age group drives more than the average for all ages. Very young and very old drivers drive the least.

TABLE 5

Automobile Travel by Age of Driver, 1969-1970

Age Group (years)	Average Annual Miles Driven per Driver	Percent of Total VMT	Percent of Total Licensed Drivers
16 – 19	4,633	4.6	9.0
20 – 24	8,260	12.6	13.1
25 – 29	9,814	13.0	11.3
30 – 34	10,274	11.3	9.3
35 – 39	9,878	10.5	9.1
40 – 44	9,833	11.1	9.7
45 – 49	9,875	11.1	9.7
50 – 54	9,447	8.8	8.2
55 – 59	9,009	7.1	6.8
60 – 64	8,112	5.0	5.4
65 – 69	5,850	2.4	3.7
70 & over	4,644	2.5	4.7
All ages	8,685	100.0	100.0

Source: U. S. Department of Transportation, Federal Highway Administration, Nationwide Personal Transportation Study, Report No. 6, April 1973.

The cost factors which influence auto travel are more interesting than the income and demographic influences since these are undergoing the most rapid changes today. Consequently, while household and demographic forecasts are used to determine miles traveled in our baseline fuel forecasts, the most detailed attention is reserved for the influences that auto travel cost will have on auto travel.

The auto travel cost factor is determined by two components – the price of gasoline and the fuel efficiency of the car used. In terms of miles traveled, a fuel price change will have a more immediate impact and this impact will tend to deteriorate over time as individuals adjust by purchasing more fuel efficient cars. However, in terms of gasoline consumed, a fuel price change will have a less immediate impact and this impact will intensify as individuals adjust by purchasing more fuel efficient cars. Expressed differently, the short term gasoline demand elasticity is lower (absolutely) than the long-term gasoline demand elasticity.

The historical price of gasoline from 1950 to 1972 has declined (in real terms) while the average fleet fuel economy has also declined. The net result has been a relative constand (real) cost of auto travel during this period with a slight decline

References p. 29.

TABLE 6

Gasoline Prices and Fuel Cost per Mile
1950 – 1974

Year	Nominal Gas Price ($/gal)	Real Gas Price 1967 Dollars ($/gal)	Real Fuel Cost per Mile[a] ($)	Real Fuel Cost per Mile Index[a] (1967=1.00)
1950	.2676	.3712	.0248	1.042
1951	.2715	.3490	.0233	.979
1952	.2756	.3467	.0236	.992
1953	.2869	.3582	.0244	1.025
1954	.2904	.3607	.0248	1.042
1955	.2907	.3625	.0250	1.050
1956	.2993	.3677	.0256	1.076
1957	.3096	.3673	.0255	1.071
1958	.3038	.3508	.0245	1.029
1959	.3049	.3493	.0244	1.025
1960	.3113	.3510	.0246	1.034
1961	.3076	.3433	.0239	1.004
1962	.3064	.3382	.0235	.987
1963	.3042	.3317	.0233	.979
1964	.3035	.3267	.0229	.962
1965	.3115	.3296	.0234	.983
1966	.3208	.3300	.0236	.992
1967	.3316	.3316	.0238	1.000
1968	.3371	.3235	.0235	.987
1969	.3480	.3169	.0233	.979
1970	.3569	.3069	.0226	.950
1971	.3643	.3003	.0221	.929
1972	.3613	.2883	.0214	.899
1973	.3882	.2917	.0223	.937
1974	.5241	.3548	.0271[b]	1.139[b]

[a] *Based on fleet fuel economy.*

[b] *Assumed the 1973 fleet fuel economy.*

Sources: Platt's Oil Price Handbook and Oilmanac for nominal price series; U. S. Department of Labor, Bureau of Labor Statistics, for Consumer Price Index

between 1967 and 1972. Data are shown in Table 6. During 1973 this trend began to reverse and, in 1974, was substantially reversed. But, what is often overlooked, in 1974 the real price of gasoline was not substantially different than existed in the 1950's. However, due to the steadily declining fuel economy through 1973 (Table 7) the average travel cost in 1974 was the highest for the available data period.

TABLE 7

Average Fuel Economy for the Automobile Fleet and New Cars

(Miles per Gallon)

Year	Average Fleet Fuel Economy (calendar year basis)	New Car Fuel Economy (model year basis)
1950	14.95	—
1951	14.99	—
1952	14.67	—
1953	14.70	—
1954	14.58	—
1955	14.53	—
1956	14.36	—
1957	14.40	13.67
1958	14.30	14.07
1959	14.30	13.85
1960	14.28	13.36
1961	14.38	13.55
1962	14.37	13.96
1963	14.26	12.62
1964	14.25	13.49
1965	14.07	12.98
1966	14.00	12.95
1967	13.93	12.86
1968	13.79	12.44
1969	13.63	12.21
1970	13.57	12.51
1971	13.57	12.21
1972	13.49	12.03
1973	13.10	11.67

Note: Motorcycle travel is included in average fleet fuel economy for 1957-64.

Sources: U. S. Department of Transportation, Federal Highway Administration, Highway Statistics, and U. S. Environmental Protection Agency, A Report on Automobile Fuel Economy, October 1973.

The net result of all of the demographic, economic and travel cost factors on gasoline demand from 1950 to 1973 is shown in Table 8 (page 16). These data are derived indirectly from tax rates and tax receipts from the individual States and the District of Columbia. They indicate a substantial increase from 24,305 million gallons in 1950 to 77,619 million gallons in 1973. During 1975 we have managed to essentially maintain our 1973 consumption levels thereby reversing the steadily increasing demand. Also, as our forecasts indicate, we should be able to eventually

References p. 29.

TABLE 8

Automobile Gasoline Consumption, 1950-1973

Year	Consumption (million gallons)	Year	Consumption (million gallons)
1950	24,305	1962	43,771
1951	26,159	1963	45,246
1952	27,969	1964	47,567
1953	29,613	1965	50,206
1954	30,915	1966	53,220
1955	33,548	1967	55,007
1956	35,326	1968	58,413
1957	36,769	1969	62,325
1958	38,095	1970	65,649
1959	40,056	1971	69,213
1960	41,169	1972	73,121
1961	42,033	1973	77,619

Note: Consumption from 1957-64 includes motorcycles.

Source: U. S. Department of Transportation, Federal Highway Administration, Highway Statistics.

reduce gasoline consumption due to improvements in fleet fuel economy. However, this will not inhibit auto mobility significantly as the improved fleet fuel economy will maintain auto travel costs at historical levels.

MODEL STRUCTURES

Only a brief overview will be provided of the model used to estimate the baseline fuel demand through 1990. More detailed documentation of this model will be available in future papers.

Four basic relationships are used:

MODEL STRUCTURE
- an estimator for market shares of new car sales,

- an estimator for new car sales,

- an estimator for scrappage, and

- an estimator for miles traveled.

A brief overview of the inter-relationships between these estimators follows.

The sales-weighted average new car characteristics are determined in the shares model given the characteristics of each market class of car (small, medium and large) available in a specific year. These characteristics include fuel economy, operating costs, and purchase price. The sales-weighed characteristics (with socio-economic forecasts) are used in the new car sales estimator to produce total sales estimates for each year. Also, the new car price data are used in the scrappage model to produce scrappage estimates for each class of car (small, medium and large) for each year. The combined inputs from these three estimators (with socio-economic forecasts) are used in the vehicle miles estimator and the resultant forecasts are incorporated in the fleet model which produces fuel demand forecasts for each year of available input data.

Market Shares Estimator — The first problem faced in developing a market shares model of auto sales is to find the most appropriate vehicle classification system. Since the data available for this study consisted of annual observations from 1962 through 1973 it is difficult to use a really homogenous market share classification scheme. The use of six to ten interdependent estimators "uses up" several degrees of freedom beyond what is adequate for 11 annual observations. This is caused by the need to specify cross price relationships between market classes in each estimator which increases the number of parameters which must be estimated.

After some experimentation, a weight based market class system was adopted using three market classes:

Small	< 3050 lbs.
Medium	3050 - 3500 lbs.
Large	> 3500 lbs.

This system produced a stable time series of observations with which to estimate parameters.

A logit probability model was employed using a single equation transformation which collapsed the three separate observation vectors into a single vector. The final form of the market share estimators is as follows:

$$S_t = \frac{1}{1 + e^{-[-4.1749 - 1.8660(X_t^S) + 3.502(X_t^M) + 5.6428(S_{t-1})]}}$$

$$M_t = \frac{1}{1 + e^{-[-4.1749 - 2.0765(X_t^M) + 3.5450(X_t^M) + 0.2589(X_t^L) + 5.6428(M_{t-1})]}}$$

$$L_t = \frac{1}{1 + e^{-[-4.1749 - 0.4299(X_t^L) + 1.8117(X_t^M) + 5.6428(L_{t-1})]}}$$

References p. 29.

Where,

S_t, M_t, L_t = Market shares of small, medium and large cars, respectively, in year t

X_t^S, X_t^M, X_t^L = An index of the real generalized price of small, medium and large cars, respectively, relative to that of all new cars in year t (1967=1.00)

New Car Sales Estimator — The model structure chosen for the estimates of new car sales is a short-run stock adjustment approach. This technique has been worked out by Chow (5), Nerlove (6), Hamburger (7) and others. A desired level (or target level) of automobile ownership is established as a non-linear function of household income. A gap between the desired level and the estimated "existing" level (from the scrappage model and "last year's" fleet and sales) generates the demand for new car purchases. This gap is combined with the cost characteristics of new cars to produce an estimate of new car sales. The "gap" is not definitional or tactological. Rather it is a function of other explanatory variables (household income, number of households and unemployment) and appears in the new car sales function with new car cost characteristics to jointly determine new car demand.

All new car cost characteristics are combined into a "generalized price" which is defined to include initial purchase price and an expected operating cost discounted to the time of purchase. In terms of model estimations, the discounted operating cost was defined over fuel economy and gasoline cost. Differential operating cost data were not available. However, for forecasting purposes, any estimated differences in maintenance cost can be incorporated in the "generalized price." For the purpose of our baseline forecast, only the expected changes in fuel economy and fuel cost were used.

The final form of the new car sales estimates was:

$$N_t = 286,721.3 \, [O_t^* - (\text{Autos}_t - D_t)]^{\,0.2178} (X_t^*)^{-1.7039}$$

Coefficient of Determination $(R^2) = 0.7807$

Standard Error of Estimate $(S.E.E.) = 0.0417$

Durbin Watson Statistic $(D.W.) = 1.177$

Degrees of Freedom $(d.f.) = 11$

where,

N_t = Total annual new car sales in year t

O_t^* = The target ownership of automobiles for the year t

$(Autos)_t$ = The stock of automobiles on hand at the beginning of year t

D_t = Scrappage of vehicles over the course of year t

X_t^* = An index of the generalized price with $1967 = 1$.

$O_t^* = (\sum_I H_I P_{It}) R_t$

P_{It} = Fraction of total households in year t having income I

H_I = Auto ownership per household of income I

R_t = Total households

I = Household income

$H_I = 0.01786 \, I^{0.4743}$ $(R^2 = 0.983)$

The relatively high apparent price elasticity of -1.7 for new car sales is misleading. First it is for the generalized price variable which includes both the cost of purchasing and operating an automobile. Second, it does not account for the feedback of the generalized price variable (X_t) on the "gap" via the endogenous scrappage estimate (D_t) which also uses X_t as an exogenous variable.

Scrappage Estimator — Overall scrappage rates have been relatively stable over time. The rate increases as vehicles age up to about 11 years of age at which time it remains constant (see Table 9, page 20). It is important to determine the influence of new car price on the scrappage of used cars. This will be important to calculate energy savings from policies which might increase the price of inefficient new cars. One expects higher new car prices to increase the value of used cars and to reduce the scrappage rate of used cars. New car price increase for large cars only, with constant or decreasing prices for small cars (however induced), will not affect the overall scrappage rate in the same manner as a price increase across all cars. However, the scrappage rate by market class would be as strongly affected resulting in substantial changes in the energy efficiency of the older vintage cars. Such changes can have substantial impacts on the fleet fuel economy.

References p. 29.

TABLE 9

Historic Scrappage of the Automobile Fleet, by Age of Vehicle
(Model Year M)

Calendar Year	Percentage of Vehicles Scrapped during Year	Percentage of Vehicles Still in Use at End of Year
M	0.00	100.00
M+1	0.20	99.80
M+2	0.55	99.25
M+3	1.05	98.21
M+4	2.01	96.23
M+5	3.47	92.90
M+6	6.02	87.03
M+7	10.16	78.43
M+8	15.70	66.12
M+9	21.46	51.93
M+10	26.00	38.43
M+11	28.89	27.33
M+12	30.15	19.09
M+13	30.00	13.36
M+14	29.17	9.46
M+15	29.29	6.69
M+16 & after	29.70	—

Sources: Automotive News, 1975 Almanac Issue, based on model years 1957 and after. Calendar year figures are computed by linear interpolation from the cumulative distribution of survivors.

The second major factor determining auto scrappage is macroeconomic conditions. Since new car sales are partly determined by the gap between desired levels of auto ownership, last year's auto stock, and auto scrappage, any macroeconomic factors which are used to estimate scrappage also impact new car demand.

Economic factors influence primarily the scrappage of marginal (i.e. older) vehicles. This was reflected in the scrappage model employed by estimating the scrappage of older vehicles (eight or nine years). Early scrappage is usually due to factors not related to the macroeconomic environment or the prices of new cars, but even severe economic and new car price changes will not generally force the scrappage of newer cars.

In using the price of new cars and the unemployment rate, the following scrappage relationship was determined.

$$SPG_t t = 0.40675 - 0.078433 \ (P_n)_t - 0.015519 \ U_t$$

$$R^2 = 0.6587$$

$$S.E.E. = 0.01544$$

$$D.W. = 1.9087$$

$$d.f. = 10$$

where

SPG_t = scrappage rate for vehicles eight or more years old, year t

$(P_n)_t$ = index of real price of new cars in year t, 1967=1.00

U_t = unemployment rate, year t

The variations of the scrappage rates so estimated are applied individually to each market class.

Vehicle Miles Traveled (VMT) Estimator — There are four important factors to include in a vehicle miles estimator:

- cost of driving,

- auto availability,

- demographic conditions, and

- available income.

Various variables have been used in many forecasting models to capture most or many of these factors. The important variables included: last year's VMT, population, number of households, number of drivers, income, gas price, fuel efficiency, vehicle cost per mile, goods and service price index, new car price, total auto stock, unemployment, and safety and emission regulations. Of course, no single model included all of these variables (usually only four or five variables could be successfully isolated). Many are redundant or partly redundant and attempts to include each would only invite severe multicolinearity problems in model estimation. After testing several of these variables in various combinations the following equation was selected to best include each of the four major influences on auto vehicle travel.

References p. 29.

$$\frac{VMT}{HHLD} = -52979.8 + 15087 \, LOG \, \frac{DI}{HHLD} + 6337.7 \, \frac{AUTOS}{HHLD} - 2204.24 \, CPM$$

$$R^2 = 0.9767$$

$$S.E.E. = 257.235$$

$$D.W. = 0.4614$$

$$d.f. = 18$$

where,

$\dfrac{VMT}{HHLD}$ = Annual VMT per household (HHLD)

$\dfrac{DI}{HHLD}$ = Real Disposable Income per household (1967 dollars)

$\dfrac{AUTOS}{HHLD}$ =Total Cars in Use (as of January 1 of each year) per household

CPM = Index of the real fuel cost per mile (price of gasoline ÷ average fleet fuel economy, deflated by CPI, all items), where 1967 = 1.00.

Model Equation Summary — In Table 10, the equations of the model are summarized as an inter-related system. Equation 4 is for the price of used cars. It was not discussed previously since it does not directly relate to the estimate of fuel consumption. Its purpose is to help indicate income impacts on households which are impacted by federal policy options which would influence the fuel economy of new cars (the price of new cars and disposable income determine the estimated price of used cars). Also, the fleet model which is used to calculate fuel consumption is not shown. This is not an estimated or behavioral model. Rather, the fleet model is an identity that follows directly from the estimated variables and historical data corresponding to the estimated variables.

Forecasts of Fuel Consumption and Related Variables — The model outlined above was used to produce a baseline fuel consumption forecast. The results are summarized in Table 11 for each year 1976 through 2000. Three sets of forecasts are provided. One set assumes constant (real) gasoline prices throughout the forecast period. Two additional forecasts assume increasing and decreasing gasoline prices to provide an outer bound with respect to the uncertainties of petroleum costs.

TABLE 10

Model Summary

1) $N_t = (286,721.3) [0_t{}^* - (Autos_t - D_t)]^{0.2178} (X_t^*)^{-1.7039}$

2a) $S_t = \dfrac{1}{1+e^{-[-4.1749 - 1.8660 (X_t^S) + 3.5092 (X_t^M) + 5.6428 (S_{t-1})]}}$

2b) $M_t = \dfrac{1}{1+e^{-[-4.1749-2.0765 (X_t^M) + 3.5450 (X_t^S) + 0.2589 (X_t^L) + 5.6428 (M_{t-1})]}}$

2c) $L_t = \dfrac{1}{1+e^{-[4.1749-0.4299 (X_t^L) + 1.8117 (X_t^M) + 5.6428 (L_{t-1})]}}$

3) $SPG_t = 0.4068 - 0.0784 (P_n)_t - 0.0155 (U_t)$

4) $(P_u)_t = (2.765 \times 10^{-5} (X_t^*)^{2.1626} \dfrac{DI}{HHLD_t}^{1.1512}$

5) $\dfrac{VMT}{HHLD} = -52979.8 + 15087 \; LOG \; \dfrac{DI}{HHLD_t} -2204.24 \; (CMP)_t + 6337.24 \dfrac{Autos}{HHLD_t}$

N_t = Total new car sales in year t

O_t^* = Target ownership of automobiles in year t

$(Autos)_t$ = The stock of automobiles on hand as of January 1 of year t

D_t = The number of autos scrapped during year t

X_t^* = An index of the real generalized price of new cars, 1967 = 1.00

S_t, M_t, L_t = Market shares of small, medium, and large cars, respectively, in year t

X_t^S, X_t^M, X_t^L = An index of the real generalized price of small, medium and large cars, respectively, relative to that of all new cars in year t, 1967 = 1.00

SPG_t = The rate of scrappage in year t of vehicles eight or more years of age

$(P_n)_t$ = An index of the real price of new cars in year t, 1967 = 1.00

$(P_u)_t$ = An index of the real price of used cars in year t, 1967 = 1.00

U_t = The unemployment rate in year t

DI_t = Total real disposable income in year t

VMT_t = Total vehicle miles travelled in year t

CPM_t = An index of the fleet real gasoline costs per miles in year t, 1967 = 1.00

$HHLD$ = The total number of households existing in year t

References p. 29.

TABLE 11

Model Results
Constant Fuel Price

Year	1976	1977	1978	1979	1980
Total Sales	11512536	12105357	10719790	11905218	12041958
Small Share	.356	.358	.360	.363	.365
Medium Share	.158	.159	.160	.161	.162
Large Share	.486	.483	.479	.476	.473
Cars In Use	95720841	98925105	100041533	102301886	104574071
Scrappage	8061625	8901093	9603362	9644865	9769772
VMT (MM miles/yr)	1065111	1103526	1129061	1162363	1196297
Gas Consumed (MM gal/yr)	79323	80246	80559	81360	82286
Avg. Gen. Price*	5995	5989	5982	5974	5966

Year	1981	1982	1983	1984	1985
Total Sales	12204908	12371009	12528804	12697365	12896472
Small Share	.368	.370	.373	.377	.380
Medium Share	.163	.164	.165	.167	.168
Large Share	.469	.465	.461	.457	.452
Cars In Use	106866276	109024159	111614876	114048909	116408853
Scrappage	9912704	10033127	10118087	10263332	10536528
VMT (MM miles/yr)	1230769	1265985	1302002	1338486	1374817
Gas Consumed (MM gal/yr)	83321	84508	85899	87504	89268
Avg. Gen. Price	5958	5949	5940	5930	5920

Year	1986	1987	1988	1989	1990
Total Sales	13112238	13350313	13600929	13859649	14118565
Small Share	.384	.388	.392	.396	.401
Medium Share	.169	.170	.171	.172	.173
Large Share	.448	.443	.437	.432	.426
Cars In Use	118670083	120881971	123060793	125240451	127443869
Scrappage	10851008	11138426	11422107	11679991	11915146
VMT (MM miles/yr)	1410883	1447004	1483421	1520300	1557898
Gas Consumed (MM gal/yr)	91140	93101	95129	97201	99320
Avg. Gen. Price	5909	5897	5884	5871	5857

Year	1991	1992	1993	1994	1995
Total Sales	14377740	14637933	14896580	15157951	14520816
Small Share	.406	.412	.417	.423	.429
Medium Share	.174	.175	.176	.177	.178
Large Share	.420	.413	.407	.400	.392
Cars In Use	129685039	131974879	134316888	136710528	139150340
Scrappage	12136570	13248093	12554571	12764312	12981004
VMT (MM miles/yr)	1596281	1635580	1675787	1716955	1759020
Gas Consumed (MM gal/yr)	101481	103684	105923	108198	110501
Avg. Gen. Price	5842	5927	5810	5793	5775

Year	1996	1997	1998	1999	2000
Total Sales	15688038	15961827	16240234	16526074	16816871
Small Share	.436	.443	.450	.458	.465
Medium Share	.180	.181	.182	.183	.184
Large Share	.384	.376	.368	.359	.350
Cars In Use	141631984	144153422	146711875	149309024	151945505
Scrappage	13206394	13440389	13681781	13928925	14180390
VMT (MM miles/yr)	1802050	1845959	1890825	1936628	1983382
Gas Consumes (MM gal/yr)	112831	115182	117554	119945	122353
Avg. Gen. Price	5756	5736	5715	5693	5670

*Includes discounted expected operating costs.

TABLE 11 (cont'd.)
Model Results
Increasing Fuel Price

Year	1976	1977	1978	1979	1980
Total Sales	11269654	11656446	10287506	11197030	11227552
Small Share	.356	.359	.361	.364	.367
Medium Share	.158	.159	.160	.160	.160
Large Share	.486	.483	.479	.476	.473
Cars In Use	95477959	98233797	98920173	100478198	101949240
VMT (MM miles/yr)	1061141	1094276	1114661	1141080	1167512
Scrappage	8061625	8900607	9601131	9639005	9756511
Gas Consumed (MM gal/yr)	79067	79660	79640	80005	80448
Avg. Gen. Price	6071	6139	6207	6274	6340

Year	1981	1982	1983	1984	1985
Total Sales	11198960	11185581	11171493	11163968	11170479
Small Share	.371	.375	.379	.383	.389
Medium Share	.161	.161	.161	.160	.160
Large Share	.469	.465	.461	.456	.451
Cars In Use	103262489	104465912	105610761	106664729	107539219
Scrappage	9885710	9982158	10026644	10110000	10295989
VMT (MM miles/yr)	1192570	1217422	1242272	1266884	1290685
Gas Consumed (MM gal/yr)	80869	81376	82031	82850	83782
Avg. Gen. Price	6432	6523	6613	6701	6788

Year	1986	1987	1988	1989	1990
Total Sales	11447102	11705911	11949158	12130010	12398268
Small Share	.394	.400	.406	.412	.418
Medium Share	.160	.160	.159	.159	.159
Large Share	.446	.441	.435	.429	.422
Cars In Use	108483766	109520611	110651443	111897385	113273548
Scrappage	10502555	10669067	10818326	10934068	11022105
VMT (MM miles/yr)	1317819	1345897	1375066	1405410	1437138
Gas Consumed (MM gal/yr)	85050	86459	87982	89592	91289
Avg. Gen. Price	6778	6768	6757	6745	6731

Year	1991	1992	1993	1994	1995
Total Sales	12585789	12767927	12945645	13123421	13302792
Small Share	.425	.432	.440	.448	.456
Medium Share	.159	.159	.159	.159	.159
Large Share	.416	.408	.401	.393	.385
Cars In Use	114763399	116365662	118067492	119845697	121669453
Scrappage	11095939	11165665	11243815	11345215	11479036
VMT (MM miles/yr)	1469926	1504016	1539298	1575682	1612942
Gas Consumed (MM gal/yr)	93042	94861	96732	98646	100584
Avg. Gen. Price	6723	6714	6703	6692	6679

Year	1996	1997	1998	1999	2000
Total Sales	13531501	13764669	14001543	14243479	14488765
Small Share	.465	.473	.482	.491	.501
Medium Share	.159	.159	.159	.160	.160
Large Share	.376	.376	.358	.349	.339
Cars In Use	123559448	125500349	127485469	129516729	131598451
Scrappage	11641505	11823768	12016423	12212219	12407043
VMT (MM miles/yr)	1651721	1691444	1732164	1773862	1816574
Gas Consumed (MM gal/yr)	102582	104603	106646	108711	110796
Avg. Gen. Price	6652	6624	6595	6565	6534

TABLE 11 (cont'd.)
Model Results
Decreasing Fuel Price

Year	1976	1977	1978	1979	1980
Total Sales	11535027	12267169	10928542	12379450	12551030
Small Share	.356	.358	.360	.362	.364
Medium Share	.158	.159	.161	.162	.163
Large Share	.486	.483	.479	.476	.473
Cars In Use	95743332	99109363	100434096	103167142	105944175
Scrappage	8061625	8901138	9603809	9646405	9773997
VMT (MM miles/yr)	1065479	1106312	1134634	1173745	1212553
Gas Consumed (MM gal/yr)	79347	80425	80917	82088	83326
Avg. Gen. Price	5988	5941	5893	5910	5762

Year	1981	1982	1983	1984	1985
Total Sales	12777766	13003457	13216366	13445051	13722936
Small Share	.366	.368	.370	.373	.375
Medium Share	.165	.167	.168	.170	.172
Large Share	.469	.465	.461	.457	.453
Cars In Use	108799495	111750161	114811349	117927570	121005385
Scrappage	9922446	10052791	10155177	10328830	10645122
VMT (MM miles/yr)	1252620	1293849	1336230	1379401	1422781
Gas Consumed (MM gal/yr)	84726	86309	88123	90174	92409
Avg. Gen. Price	5707	5652	5596	5540	5484

Year	1986	1987	1988	1989	1990
Total Sales	13799240	13975909	14212912	14484393	14766342
Small Share	.373	.381	.384	.387	.391
Medium Share	.174	.176	.178	.180	.181
Large Share	.448	.444	.439	.433	.428
Cars In Use	123784714	126380364	128847772	131249482	133625830
Scrappage	11019910	11380260	11745503	12082683	12389994
VMT (MM miles/yr)	1462074	1500566	1538745	1576958	1615573
Gas Consumed (MM gal/yr)	94511	96649	98818	101007	103225
Avg. Gen. Price	5482	5480	5477	5473	5468

Year	1991	1992	1993	1994	1995
Total Sales	15083687	15396549	15700308	16000329	16293891
Small Share	.395	.399	.403	.408	.413
Medium Share	.183	.185	.187	.189	.191
Large Share	.422	.416	.410	.403	.396
Cars In Use	136035357	138494414	141012045	143595755	146247798
Scrappage	12674160	12937492	13182677	13416620	13641848
VMT (MM miles/yr)	1655290	1695930	1737522	1730170	1823858
Gas Consumed (MM gal/yr)	105507	107835	110205	112619	115072
Avg. Gen. Price	5456	5444	5430	5416	5401

Year	1996	1997	1998	1999	2000
Total Sales	16585672	16880345	17177182	17481926	17792427
Small Share	.419	.424	.430	.436	.443
Medium Share	.193	.195	.197	.199	.201
Large Share	.389	.381	.373	.365	.356
Cars In Use	148965695	151743369	154569353	157437152	160339809
Scrappage	13867775	14102671	14351198	14614127	14889771
VMT (MM miles/yr)	1868665	1914480	1961326	2009131	2057861
Gas Consumed (MM gal/yr)	117567	120093	122649	125227	127823
Avg. Gen. Price	5386	5369	5352	5334	5316

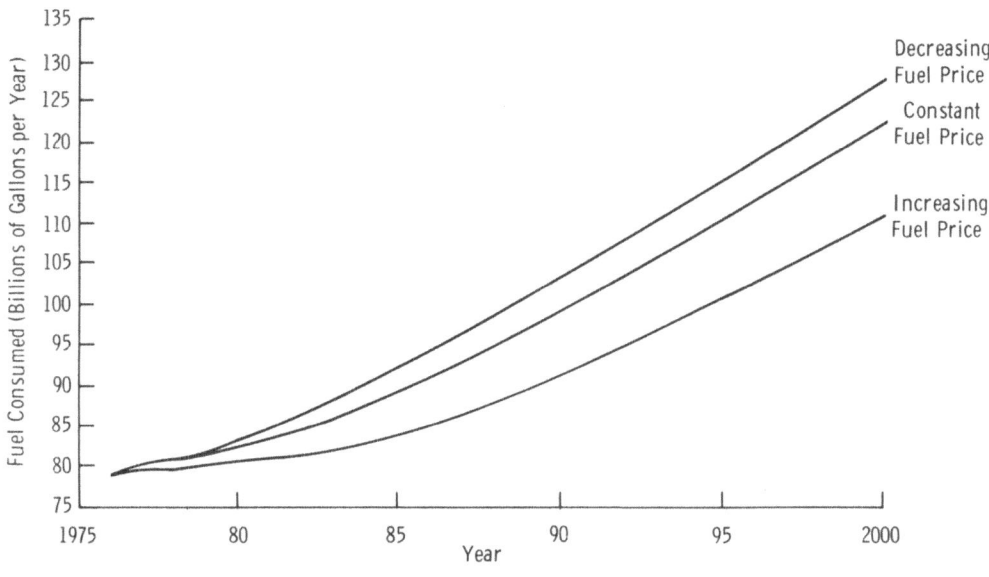

Fig 4. Projected fuel consumption.

The input (exogenous) variables were derived from secondary sources. These inputs are documented in Table 12 (page 28).

SUMMARY

Auto and highway travel demand demonstrate regular demographic and economic patterns. These patterns appear to be relatively unaffected by regional characteristics of the availability of alternative modes of transportation. This is not to say that these factors have no impact on highway travel demand, but simply that long-term aggregate forecasting cannot benefit from these relationships. They are

- too marginal to statistically quantify in a useful way, and they

- require forecasts of exogenous variables that are less reliable than the endogenous variables.

Given this aggregate framework, a multi-equation model is used to relate forecasts of population, income, auto characteristics, economic conditions, and fuel costs to estimates of auto and travel demand and the fuel economy characteristics of the future fleet. Using forecasts of input variables which reflect normal economic growth expectations and relatively stable fuel prices, highway travel and fuel consumption increase substantially through the year 2000. Under the decreasing fuel price

References p. 29.

TABLE 12

Exogenous Variables

	Fuel Prices*			Number of Households†	Disposable Income††	Unemployment †††	Fuel Economy •, mpg			Vehicle Prices**		
Year	Constant	Rising**	Falling***				Small	Medium	Large	Small	Medium	Large
1975	$.610	$.610	$.610	69840530	$10,074/ household	8.9%	22.3	15.3	12.6	$3,000	$3,598	$4,654
1976	.610	.632	.608	70976148	10,275	7.5%	22.7	15.7	13.0	3,000	3,598	4,654
1977	.610	.654	.596	72130232	10,481	6.1%	23.0	16.1	13.3	3,000	3,598	4,654
1978	.610	.676	.584	73303082	10,691	4.7%	23.4	16.5	13.7	3,000	3,598	4,654
1979	.610	.698	.562	74995002	10,904	4.7%	23.8	16.9	14.1	3,000	3,598	4,654
1980	.610	.720	.550	75706303	11,122	4.7%	24.2	17.3	14.5	3,000	3,598	4,654
1981	.610	.750	.536	76937300	11,345	4.7%	24.5	17.7	14.8	3,000	3,598	4,654
1982	.610	.780	.522	78188313	11,572	4.7%	24.9	18.1	15.2	3,000	3,598	4,654
1983	.610	.810	.508	79459668	11,803	4.7%	25.3	18.5	15.6	3,000	3,598	4,654
1984	.610	.840	.494	80751695	12,039	4.7%	25.7	18.9	16.0	3,000	3,598	4,654
1985	.610	.870	.480	82064731	12,280	4.7%	26.0	19.3	16.3	3,000	3,598	4,654
1986	.610	.872	.482	83399117	12,526	4.7%	26.0	19.3	16.3	3,000	3,598	4,654
1987	.610	.874	.484	84755200	12,776	4.7%	26.0	19.3	16.3	3,000	3,598	4,654
1988	.610	.876	.486	86133334	13,032	4.7%	26.0	19.3	16.3	3,000	3,598	4,654
1989	.610	.878	.487	87533876	13,292	4.7%	26.0	19.3	16.3	3,000	3,598	4,654
1990	.610	.880	.488	88957191	13,558	4.7%	26.0	19.3	16.3	3,000	3,598	4,654
1991	.610	.884	.490	90403650	13,829	4.7%	26.0	19.3	16.3	3,000	3,598	4,654
1992	.610	.888	.490	91873628	14,106	4.7%	26.0	19.3	16.3	3,000	3,598	4,654
1993	.610	.897	.490	93367509	14,388	4.7%	26.0	19.3	16.3	3,000	3,598	4,654
1994	.610	.896	.490	94885680	14,676	4.7%	26.0	19.3	16.3	3,000	3,598	4,654
1995	.610	.900	.490	96428537	14,969	4.7%	26.0	19.3	16.3	3,000	3,598	4,654
1996	.610	.900	.490	97996481	15,269	4.7%	26.0	19.3	16.3	3,000	3,598	4,654
1997	.610	.900	.490	99589919	15,574	4.7%	26.0	19.3	16.3	3,000	3,598	4,654
1998	.610	.900	.490	101209268	15,886	4.7%	26.0	19.3	16.3	3,000	3,598	4,654
1999	.610	.900	.490	102854947	16,203	4.7%	26.0	19.3	16.3	3,000	3,598	4,654
2000	.610	.900	.490	104527386	16,527	4.7%	26.0	19.3	16.3	3,000	3,598	4,654

* 1974 dollars.
** Rising case based on low domestic oil and gas availability and high import prices ($18/bbl in 1990).
*** Falling case based on expected domestic oil and gas availability and import prices ($7/bbl).

† Household forecasts based on Census Series II Population Forecasts (Current Population Reports, Series P25, No. 541) and Census household-population factors (Current Population Reports, Series P23, No. 40).
†† 1967 dollars. (Assumes 2 percent growth rate.)
††† Transition from current unemployment level to equilibrium unemployment of 4.7% by 1978.
Jack Faucett Associates, Project Independence and Energy Conservation, Transportation Section. Council On Environmental Quality, 1974.

• Based on expected engine improvements and weight reductions. Provides 1980 goal of 40% improvement (over 1975) only with market shifts or 40% goal in 1985 if market shares remain constant.
** 1975 dollars. Automotive News, 1975. Future (rear) prices assumed constant.

assumption, auto fuel consumption increases 85 percent over 1976 levels by the year 2000. Under the increasing fuel price assumption this growth is reduced to 40 percent for the same period.

These forecasts could be reduced substantially if more fuel efficient auto supply scenarios were introduced. The assumptions used were the most conservative. Given technological advances in engine and chassis design, substantially lower fuel consumption can be expected.

REFERENCES

1. F. C. Wykoff, "A User Cost Approach to New Automobile Purchases," Review of Economic Studies, Vol. XL(3), No. 123, July 1973, pp. 377-390.
2. U. S. Department of Transportation, Federal Highway Administration, "Nationwide Personal Transportation Study," 1974.
3. "Survey of Consumer Finances," Survey Research Center, Institute for Social Research, University of Michigan.
4. U. S. Department of Commerce, "Current Population Reports," Series P-65.
5. G. C. Chow, "Demand for Automobiles in the United States," North Holland Press, Amsterdam, 1957.
6. M. Nerlove, "A Note on Long-Run Automobile Demand," Journal of Marketing, Vol. 22, July 1957.
7. M. J. Hamburger, "Interest Rates and the Demand for Consumer Durable Goods," American Economic Review, Vol. 57, December 1967.

BIBLIOGRAPHY

1. L. J. Atkinson, "Consumer Markets for Durable Goods," Survey of Current Business, April 1952.
2. Business Week, December 14, 1974, p. 62.
3. Chase Econometrics Associates, "The Effects of Tax and Regulatory Alternatives on Car Sales and Gasoline Consumption," prepared for the Council on Environmental Quality, NTIS PB 234 622, May 1974.
4. T. R. Dyckman, "An Aggregate Demand Model for Automobiles," Journal of Business, Vol. 38, July 1965.
5. H. S. Houthakker and L. D. Taylor, "Consumer Demand in the United States 1929-1970," Cambridge, 1966.
6. C. F. Roos and V. von Szelski, "Factors Governing Changes in Domestic Automobile Demand," The Dynamics of Automobile Demand, New York, 1939.
7. R. E. Mellman, "A Critical Analysis of Automobile Demand Studies," Transportation Systems Center Report No. WP-210-U2-84, 1975.
8. R. Schuessler and R. Smith, "Working Models of Fuel Consumption, Emissions and Safety Related to Auto Usage and Purchasing Behavior," Transportation Systems Center Report WP-230-U2-52, 1974.
9. D. B. Suits, "The Demand for New Automobiles in the United States, 1929-1956," Review of Economics and Statistics, Vol. 39, November 1958.
10. L. White, "The Automobile Industry Since 1945," 1971.
11. S. Wildhorn, B. K. Buright, J. Enns and T. Kirkwood, "How to Save Gasoline: Public Policy Alternatives for the Automobile," Santa Monica: The Rand Corporation, 1974.
12. D. B. Suits, "Administered Prices," Part 7, Senate Antitrust Subcommittee Staff Report, 1958, pp. 3998-3999.

DISCUSSION

E. W. Beckman (Chrysler Corporation)

I'd just like to ask you to repeat those fleet mpg figures that you cited just a moment ago. I assume by fleet, you mean the car population that is on the road at this time.

Hemphill

Yes, that's correct. Our estimate for the base year, 1976, is 13.3 mpg. For 1980 it is 14.3, and for 1985 it is 17.3.

Unidentified

You referred to real cost. Does that include cost of development?

Hemphill

Yes.

J. L. Beckham *(Cadillac Motor Division, General Motors)*

I was just wondering what considerations, if any, were given to the continuance of emission standards, or lowering of them, in your modeling for fuel economy?

Hemphill

I've been raising that issue, fairly unsuccessfully, on The Hill for about six months. There is absolutely no consensus, even among all the experts, as to how much different emission standards will cost in fuel economy. We've made our own estimates of the regulatory climate and we've assumed the Rogers subcommittee bill, as currently written. We've made some assumptions about what could be done based on all the work that's been done on fuel economy and we can plug those into the model. Now, the model is sufficiently flexible, that you can plug in a different input and you'll clearly get different conclusions. The model itself doesn't need any internal modifications to take account of emission factors. This stuff is so uncertain and there's so much controversy that we've had to plug that in externally.

S. S. Penner *(University of California, San Diego)*

You mentioned that the existence of a rapid transit system has essentially no significant effect on private car ownership. I think you gave a differential of 14% if you exclude New York City. Is your data sufficiently detailed to show whether or not there is a change in this pattern over the last couple of years?

Hemphill

There have been some perturbations over the past couple of years in terms of mass transit ridership. Mass transit ridership has been steadily declining since about 1946 in terms of passenger-trip-miles. That decline has been slow, but it's a decline nonetheless. It turned up, not surprisingly, during the embargo. There seems at this point to be a lot of confusion. The month-to-month data are very jumpy, but everything we have seen indicates that availability or nonavailability of transit does not have an overwhelming impact on car ownership. The problem with the data, of course, is that there really aren't very many places where mass transit is sufficiently available to make it a worthwhile substitute. Roughly, you have to go with the six cities I mentioned, and when using data from only six places it's a little suspect. The real uncertainty on mass transit is that we have no idea how much there's going to be in the future. So, to try and predict new car demand based on a clearly-unknown projection of mass transit is a rather reckless exercise.

S. S. Penner

Do you have any information on the relationship between the number of cars per family and number of miles driven?

Hemphill

I can't answer that right now. I'd have to go back and check it.

J. W. Davison *(Phillips Petroleum Company)*

You commented that your base-line case had a constant fuel price. Does your model have a provision for changing the price of the fuel?

Hemphill

Yes it does. The model wouldn't be terribly useful to us if it didn't. You can kind of sketch your own scenario as to what you think gasoline prices will be, whether they're going up or down, and then run through and see what that does to the market share and the fuel economy of the fleet.

W. T. Lyn *(Cummins Engine Co., Inc.)*

Did you try out your model on historical data to see if it would predict present day conditions?

Hemphill

Yes. That's basically how the various constants and relationships which determine the equations were derived.

Lyn

How far do you go back?

Hemphill

To 1950 in some cases, 1960 in other cases. It depended on how much good, solid time series data we have for that particular area.

R. E. James *(General Motors Corporation)*

Have you used your model to evaluate the effects on gasoline consumption of the auto industry's voluntary program to improve fuel economy by 1980? A second question is, have you evaluated the effect of the two bills before the Senate that mandate 28 mpg by 1985?

Hemphill

The first answer is, we've done a lot of evaluation of a 40% improvement in fuel economy by 1980, and I think we've got some estimates, depending on the assumptions you make as to price changes between now and then, of savings somewhere between 400,000 and 600,000 barrels a day. As to estimating the impact of the 28 mph standard by 1985, we've got some real difficulties here. You can put into this model the miles-per-gallon of the new car fleet, broken down by car class, and stipulate that it's going to be 28 miles per gallon. Then you have to make some assumptions as to what price changes that would imply, and then you can come out with a set of market shares. The problem is that it all has to be done exogenously and that's kind of similar to the emissions case. It's based so much on engineering judgement and probabilities that I'm very uncomfortable doing that. I'm not convinced from anything I've seen that you can get 28 mpg in 1985, assuming statutory emission standards, with anything less than a 100% sub-compact and compact cars mix, and I don't think that's going to occur. So, if you went ahead and plugged that in, you would find it saves a lot of fuel. But the likelihood of that occurring is remote.

F. A. Williams *(University of California, San Diego)*

There are always uncertainties in forecasts of this kind. I wonder what the uncertainty limits are in these projections?

Hemphill

We've got that information and it can be provided upon request. We just haven't included it in this paper because of time and space limitations.

J. M. Colucci *(General Motors Research Laboratories)*

How did you account for changes in scrappage rate due to shifts in the economic conditions? We know that during a recession people don't buy as many cars and that they keep their old car a lot longer. How did you handle that?

Hemphill

We used two variables which we found are the best indicators of scrappage rate. We used the price of new cars and the unemployment rate. In making our projections for the future, we used a Pollyanna kind of estimate of 3.5% GNP growth, and an unemployment level sinking, by 1978 to 4.5% and staying constant from then on. If these parameters are changed, the scrappage rate will change. I think that's one of the two or three things about this approach that makes it a step forward compared with previous models. Almost every other model I've seen just says, 10,000,000 cars a year go off the road, period. Everybody knows that's nonsense, and it has a big impact on your car sales.

THE U. S. ENERGY OUTLOOK THROUGH 1990

S. J. BEAUBIEN

Shell Oil Company, Houston, Texas

We should all be concerned about the future because
we will have to spend the rest of our lives there.
    − *C. F. Kettering*

ABSTRACT

Each sector of the energy market tends to see itself as the center of events, with only vague connections with other energy-consuming sectors. The truth, however, is that all sectors are highly interrelated, and the quantity and type of fuel available to one depends greatly on the demands and the relative importance of each of the other consuming sectors. Thus, it is necessary to establish a total energy balance before a credible forecast for fuel can be made, for example in the transportation sector. Once this concept is accepted, it becomes apparent that a still wider scope of inquiry is needed to provide estimates for social, political, and economic factors which impact on the various fuel demand levels. This paper draws upon independent social/political/economic studies to provide the required input premises for estimating total U. S. energy balances through the year 1990.

The results are given in terms of total energy demand by market sector, and total energy supply by primary fuel. One important finding is that in spite of a decrease in the rate of growth of total energy demand, by 1990 we will require an additional 19 million barrels/day oil equivalents.* Half of this new demand will be met with nuclear

---

In this paper, except where otherwise noted, barrel of crude oil equivalent (COE) is used as a yardstick for comparing amounts of different forms of energy. Energy can be conveniently measured in terms of heat produced and a barrel of crude oil produces, on average, 5.8 million British Thermal Units (Btu's).

power, while coal and oil will provide most of the remainder. Also provided are details of supply and demand for each major fuel source. It is found that while the domestic supplies (including Arctic) of all fuels except natural gas increase, the supply/demand gap continues to grow through 1985 and can only be met by increasing oil imports. A further analysis of the transportation sector by specific oil products (fuels) and by mode (e.g., autos, trucks, aircraft, etc.) is also presented. Here we find that transportation fuel demand, which has been growing at a rate of nearly 5% per year in the recent past, will decline to about 1% per year over the forecast period. Much of this decline is due to a reduced demand for passenger car motor gasoline.

One of the conclusions reached is that the U. S. capability to attenuate demand through conventional conservation measures is insufficient to counter-balance domestic supply deficiencies. Further, to effect a substantial decrease in oil imports before 1985 would require a national program that would impact in a major way upon U. S. patterns of consumer use and lifestyle.

INTRODUCTION

The United States is in the midst of a critical evaluation of its current and long-term energy opportunities and options. This evaluation is greatly affected by estimates of actions taken or anticipated in the U. S. and world political arenas — more so today than at any time in the past.

It is in this more uncertain climate of projecting the actions of men rather than markets that these forecasts were developed.

A PERIOD OF TRANSITION

Both the U. S. and world energy sectors are undergoing a period of transition. Changes, both expected and unexpected, have affected both the cost of energy and the way energy is seen by its many users.

The 1973/74 oil embargo and mandated price increases by the Organization of Petroleum Exporting Countries (OPEC) had the consequence of driving up world crude oil prices. Both the price increase and the embargo came at a time when U. S. dependence on foreign crude oil was rising.

The consumption of energy in the United States, which had been growing in all sectors of the economy, reached 35 million barrels per day crude oil equivalents (COE) in 1974. Domestic production of energy however reached a plateau at about 29 million barrels per day COE during the 1970-73 period. The difference between these two values created the energy gap — which was filled by imported oil. Thus, while in 1970 imported oil accounted for slightly less than one quarter of total U. S.

oil supply, by 1974 oil imports amounted to more than a third.

One of the main causes of the energy gap is the historical price of energy. Over a period of years prior to the embargo, U. S. energy prices actually declined relative to the price of other commodities. While consumption was thus encouraged, government price controls on natural gas and attractive world prices for other fuels kept prices in the United States too low for domestic supply to keep up with demand. The energy gap that developed was most economically satisfied by imports of formerly cheap foreign oil.

A weak economy, government inaction on energy policy, litigation concerning environmental conservation, and a general feeling of uncertainty caused delays in critical energy projects. Examples of projects that have met with obstacles are the Alaskan pipeline and exploration for oil and gas in offshore areas. Another example was the suspension in the Fall of 1974 of the Colony Development oil shale project in Colorado because of rising costs and uncertainties in U. S. energy policy. That suspension put in limbo the date for realization of U. S. expectations for energy from the vast western shale oil deposits.

In face of these delays we note that U. S. oil production peaked in 1970 and natural gas production peaked in 1973. National coal production, which had declined appreciably in the 1950s has only recently come back to about 600 million tons per year, a level that it held in the late 1940s.

In sum, U. S. energy requirements have continually increased, but in recent years our domestic production capability — bogged down by uncertainty and internal delays — has not expanded. The resulting increase in imports has awakened the U. S. to both the insecurity and economic hazards of foreign dependence.

The nation is in the midst of a critical evaluation of its current and long-term energy opportunities and options. Some of the decisions which will ultimately evolve from this national debate must certainly take into account estimates of our technological capabilities. But, at least as important will be estimates of anticipated institutional actions bearing upon the economic and social climate in which technological progress operates.

PREMISES AND PROJECTIONS

Shell Oil Company has prepared forecasts of U. S. energy supply and demand for many years. As interest in energy increased, Shell took the forecasts to various public groups. Beginning in 1970, Shell forecasts were presented to federal and state governments and citizens organizations.

In late 1975, Shell prepared this forecast of the 1980 to 1990 period. It considers not only potential technological advances, but also assesses major energy-related economic and social issues. A framework was thus developed to insure a uniform and consistent viewpoint during the actual development of the energy supply and demand projections. Not all of the elements of this framework can be covered in this paper, but some of the more important components are discussed below.

The forecast reflects a premise of a weak economy and high unemployment through most of 1975, with movement into a moderate cyclical recovery in 1976. The upturn is predicated upon the end of inventory liquidation and upon increasing real incomes as the rate of inflation falls. As shown in Fig. 1, this is followed by several years of sustained growth as the economy gradually narrows the gap between actual production levels and full-employment potential output. Due to the severity of the recession, full-employment production potentials are not reached until the early 1980's. Shell expects that potential economic growth will be slightly lower in the future than in the recent past because of lower population growth, more stringent environmental standards and higher energy prices.

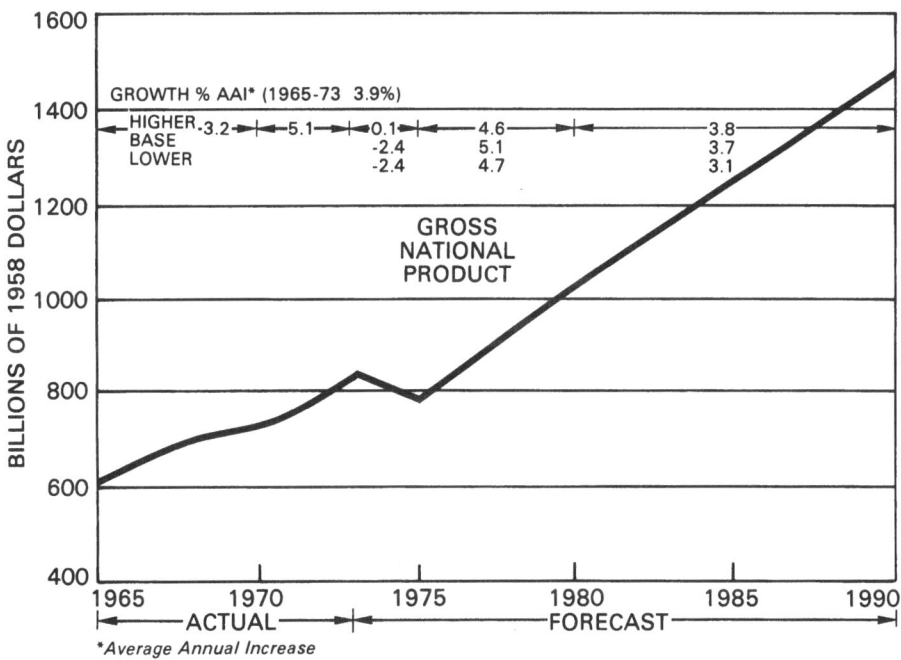

Fig. 1. Real gross national product.

It is further premised that the inflation rate will decline significantly in the next several years (Fig. 2), because of the severe recession and the likelihood that a number of factors which contributed importantly to the recent acceleration in inflation will not recur. These special factors include the worldwide commodity boom, worldwide declines in food production, the deterioration of the dollar exchange rate and multifold oil price increases. The recent experience will, however, have some spill-over

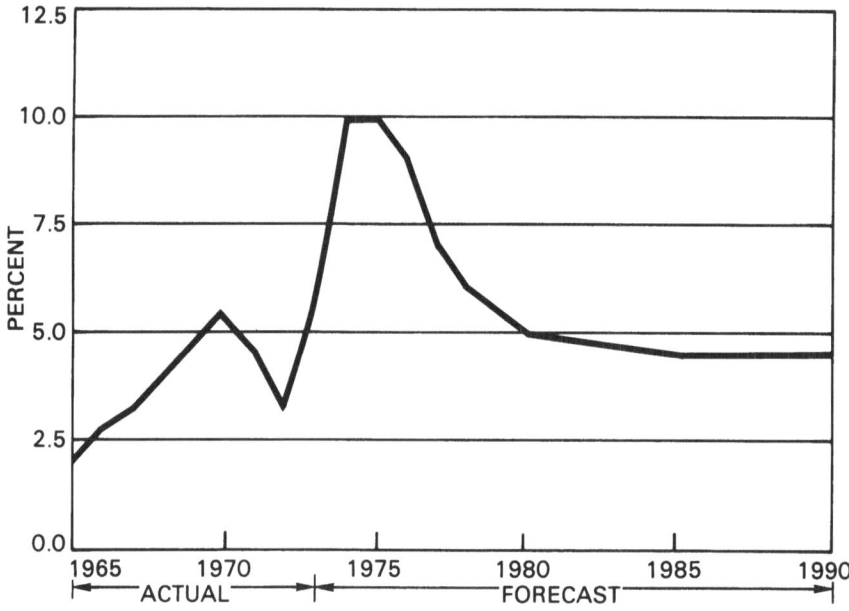

Fig. 2. Inflation rate.

effects on inflation rates in the remainder of the period. Thus we project inflation rates considerably above the postwar norm for the remainder of the period.

The political arena in the near term is seen characterized by strong vocal support for measures designed to decrease reliance on foreign oil through demand suppression, combined with reluctance to impose the harsh measures required to significantly diminish imports. As a result, government initiatives are limited.

During the near term, public concern for environmental protection is expected to remain high, but there is some willingness by Congress to consider interim compromises on air quality regulation and offshore leasing. As the economy continues to recover through 1977 and 1978, the associated increases in oil imports and resulting public concern are expected to provide the necessary political impetus to remove some of the barriers to increased domestic energy supply. Legislation is seen being passed to reduce regulatory delay in nuclear plant construction, further remove barriers to early leasing of promising offshore tracts, and encourage production of shale oil and coal derived synthetic hydrocarbons. Still, even in this time frame, a high degree of environmental protection is expected to be required by federal law, with restrictions in some western states being more severe.

In addition to the growing costs to be faced by the oil industry, it is considered likely that the burden of increased taxes will inhibit capital formation, thereby constraining energy investment.

Prices for both oil and gas are premised to increase in the future. This should provide considerable incentive for both energy-saving investment and prudent energy use by consumers. Thus, throughout the next decade, the U. S. is expected to pursue a course designed to slow the growth in demand, but in absolute terms demand will increase.

The forecast assumes that the much discussed Mackenzie Valley gas pipeline will not be built during the forecast period, but the Trans Alaska Pipeline System is expected to be operating by 1978 and a parallel gas pipeline by 1982. Further, a second Arctic oil pipeline is projected for 1983.

The technological forecast also anticipates gradual introduction of new technologies with no dramatic surprises.

By 1990 it is expected there will be perhaps as many as 6 million electric cars in the U. S. automobile fleet, 10 percent of them powered by fuel cells.

The first nuclear breeder reactor is not expected to become commercial before 1990, with nuclear fusion not a reality for about another decade after then. Their impacts are therefore outside of the forecast period.

By 1990 solar power, to supply part of the space heating and water heating requirements, is projected for as much as a fifth of new residential construction.

It is premised that there will be enough oil in world markets to satisfy U. S. and foreign demand through the forecast period. However in the 1985-1990 period world oil demands are expected to strain production limits set by exporters.

U. S. ENERGY DEMAND

U. S. energy demand by market for the period from 1965 to 1990 is shown in Fig. 3.

During the period 1965 to 1970, total U. S. energy demand grew at nearly a 5 percent annual rate, slowing in recent years to 3 percent. Shell believes that for the forecast period, U. S. energy demand will continue to grow, but at a lower than historic rate. Annual growth will be slightly larger than 2 percent for the remainder of this decade, and will increase to slightly less than 3 percent through the 1980s.

The decline in growth from the historic rate reflects the effects of both domestic energy saving measures and increased future energy prices, along with lower population growth.

The transportation market, whose share of the total energy market decreases through the forecast, remains predominately dependent on petroleum. This conclusion stems from Shell's belief that most transportation developments will be in internal combustion engines and that increasing improved fuel utilization will modify demand growth. The transportation market has traditionally had about a 23 percent

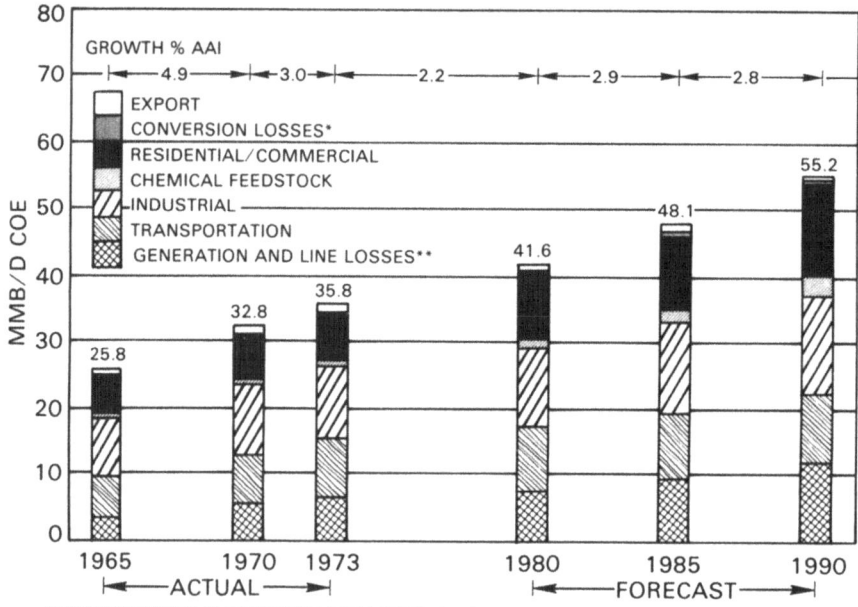

Fig. 3. U. S. energy demand by market.

share of total energy demands, but by 1990 it is projected to drop to about 18 percent of a larger energy base.

The chemical feedstock market is now at about 3 percent of total energy demand, but is expected to double to about 6 percent over the forecast period. This growth reflects the growing tendency to replace natural products with synthetics made from petrochemicals as the resource base of natural products declines.

In the residential-commercial market, which summarizes the demand of residences, stores and office buildings, natural gas remains the largest single primary fuel. While this market is expected to grow, the share of the total energy market will remain steady at 22 to 23 percent.

In the industrial market, coal, and to a lesser extent oil, must replace natural gas as the primary fuel for industrial boilers. This market is expected to decrease its share of total demand from 30 percent today to 27 percent by the end of the forecast.

Aside from the chemical feedstock sector which grows from a small base, the utility market is expected to have the fastest growth rate. Its growth is predominantly achieved through the rapid increase in nuclear energy. This market is forecast to grow from 26 percent to 32 percent of total energy demand by the end of the period.

U. S. ENERGY SUPPLY

As outlined in the previous section, Shell expects total U. S. energy demand to rise from 35.8 million barrels per day COE in 1973 to 55.2 million barrels per day COE in 1990.

Fig. 4 summarizes U. S. energy supply projections through the forecast period.

Through 1990, oil, gas and coal continue to provide the bulk of U. S. energy supply. Nuclear energy however provides over half of the incremental growth in total supply between 1973 and 1990.

While oil supply — including imports — continues to grow slightly, its share of total energy supply decreases somewhat from 46 percent in 1973 to 42 percent in 1990.

The natural gas share of total supply is reduced by half, decreasing from 30 percent in 1973 to 15 percent in 1990.

Coal supply grows slowly through the period and increases from 19 percent of total energy supply in 1973 to 21 percent in 1990.

Nuclear increases its share of total energy supply from around one percent in 1973 to 20 percent in 1990, thus becoming as important as coal.

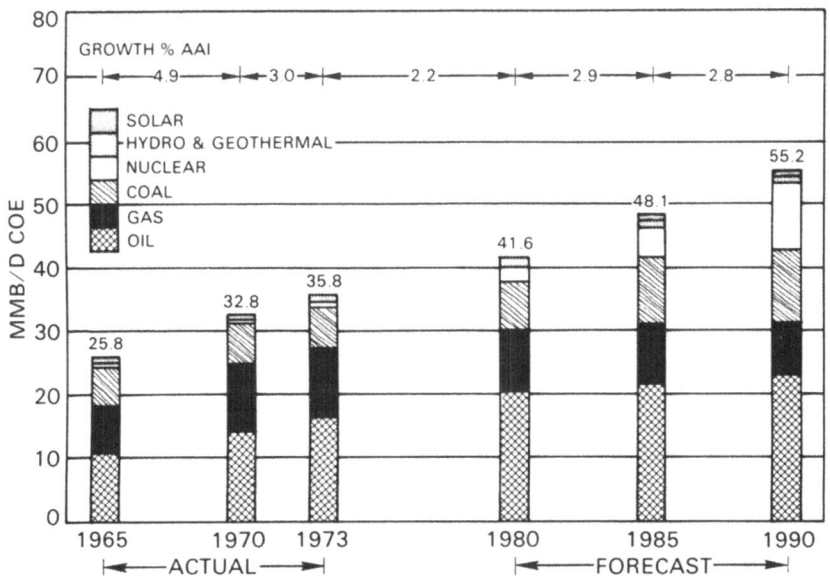

Fig. 4. U. S. energy supply by primary fuel source.

U. S. OIL DEMAND

Although total U. S. oil demand is forecast to increase, the rate of growth declines

from an historical annual 5 percent to around 1 percent toward the end of the
forecast period, Fig. 5.

Shell sees the residential-commercial sector increasing in total oil demand primarily
because of the scarcity of gas and the fact that the gas is partly replaced with
electricity and oil for space heating. However, its share of the oil market will remain
constant at about 16 percent.

In the industrial sector, we again see growth in oil demand as natural gas is diverted
to other sectors. Industry will consume 25 percent of oil supply in 1990 compared to
16 percent in recent years.

In the utilities market, we see a slight increase in oil use and then a decline as coal
and nuclear begin to replace oil for electricity generation. While utilities consumed 10
percent of total oil in 1973, by 1985 their share will decline to 8 percent and to 4
percent in 1990.

The transportation market, which is made up of autos, trucks, railways, airplanes
and other mobile equipment, depends heavily on oil. Its growth rate slows so that its
share of the total oil supply declines from 52 percent in 1973 to about 42 percent in
1990.

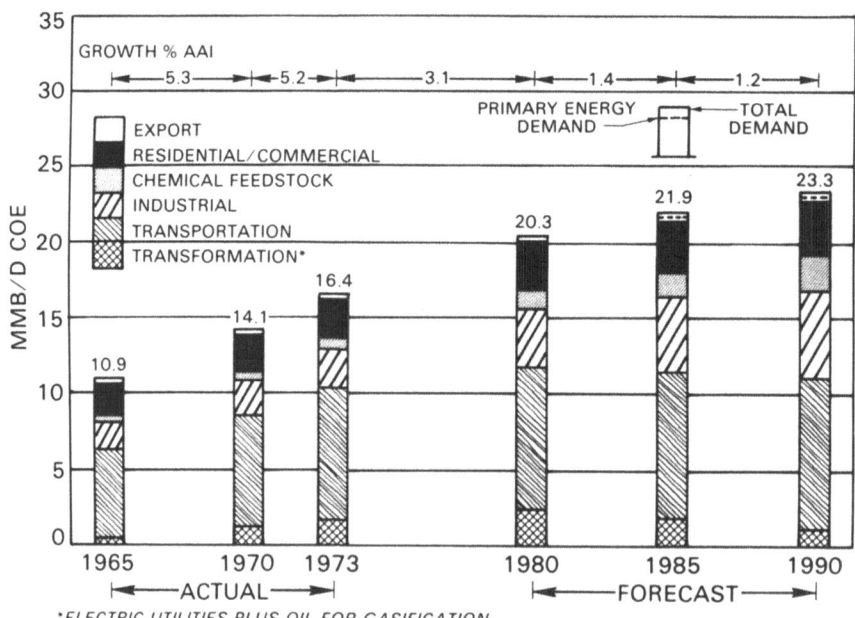

Fig. 5. U. S. oil demand by market.

The reduced growth is due to improved fuel utilization in both automobiles and
airplanes. Cars are projected to be smaller and lighter and to yield more miles per

gallon. In addition, car pools and mass transit developments will retard the growth of miles traveled per vehicle, and airplanes are projected to become more efficient and to utilize higher seat load factors.

Because of the special importance of oil to the transportation sector, this market has been further analyzed by specific product and by transportation mode, Fig. 6. Motor gasoline for autos declines over the forecast period, but this is offset by increasing consumption for trucks and busses. The net result is about level demand for motor gasoline. Diesel fuel for trucks and busses is the fastest growing sector, amounting to about 5-1/2% per year, and jet fuel grows at a rate of about 1.7% per year.

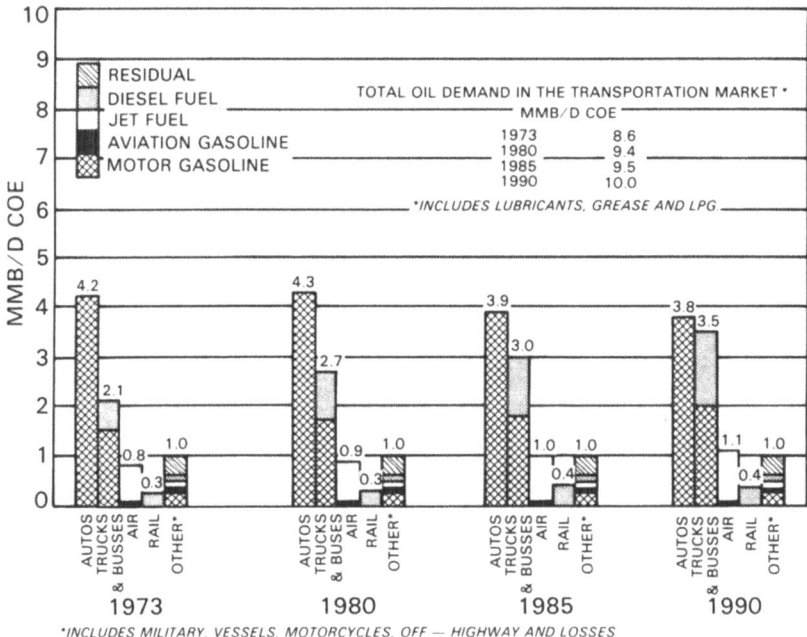

Fig. 6. Oil demand in the transportation market by product and mode.

U. S. OIL SUPPLY

In 1974, U. S. oil supply totalled 15.8 million barrels per day, of which net imports amounted to 5.9 million barrels per day.

As shown in Fig. 7, Shell expects the growth rate of U. S. oil supply to decline from its historic 5 percent per year to about 3 percent per year for the rest of this decade. A still further decline to about 1 percent per year is expected for the 1980-1990 period.

In 1990 a little more than half of the total oil supply will be domestic and the

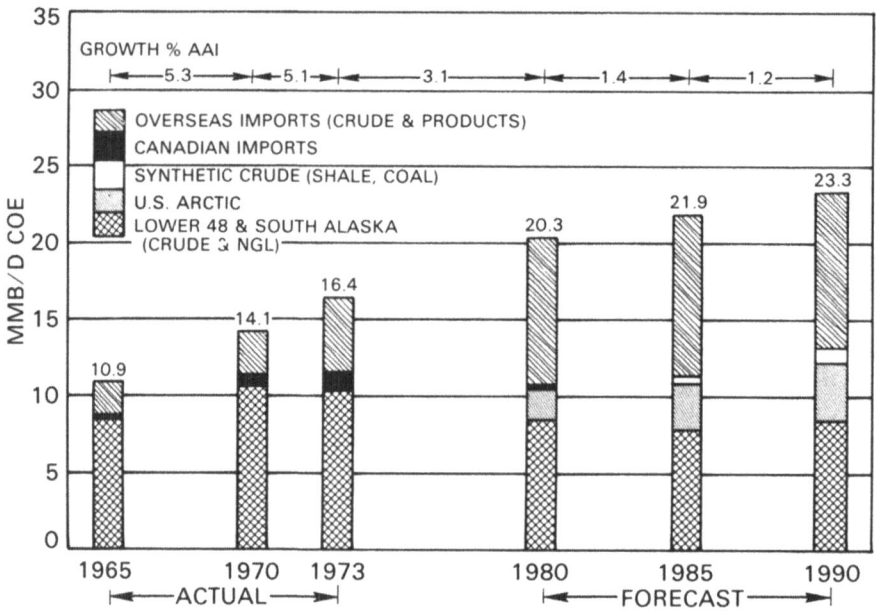

Fig. 7. U. S. oil supply.

remainder imports. Production in the lower 48 states and South Alaska is currently declining; Shell expects future U. S. energy initiatives to do no more than arrest the decline at around 8 million barrels per day by 1990. However, we do expect the U. S. Arctic to contribute another 4 million barrels per day. Moreover we see synthetic crude oil from unconventional sources (Fig. 8) to begin to appear in 1985 and amount to 1 million barrels per day by 1990.

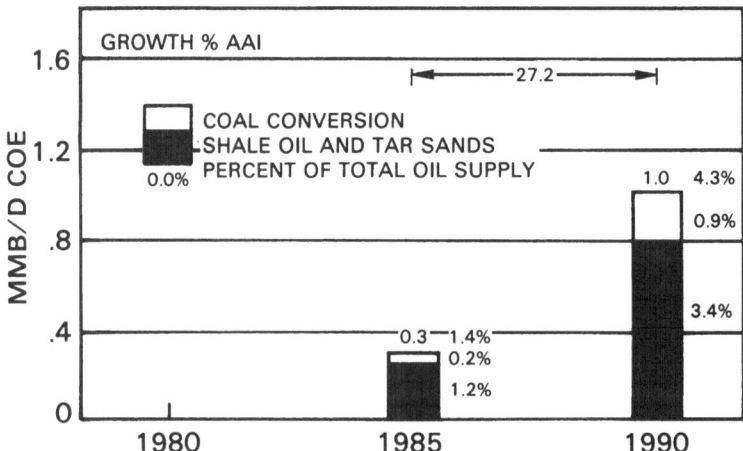

Fig. 8. Unconventional oil supply.

U. S. GAS DEMAND

Supplies of domestic natural gas are extremely limited, which makes a discussion of natural gas demand somewhat misleading; i.e., the term demand here is really a misnomer. Because of supply limitations, a much better way of viewing the situation is how the available supplies are expected to be allocated over the forecast period.

Reflecting the projected decrease in availability of U. S. natural gas, Fig. 9 illustrates the future decline in gas use.

Although in the past natural gas use rose at an annual rate in excess of 6 percent, Shell forecasts future U. S. natural gas usage to decline at around 1 percent per year.

Most of the natural gas will be diverted from utility and industrial boilers to the residential-commercial market, which will receive preferential allocation of limited natural gas supplies.

The use of natural gas as a chemical feedstock is expected to grow, but at a lower rate than other feedstocks. This is mainly because natural gas is used to manufacture fertilizers, which are projected to grow more slowly than other petrochemical products.

In the utilities market, Shell expects natural gas to be all but phased out, except for peak power use, by the end of the forecast period.

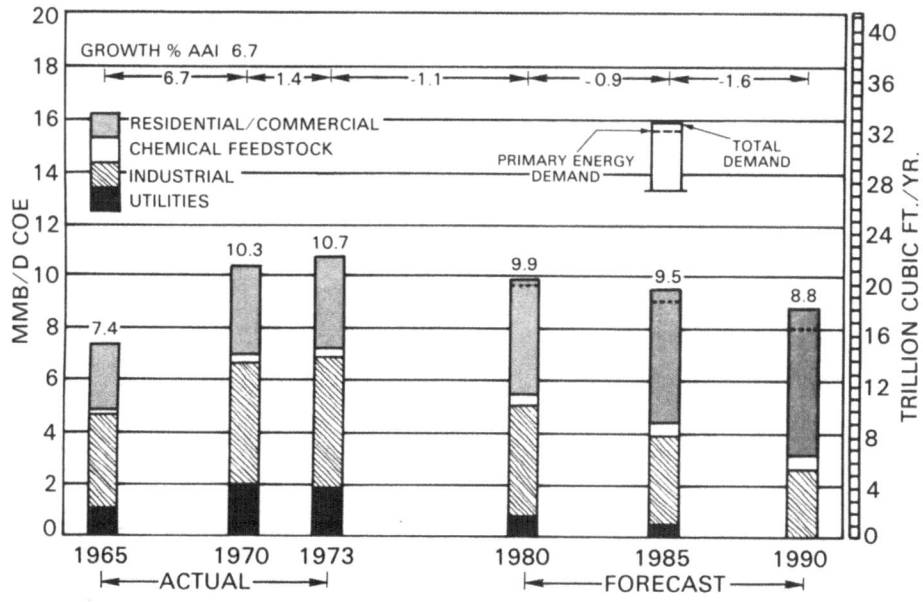

Fig. 9. U. S. gas demand by market.

U. S. GAS SUPPLY

As suggested earlier, Shell expects U. S. natural gas supply to decline through the forecast period. The largest component of domestic supply through 1990 remains gas from conventional sources, primarily in the lower 48 states and South Alaska. Despite expected new discoveries, production from these sources steadily declines, reaching slightly less than 12 trillion cubic feet a year by 1990. This picture differs from the U. S. oil supply picture where a declining production trend was reversed. Despite natural gas supplies from the U. S. Arctic, amounting to 1.7 trillion cubic feet per year in 1990, and supplemental imports of LNG amounting to 2.3 trillion cubic feet per year, the total supply of natural gas declines from about 22 trillion cubic feet per year in 1973 to about 18 trillion cubic feet per year in 1990, Fig. 10.

Imports from Canada will also decline through the period. The modest increases in supply from such sources as synthetic natural gas from oil and from coal gasification (Fig. 11) fail to arrest the decline, which will reach an average annual rate in excess of 1 percent for the period between 1985 and 1990.

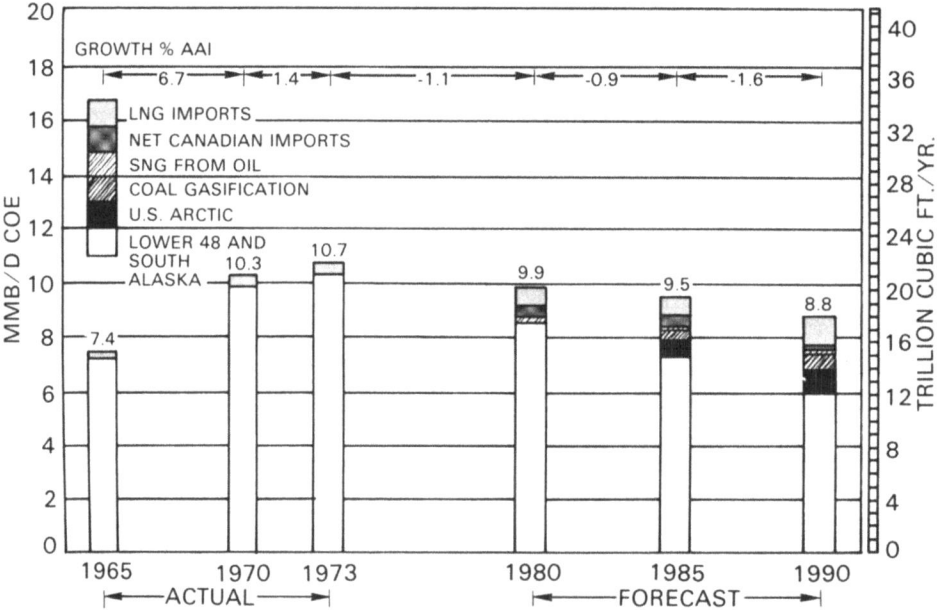

Fig. 10. U. S. gas supply.

U. S. COAL DEMAND AND POTENTIAL SUPPLY

Demand – Coal is the United States' most abundant resource. Fig. 12 illustrates the substantial increase in U. S. coal demand stemming from anticipated changes in

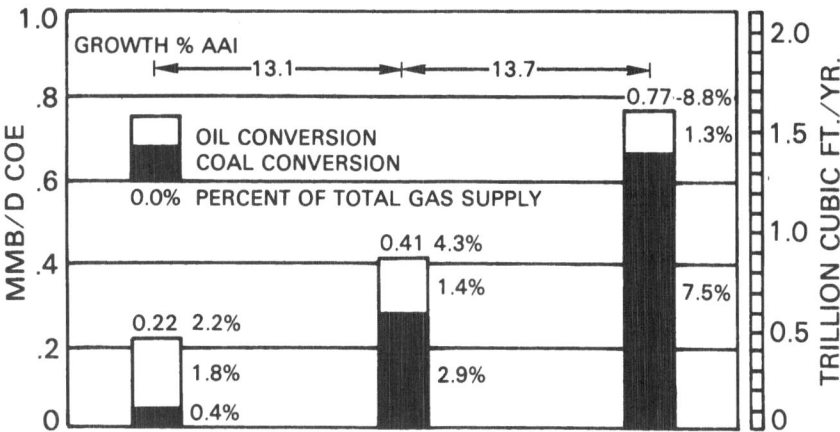

Fig. 11. Synthetic gas supply.

U. S. policy to develop domestic energy resources.

Shell forecasts coal consumption will grow materially through the forecast period. The largest growth is expected to be in the industrial market where coal replaces natural gas as boiler fuel.

Through 1985 coal use is expected to grow in the combined utility and industrial market, but after 1985 growth of nuclear capacity will reduce the contribution of coal in utility use. Coal gasification increases in importance after 1980, reflecting the

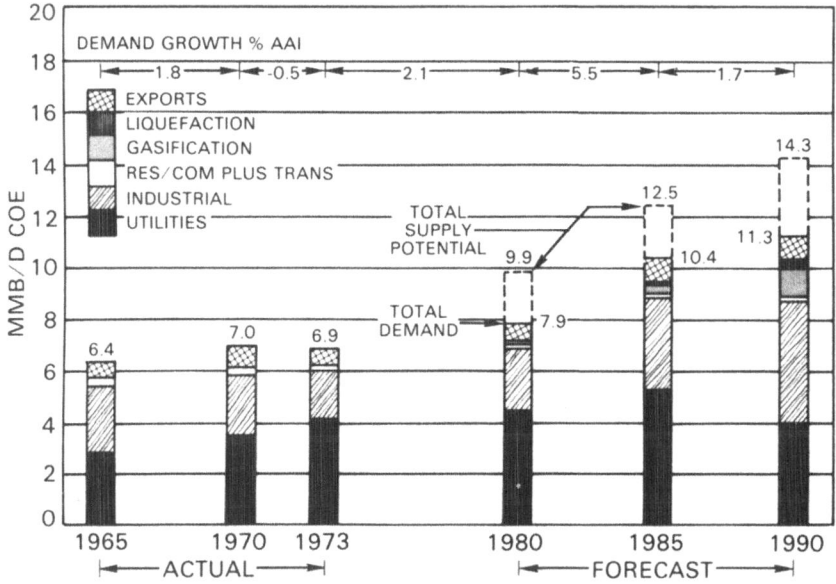

Fig. 12. U. S. coal demand and potential supply.

fact that this technology is already available. In contrast, coal liquefaction technology is less advanced and its contribution will be small.

Supply — Unlike the case of other fuels, coal consumption in the short to intermediate term is projected to be limited primarily by demand rather than by supply. Important secondary limitations do show up, however, in the supply side, such as environmental constraints, restraints on mining, and equipment and transportation limitations.

The area of the chart showing potential supply in excess of demand reflects the coal which could be available if mines were operated longer hours, if mines which were closed because of marginal economics were reopened, and if problems associated with transportation and mining equipment supply were resolved. This incremental potential supply amounts to some 2 million barrels per day COE in 1985 and 3 million barrels per day COE in 1990.

One of the challenges over the next 15 years will be the development of western coal which is located primarily in areas of low population density and remote from energy consuming areas.

The U. S. has, within the lower 48 states, about one-third of the known economically recoverable coal reserves in the world. Fig. 13 summarizes the distribution of the U. S. coal by geographical region, mining method and sulfur content.

Total coal in place, if calculations are restricted to economically recoverable seams, located at depth of 1,000 feet or less, is estimated to be about 258 billion tons. Roughly half is surface mineable and that is primarily in the West. About half of the total is relatively low in sulfur and that also is located chiefly in the West.

AREA	MMBTU PER TON	ECONOMICALLY RECOVERABLE RESERVES BILLION TONS					
		MINING METHODS			% SULFUR		
		SURFACE	UNDER-GROUND	TOTAL	≤ 1.0	1.1 TO 1.5	> 1.5
WESTERN	18.9	75.4	59.2	134.6	100.2	33.3	1.1
GULF	16.4	3.5	–	.5	–	3.5	–
MIDWESTERN	23.8	18.6	40.7	59.3	.6	.6	58.1
APPALACHIA	26.4	11.8	48.8	60.6	20.5	9.9	30.2
TOTAL	21.7	109.3	148.7	258.0	121.3	47.3	89.4

Fig. 13. U. S. coal reserves.

ELECTRICITY SALES

Electricity sales, Fig. 14, double between 1973 and 1990. However, the projected growth rate does not achieve the historical levels of 7 to 8 percent, but holds at about 5 percent per year after 1980. The decline in growth rate stems from increased electricity cost, resulting primarily from fuel price increases. While all markets for electricity grow through the forecast period, the largest growth is expected to be in the residential/commercial market. Shell's expectation that electric cars will enter the automobile fleet late in the forecast period is illustrated by a forecast demand of 70,000 barrels per day COE in 1990 — about 1 percent of the total electricity market.

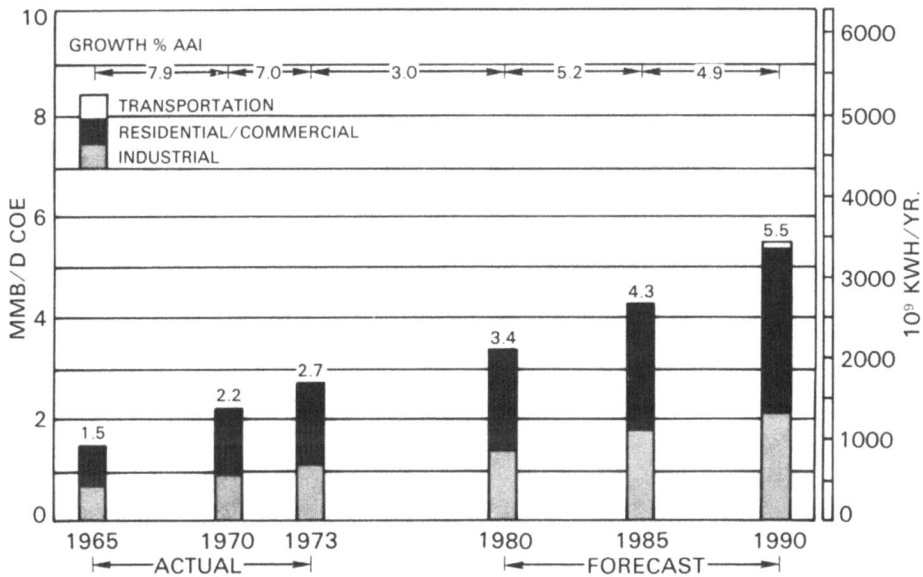

Fig. 14. Electricity sales.

ELECTRICITY GENERATION

Fig. 15 shows the energy input to the utilities to generate electricity. Note that energy input is roughly 3 times as large as the energy received as electricity (previous Fig. 14). The difference between energy input and electricity sales is energy lost in generation and line losses.

Growth in the electricity market is provided predominantly by nuclear energy. Coal, which historically has accounted for around half of the energy for electricity generation, increases in volume slightly through 1985 but subsequently begins to lose its share of the market. Natural gas and oil usage decline through the period as well. Hydroelectric power (including geothermal) remains nearly constant and thus also loses market share.

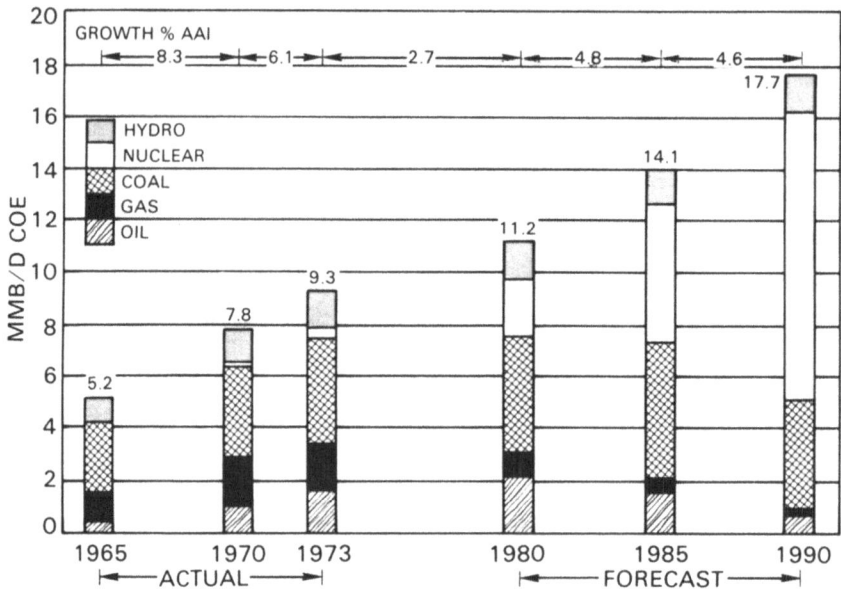

Fig. 15. Electricity generation — input energy.

Nuclear energy while contributing less than 5 percent of the total input energy in 1973 is forecast to grow by 1990 to almost two-thirds of total energy supply used to generate electricity. We are reasonably confident that sufficient U_3O_8 will be available to support the projected nuclear growth. However, it will be necessary to begin immediately to expand the nation's uranium enrichment capability.

U. S. ULTIMATE ENERGY RESERVES

Ultimate energy reserves are defined as reserves which are economically recoverable under circumstances reasonably likely to transpire. Coal is the most abundant U. S. energy resource as Fig. 16 illustrates. Crude oil and natural gas liquids represent about 13 percent of total reserves or about 180 billion barrels. Natural gas represents about 9 percent of the total and shale oil some 6 percent. Uranium reserves of 525,000 tons, when used in light water reactors, represent 3 percent of our total energy reserves.

While there are abundant reserves of energy in the U. S., it is not possible to produce sufficient amounts in usable forms to satisfy our energy demands in the immediate future. As a consequence, we will have to import energy, and the most practical form for the imports will be oil.

IMPORTS/EXPORTS

Even with projected reduced energy growth rates, domestic energy supply will be

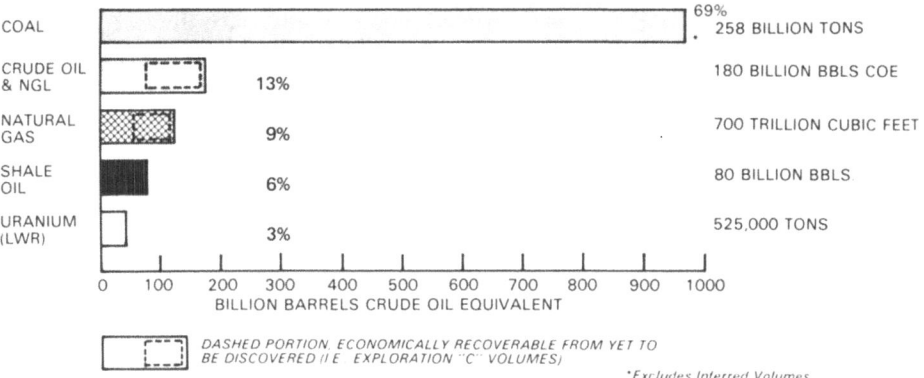

Fig. 16. U. S. "ultimate" energy reserves.

insufficient, and Shell's forecast shows that the U. S. must continue to depend on imported oil to make up for the imbalance between supply and demand.

Oil imports, Fig. 17, are expected to grow to about 10 to 11 million barrels per day by 1985 and then level off.

The projected requirements for U. S. oil imports are expected to be available from interregionally-traded oil.

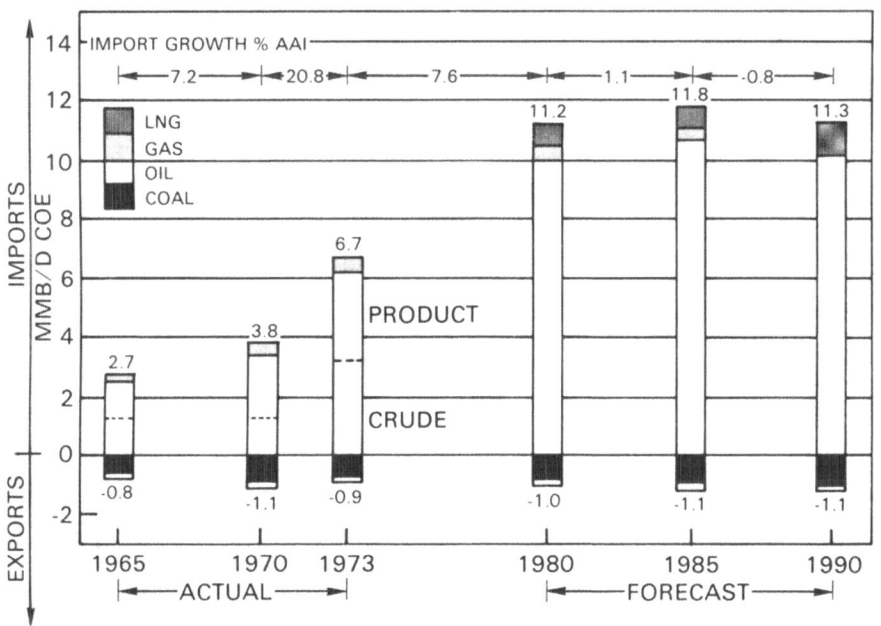

Fig. 17. U. S. energy imports/exports.

In 1974, Canada was the largest single supplier of crude oil to the U. S., and of the 5 largest suppliers (Canada, Nigeria, Iran, Saudi Arabia and Venezuela) only Saudi Arabia participated in the 1973/74 boycott. Canada has indicated her intention of phasing out oil imports to the U. S. by 1983, and it seems likely that Venezuelan production has peaked. Its exports to the U. S. will therefore probably not increase. Accordingly, imports are likely to come increasingly from Eastern Hemisphere nations, most of which are members of OPEC.

In the near future, LNG is expected to be imported from Algeria only. Because of technical problems, significant increases in volume will not occur before 1977. Imports from the Persian Gulf or the Far East are not expected before the 1980's.

Throughout the forecast period, the U. S. is seen continuing to export metallurgical coal and oil products which are mainly coke, waxes and lubricating oils.

WORLD OIL PRODUCTION

Having determined the level of imports needed by the U. S., the question is: will these imports, as well as imports required by the rest of the free world, be available? Fig. 18, showing the regional distribution of World oil production, indicates that the major areas of production are the Middle East and the Communist countries. Of all regions, the Middle East will be the principal exporter.

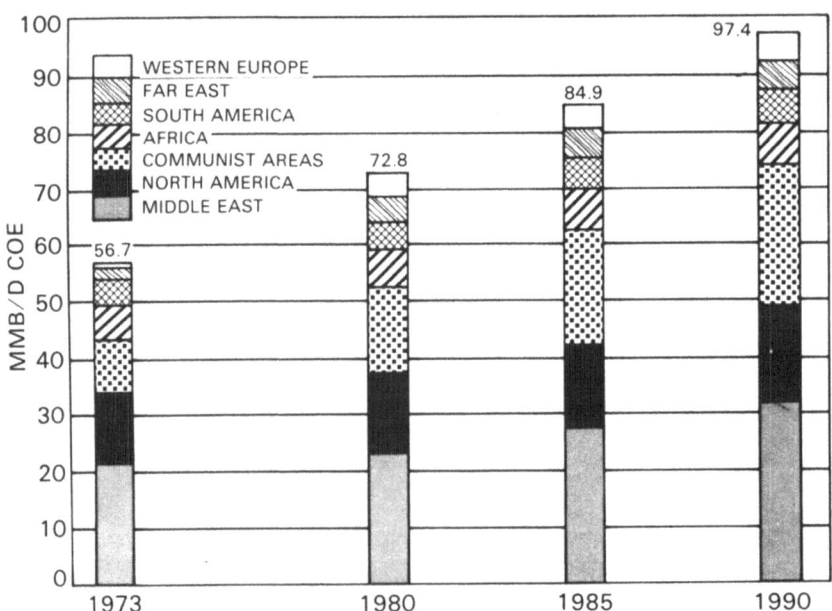

Fig. 18. World oil production.

INTERREGIONALLY TRADED OIL

After making allowance for exports from Communist countries, and after allowing for internal consumption in the producing areas in the rest of the world, the net volume of interregionally traded oil in the world outside Communist countries (WOCA) can be estimated.

Because of anticipated increased world oil prices, and because of increased oil production in consumer countries, Shell believes that the demand for interregionally traded oil will level off through the forecast period at around 30 million barrels per day, Fig. 19. This volume is sufficient to satisfy the requirements of WOCA. To reach these levels, it is premised that Saudi Arabia, having the largest production base, will produce at a rate of 13 million barrels per day in 1990. This premise is consistent with our projections for a healthy world economy which will provide an opportunity for the Saudis to invest their surplus funds.

In the U. S., growth in total energy consumption will slow; moreover, our mix of supply is projected to shift as we begin to emphasize nuclear and coal development. Nevertheless, in spite of all our efforts, U. S. oil imports are expected to increase from about a fourth of interregionally traded oil in 1975 to about a third in 1990.

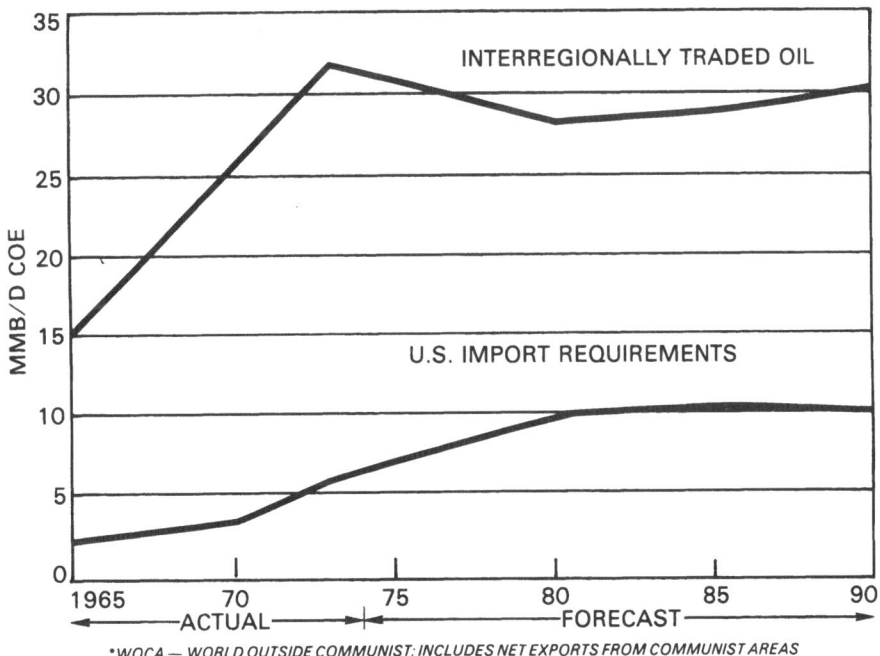

Fig. 19. Interregionally traded oil in the world outside Communist countries.

SUMMARY

This is a digest of the more significant findings derived from the energy forecast.

Total Energy — Demand will continue to increase, but the rate of growth is expected to decline from about 5 percent (1965-1970) to 2 to 3 percent average annual increase (AAI) from 1970 to 1990.

About half of the incremental supply required to meet these new demands is projected to come from nuclear power, with coal and oil (including imports) providing most of the remainder.

Oil — Demand will continue to increase, but the rate of growth will decline from about 5 percent (1965-1970) to 1 to 2 percent AAI from 1980 to 1990.

Gas — Domestic supply, including synthetic gas, will decline. Even when supplemented by imports, total supply will decline.

The declining domestic supply, coupled with the small base of coal gasification and limited imports availability, will lead to intensified competition among consumers for gas.

Coal — Coal is demand limited. However, the growth rate is expected to increase from about 2 percent AAI (1965-1970) to 3 to 4 percent AAI from 1973 to 1985. After 1985 coal growth is expected to slow.

Uncertainties impacting on the long-term market for coal will impede supply development, e.g., oil import and environmental policies, coal conversion economics, transportation and equipment constraints, and rate of growth of nuclear power.

Nuclear — Strong nuclear growth is projected, but its small present base will not result in a decline in the utility market's demand for fossil fuels until 1985.

Electricity Sales — Demand will continue to increase, but the rate of growth will decline from an historic 7 to 8 percent AAI to 4 to 5 percent AAI in the 1980-1990 period.

Oil Imports — Even with reduced energy demand growth rates, domestic energy supply will be insufficient and the U. S. must continue to depend on imported oil.

Oil imports are expected to peak at a level of 10.7 million barrels per day by 1985.

The required U. S. oil imports are expected to be available from interregionally-traded oil. However, to satisfy the free world needs in 1990, Saudi Arabia is premised to produce at a level of approximately 13 million barrels per day.

Environmental Constraints — Domestic supplies from all major sources are severely restricted.

The substitution of coal for oil and natural gas is impeded.

Conservation – U. S. capability to attenuate demand is insufficient to counter-balance domestic supply deficiencies.

CONCLUSIONS

Throughout the forecast period, most U. S. energy will be domestically produced. However, while in 1974 the nation was about 16 percent dependent on imported energy, we forecast that in 1990 about 20 percent of total energy will be imported. Of these imports most is oil (crude and products) with some liquified natural gas. While U. S. dependence on foreign sources of energy will increase somewhat, the nation will be relatively less dependent on imports than other major consuming blocs.

The U. S. will still have a predominately fossil fuel economy with oil, natural gas and coal making up 77 percent of total energy in 1990.

Although new sources of energy such as oil from shale, gas and oil from coal, and energy from the sun will begin to be utilized, they will make only a very small contribution to U. S. energy supply by 1990. Sources such as fusion, and oceanic and tidal energy will not yet have appeared, for it is felt that these involve longer-term technologies that will manifest themselves only in the 21st century.

Passenger automobiles will be smaller and more efficient. Electric cars will begin entering the fleet significantly in about 1985, probably as second cars or special purpose vehicles.

Energy consumers will have become more conservation conscious; home insulation and appliance efficiency will have become increasingly important considerations in personal consumption.

There will be adequate energy sources available, both foreign and domestic, to support continued U. S. and world economic growth. Moreover, we believe that the energy supply expectations in the forecast can be achieved without paying an unacceptable price in damage to our air, water and land.

The Shell energy forecast does present us with one predominant conclusion. To effect a substantial decrease in oil imports for the United States before 1990 would require both a comprehensive and exhaustive program. Although impacting upon patterns of supply, this program would affect demand with such force as to produce a dramatic change in the entire United States pattern of consumer use and lifestyle. Is the United States willing and able to embark upon such a program? Do we know how to formulate a successful program? If not what are the other alternatives?

This forecast will not attempt to answer those questions. It merely presents what Shell considers to be a reasonable description of where we are today and where we could be in the future. The final answer will depend on the cooperative actions and the will of all our people, as represented by government, industry, consumer and citizen. It will be formulated progressively over the time span of the forecast. On the appropriateness of this answer will depend the future well-being of the United States.

DISCUSSION

E. M. Dickson *(Stanford Research Institute)*

I'm interested in relating your scenarios to Mr. Hemphill's scenarios. Presumably you iterated the global projections to find when supply and demand were in balance. Can you tell us at what relative price that balance was achieved?

S. J. Beaubien

I made a firm commitment not to discuss price of energy before I came here. The only thing I feel free to say to you about price is that oil and gas prices, as well as other energy prices, will increase.

S. S. Penner *(University of California, San Diego)*

It seems to me that your prognosis for synthetic fuel development is very conservative, particularly on shale. You said nothing about possible recovery from waste, at least I didn't recognize it. You had solar energy utilization almost the same in 1985 and 1990, as near as I can read your scale. I would have regarded your 1985 estimate as optimistic and your 1990 estimate as pessimistic. There are many uncertainties in this kind of scenario, and it is, therefore, very difficult to make a long-range prediction that is comfortable. Now, what I find particularly uncomfortable is the final outcome, with the U. S. importing 33% or so of the interregionally traded oil. Do you really regard this as an economically viable long-term goal? Perhaps you could comment?

Beaubien

The long-term situation is economically viable to 1990, which is as far as our forecast goes. We will need one-third of the interregionally traded oil, but our share, in 1990, will not be growing. Interregionally traded oil peaks out at around 30,000,000 barrels a day. So getting one-third of that is not excessive, because built into this is the concept that the rest of the free world also has equal access to that oil. What we have done is to look at the supply and demand in other countries, and allocated as much of that oil to the other countries as their economies will support. We have no preferred position, and in other words, we're not pushing anybody out. The oil is available to the United States.

M. C. Hardin *(Detroit Diesel Allison Division, General Motors Corp.)*

In your assumptions you show a sharp increase in nuclear energy in the time period to 1990. Do you have any serious doubt as to the attainability of that level within the time frame scheduled?

Beaubien

Our value is lower than values commonly reported in the literature, lower than the government has predicted, for example. Only recently is ERDA coming to a level where we are. That's point one. Point two is we see difficulties now, like getting siting for nuclear plants, and also we find a great delay period, somewhere in the neighborhood of 10 years or longer, from concept to completion. In other countries this takes only 5 or 6 years, and we see some concern growing in our Congress as they begin to recognize that we are falling further and further behind in self-sufficiency. Our imports go up and up. As a consequence of their concern, possibly they will make regulations which will permit a more rapid introduction of nuclear power so that we can get rid of these 10 to 12 year delays and get down to around 6, 7, and 8 year delays.

J. W. Davison *(Phillips Petroleum Co.)*

The last question recalls an issue that I want to raise. If you consider all energy resources, and this is a personal opinion, self-sufficiency will slow down. I think many of you may have seen the article in the last issue of BUSINESS WEEK that pointed out that numerous delays and stretch-outs are occurring. I also would like to mention Arctic oil which you also discussed. You made your point very well, showing that we are dependent on foreign oil and that this will continue. The article reinforces what you said. Arctic oil was very slow to come on. The enormous capital required to develop Arctic oil and nuclear power means just that much more dependence on foreign oil.

Beaubien

Thank you. I might say that we have considered capital costs also, and I agree that it's a very difficult issue.

S. L. Meisel *(Mobil Research and Development Corp.)*

I think Prof. Penner raised some interesting points about the difficulty of predictions. We also have made some predictions. We made every possible optimistic assumption we could make with regard to shale oil, with regard to nuclear energy, and with regard to off-shore oil. Using these optimistic assumptions we could get your figure of one-third of interregionally traded oil down to just somewhat less than 25% by 1985. So, this is the sort of thing we're going to be confronted with. The amount is not precisely predictable, but it will be sizeable.

M. A. Elliott *(Energy Consultant)*

This is just a comment relative to Prof. Penner's remark on synthetics. I just retired from a company in Houston that's been getting nowhere for 10 years and 7 months trying to put a coal gasification plant in the four-corners area. The institutional constraints are such that it has taken us 10-1/2 years to get an environmental impact statement. So I think that maybe Prof. Penner is too optimistic as far as coal gasification is concerned.

E. E. Hughes *(California State Air Commission)*

What assumptions did you make regarding the substitution of oil from utilities and other uses in the transportation sector? How large is this contribution by 1990?

Beaubien

I don't think I can give you that directly because you just can't say all the utility oil will wind up in transportation. It goes into an available pool from which, if transportation demand is sufficient, it can get some of that oil.

ENERGY CONSERVATION AND FUEL-VEHICLE OPTIMIZATION

H. F. SHANNON

Exxon Research and Engineering Company, Linden, New Jersey

ABSTRACT

A recent study done by Exxon Company, U.S.A. and Exxon Research and Engineering Company has attempted to define the relationships between unleaded octane levels, compression ratio, and fuel economy. This paper will review this work which indicated a net conservation of crude oil for octane levels as high as 87 Motor Octane under certain conditions. It also tries to show that under today's or the foreseeable future's capital constraints, this is not an attractive route to energy independence.

When considering the efficient use of petroleum resources for vehicular transportation, one must consider the total fuel-vehicle relation. This paper will review studies of gasoline/distillate ratios carried out by Exxon Research and Engineering for the EPA.

In any study dealing with refinery economics, a large grid of assumptions is made. Without wishing to denigrate these exercises in logic, the more obvious pitfalls will be examined. These assumptions which are normally buried deeply in the cost and yield calculations are critical to the outcome of the study. Obvious examples include economies of scale, cost of refinery fuel, credits, and disposition of light products resulting from more severe processing. Unrealistic assumptions of this nature limit our ability to forecast future opportunities or problems.

In looking ahead, it is imperative that both the petroleum and the automotive industries use such tools cautiously and with the utmost concern for their technical validity.

INTRODUCTION

With the dramatic rise in crude oil and investment costs, a restudy of the relation between increased unleaded gasoline octane levels and higher compression ratios

References p. 69.

became imperative. Such a study should consider the automotive fuel economy benefit that could be gained from increased compression ratio made possible by higher octane unleaded fuel, and the refinery investment costs and additional energy consumption that would be incurred in producing the higher octane unleaded gasoline. The net energy effect of increasing automotive compression ratios and unleaded gasoline octane levels is the difference between these two offsetting effects. This net energy effect is presented in this paper along with some additional subjects pertinent to the subject of energy conservation by fuel-vehicle optimization. These additional subjects are: 1) the effects on refinery energy consumption of changing the proportions of gasoline and distillate; 2) engine modifications to reduce octane requirements.

OPTIMUM OCTANE NUMBER OF UNLEADED GASOLINE

Relationship Between Compression Ratio and Fuel Economy — A 1971 study (1) developed, from published data, a relationship between compression ratio and fuel economy at conditions of constant vehicle performance. To do this, it was assumed that vehicle acceleration capability would be held constant either through reductions in engine displacement or by increases in rear axle gear ratio. It should be noted that, if these steps are not taken, increasing compression ratio will provide increased vehicle performance at the expense of some of the potentially available fuel economy improvement. The upper curve in Fig. 1, which represents a car operating on a level road at a constant speed of 40 mph, shows this relationship. It was derived from data on the General Motors Research engine, published by Roensch in 1949 (2), and verified by Exxon Research and Engineering Co. using data obtained by ten investigators encompassing twenty different engines, all operating at constant speed conditions. These data were developed using both dynamometer and actual road measurements.

Because the average motorist does not always drive at 40 mph on a level road, the fuel economy gains realized from increased compression ratios in typical driving patterns will be less than those developed for constant speed operation. Based upon available data, the fuel economy effect for these typical driving patterns was estimated to be about 80% of the constant speed effect. The lower curve in Fig. 1 shows this discounted effect and it was used as the basis for this study.

Gasoline Grades versus Optimum Octane Number — Any study considering the optimum use of octane must also deal with the benefits derived from splitting the gasoline pool into two or more grades. In Fig. 2, the unleaded MON level of the total gasoline pool is plotted against compression ratio for one, two, and three grade marketing systems. In each case, 98% customer satisfaction has been assumed; that is, the octane requirements perceived by 98 out of every 100 drivers will be satisfied

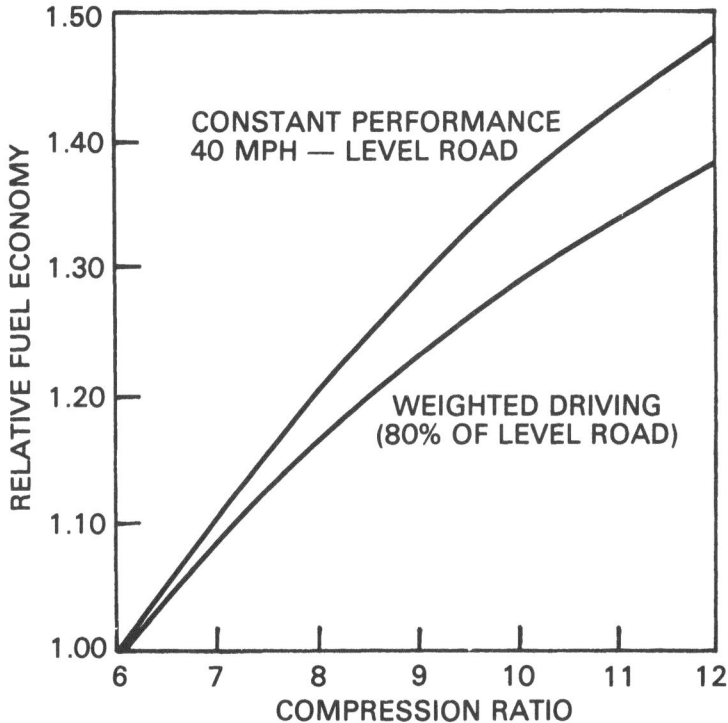

Fig. 1. Effect of compression ratio on relative fuel economy.

with one or more of the grades produced. Motor Octane Number (MON) has been used because 1972-1974 vehicles respond predominantly to MON rather than to Research Octane Number (RON).

These curves are based on historical Coordinating Research Council Equipment Survey Data modified to reflect the slightly higher octane requirement increase estimated to result from the use of unleaded fuels. With a single grade system it is estimated that a pool MON of 88 is needed to provide 98% satisfaction to a vehicle population that averages 8.2:1 compression ratio, approximately that of 1972-1975 model year cars. However, if the pool is split into two grades, a pool MON of only 84 will provide this same customer satisfaction. This effect can be extended by splitting the pool into three grades, where an 83 MON pool provides 98% customer satisfaction. The grade ratios and octane levels in this three grade case are: premium, 20% at ∿88 MON; intermediate, 38% at ∿83 MON; and regular, 42% at ∿80 MON.

If the concept of a multiple grade pool is combined with the benefits derived in a typical driving regime from higher compression ratios, a curve can be developed which shows the relationship between relative fuel economy and pool MON or compression ratio. This curve is shown in Fig. 3 and it was used to predict the vehicle fuel economy effects in the current studies.

References p. 69.

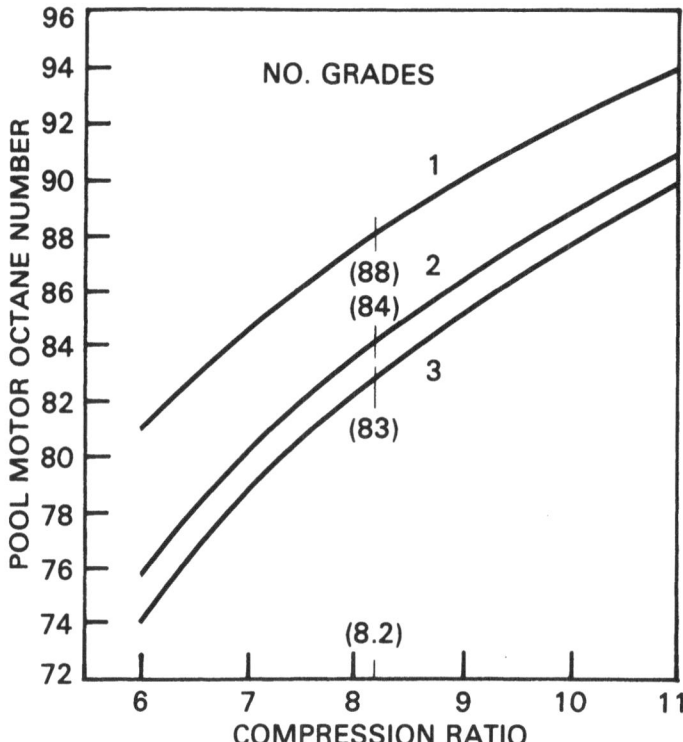

Fig. 2. Effect of multiple grades on pool motor octane number.

Refinery Energy Consumption for Increased Octane – After defining the fuel economy benefits that could be gained from increased compression ratios, the increased energy consumed by the refinery in producing the required higher unleaded octane gasoline was calculated. The refinery model used in this study was designed to depict a generalized 150,000 barrels per day (BPD) Gulf Coast refinery in which the capacities of all major units were fixed. As unleaded octane level was increased, the model had the flexibility to add new units in the gasoline processing area, but the sizes and operating parameters of all base case processing units were held constant.

In addition to studying the hypothetical refinery (ER&E), it was decided to define the investment, operating cost, and energy consumption that would be incurred in raising unleaded gasoline octane levels at each of three actual refineries operated by Exxon Comapny, U.S.A. at Baytown, Texas (BT); Baton Rouge, Louisiana (BR); and Bayway, New Jersey (BW).

Assuming a hypothetical environment in which all gasoline would be unleaded,

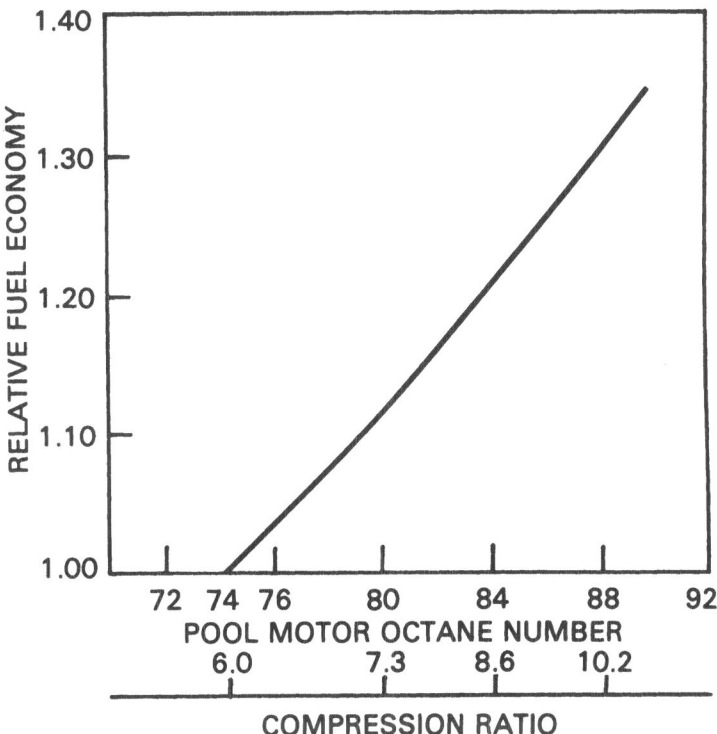

Fig. 3. Relationship between pool motor octane number and fuel economy.

Linear Program simulations were used to predict the base case unleaded MON producible for each of the four refineries studied. These values are shown in Table 1, along with the base case crude run and gasoline production for each refinery. This table shows the significant variation found in the MON producible at each refinery. This variability is a function of several factors, including crude type and base case processing unit characteristics at the individual refineries.

The base case producible MON includes the effects of debottlenecking the existing catalytic reformers at each of the refineries. Therefore, for gasoline octane levels above these base points, it was necessary to install grass roots octane generation facilities. The grass roots facilities utilized in this study are those that are commercially available today, including single-pass light virgin naphtha isomerization, catalytically cracked naphtha reforming, recycle virgin naphtha isomerization, plus an Exxon proprietary process. Both catalytic isomerization and catalytic reforming require feed hydrofining for removal of sulfur and/or nitrogen and olefins.

References p. 69.

TABLE 1

Base Case Refinery Operations

	Baytown	Baton Rouge	Bayway	Hypothetical
Crude Run (MBBL/D)*	570	460	300	150
Gasoline Production (MBBL/D)	225	238	140	73
Motor Octane Productibility	82.8	83.6	81.9	82.9

* *Thousands Barrels per day*

As these units were added to each refinery model to increase unleaded octane producibility, the overall severity of the refining operations increased, resulting in increased energy consumption and decreased gasoline yield per barrel of crude oil input. However, since automobile fuel economy also increases as compression ratio is increased, total gasoline production was not readjusted to the base level throughout the range of unleaded octanes studied. Instead, gasoline production was adjusted at each octane level to maintain constant total consumer miles driven from case to case. On this basis, the incremental refinery energy consumption was calculated for each octane increase above the base case level. These data are plotted in Fig. 4.

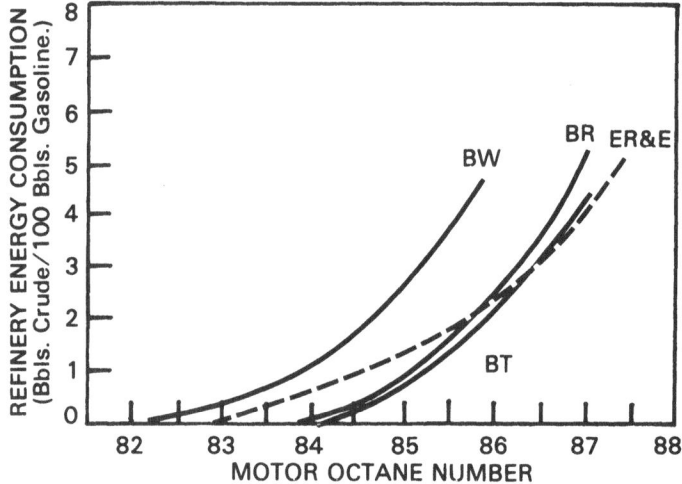

Fig. 4. Energy consumption in refining.

In order to establish a common base from which to express the net energy effects of increased compression ratio, it was necessary to adjust each refinery to the same initial octane level. This was accomplished by first assuming that compression ratios remain constant at their current level of about 8.2 to 1. As discussed earlier, assuming an optimized three grade system would be employed, an average gasoline pool octane level of 83 MON would be needed to provide 98% customer satisfaction for a car population with this compression ratio. Adjusting each of the four refineries to this base point resulted in the energy consumption curves shown in the lower half of Fig. 5. By combining these results with the automobile savings curve developed earlier and replotted at the top of Fig. 5, the net energy savings resulting from increased compression ratios were calculated. These are shown by the four net savings curves in the center of Fig. 5. Looking first at the Baytown, Baton Rouge, and hypothetical refinery (ER&E) curves, the maximum net energy savings would be predicted to occur at a pool octane number between 86 and 87 MON, corresponding to a compression ratio of about 9.4 to 1. This is in essential agreement with the results obtained in a 1971 ER&E study (1), if the 97 RON determined in that study is converted to the corresponding MON. The net energy savings generated in moving to this MON level would vary from 2.8 barrels of crude per 100 barrels of gasoline at Baton Rouge to 3.3 barrels per 100 barrels at Baytown.

As shown, the Bayway inflection point is significantly lower than those at the other refineries; it occurs at about 84.5 MON, corresponding to about 8.8 to 1 compression ratio, and it yields a net energy savings of only 1.3 barrels of crude per 100 barrels of gasoline. This results because Bayway started at a lower pool MON producibility than the other refineries. Therefore, just to increase its gasoline pool to the 83 MON base point, it was necessary to install once-through isomerization, the least energy intensive octane generating unit studied. Since this unit was used in the base case, only the other three, more energy intensive octane generating processes were available to increase the Bayway pool MON above 83.

Several points should be made about the net savings curves shown. The first is that they tend to decline rapidly once the inflection points are passed. Secondly, the curves highlight the differences in the maximum MON levels attainable at each of the refineries. If octane demands above these maximum levels were placed on the individual refineries, they would be incapable of blending motor gasoline. The third significant point about these net savings curves is the large differences shown among the net energy savings for only four refineries. Consideration of these three factors suggests that the establishment of a single optimum MON or compression ratio for the entire nation would be an extremely complex task; and if done improperly, could result in some refiners being unable to blend motor gasoline at the octane level selected.

The net energy savings curves calculated in this study are subject to several sensitivities, both positive and negative. On the positive side, an increased response to

References p. 69.

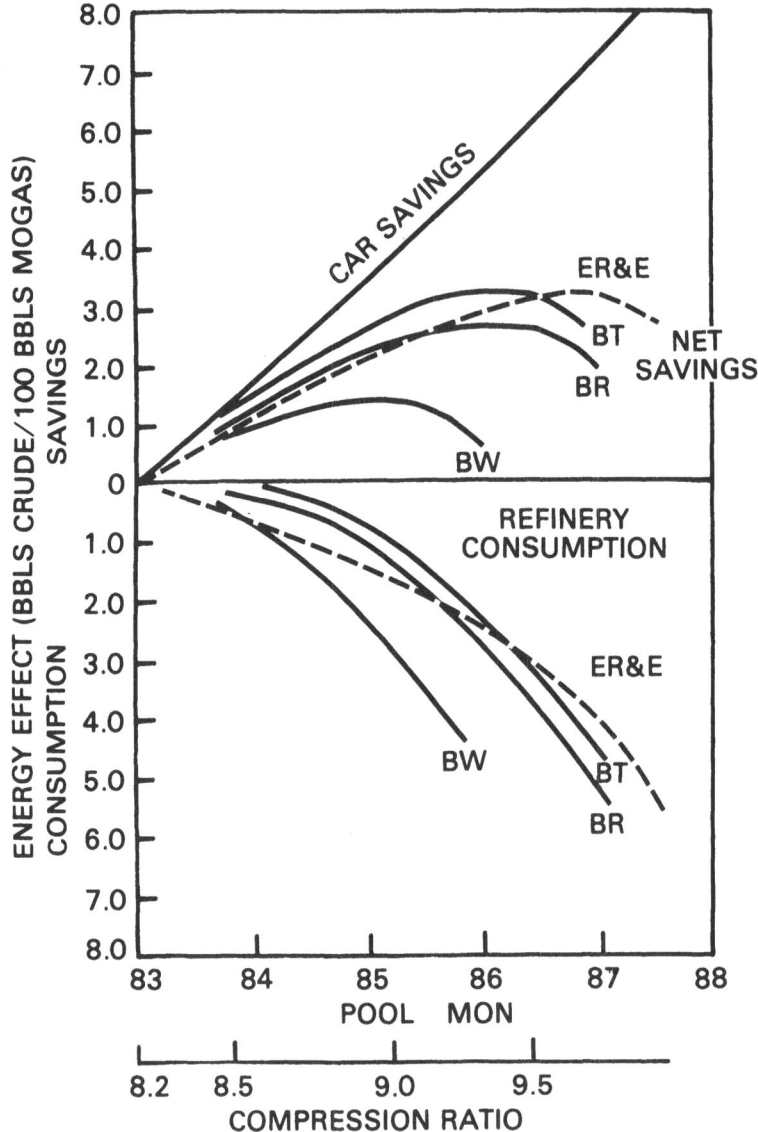

Fig. 5. Potential energy savings.

Research Octane Number by 1975 and later cars would tend to increase the realizable maximum net energy savings shown. For example, if the road octane number requirement for new cars were to include a 25% dependency on RON (rather than the zero dependency used in this study), the net energy savings predicted for Baytown in the above calculation would be increased from 3.3 to about 4 barrels of crude per 100 barrels of gasoline. On the negative side, fuel economy debits associated with the control of increased HC and NO_x emissions from automobiles at the higher compression ratios could decrease or eliminate the net energy savings.

Costs of Optimizing Octane Number — After determining the net energy savings possible with increased compression ratio, the effective cost of the equivalent crude oil saved was calculated. This was done to compare the relative attractiveness of increased compression ratio to typical crude oil prices. These costs were calculated by dividing the non-energy related refinery operating costs, including a 10 percent discounted cash flow rate of return on the capital employed, by the net crude oil savings predicted for each refinery. Expressed in terms of 1975 dollars, they were found to vary from the $18-20/BBL level for the hypothetical, Baytown, and Baton Rough refineries to about $30/BBL for Bayway. This is about 1.5 to 2.5 times the current prices of both imported and new (non-price controlled) domestic crude oil.

Summary — Increasing compression ratio and unleaded gasoline octane can improve automobile fuel economy, but there is a partially offsetting increase in refinery energy consumption. Based on this study of one hypothetical and three actual refineries, possible net crude oil savings vary between 1 and 3 percent of gasoline consumption. However, considering the high cost and low volume of crude oil generated, and the significant uncertainties involved, increasing compression ratio is not economically attractive relative to the current prices of either imported or new domestic crude oil.

ENERGY CONSERVATION BY CHANGING PROPORTIONS OF GASOLINE AND DIESEL FUEL

In July 1974, the Government Research Laboratory of Exxon Research completed a study on the changing proportions of automotive distillate and gasoline (3). The study, carried out for the Environmental Protection Agency (EPA) assumed a grass roots refinery and concerned itself with process energy conserved as increasing volumes of automotive distillate are made. Fig. 6 shows one of the plots developed; it indicates a 2% savings in crude due to reduced process energy requirements at a 1:1 ratio of auto distillate to gasoline. In highlighting this energy savings, we should not assume that auto distillate volume will grow to that of gasoline. Table 2 shows Exxon's projections of U.S. fuel demand. The distillate volumes shown here include auto diesel, residential and commercial heating oil, railroad and marine transportaation diesel, electric utility distillate fuel, and displacement of natural gas. The potential growth of auto distillate can be summarized as follows: ratio of gasoline to auto distillate is now 10:1 on a BTU basis; it could change in a future grass roots refinery to 1:1, but more realistically it could be 5 or 6:1 by 1990.

REDUCTION OF ENGINE OCTANE REQUIREMENTS

Exxon has an EPA contract to raise the compression ratio of a vehicle without
References p. 69.

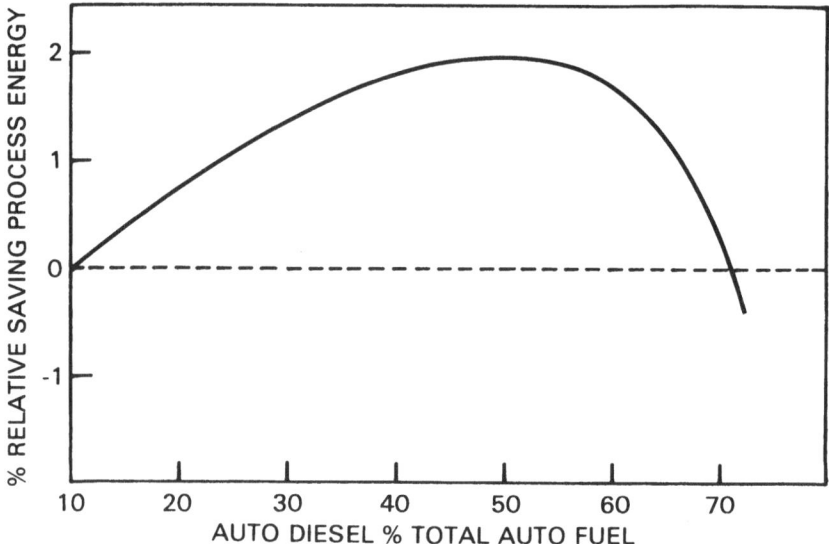

Fig. 6. Effect of automotive diesel fuel proportion on refinery process energy savings.

raising its octane requirement. We are working on three approaches: an on-board knock sensor which retards timing when knock is detected; more squish area for greater turbulence; and dual ignition to reduce the time of flame propagation.

The knock sensor coupled to a special distributor that Delco-Remy is building for us looks like the best bet. The possibility of higher emissions with higher compression ratio could rule out such an approach.

We are also looking for ways to raise vehicle satisfaction levels with respect to octane without raising octane. That is, lower the requirement or change the relative importance of Research and Motor Octane. We are looking at coolant temperature, spark timing, A/F ratio, EGR trade-offs and at transmission modifications that could reduce knock.

TABLE 2

Projected U.S. Fuel Demand

MBBL/D

	1970	1980	1990
Gasoline	5800	7200	8100
Distillate	2500	3700	4700

SUMMARY

Our study of higher unleaded gasoline octane levels and higher compression ratio indicates that octane investment is not an attractive way to decrease our foreign energy dependence. We suggest that some of the changes made as part of engine modification for exhaust emission control should be reexamined. We hope to define some of these soon. Lastly, in our view it is reasonable to project automotive distillate doubling its share of total automotive fuel by 1990.

REFERENCES

1. *E. S. Corner and A. R. Cunningham, "Value of High Octane Number Unleaded Gasolines in the U.S.," presented before the Division of Water, Air and Waste Chemistry of the American Chemical Society, Los Angeles, California, March 28-April 2, 1971.*
2. *M. M. Roench, "Thermal Efficiency and Mechanical Losses of Automotive Engines," Society of Automotive Engineers Journal, Vol. 51, 1949 pp. 17-30.*
3. *F. H. Kant, et. al., "Effects of Changing the Proportions of Automotive Distillate and Gasoline Produced by Petroleum Refining," Environmental Protection Agency Report 460/3-74-018, July 1974.*

DISCUSSION

E. E. Spitler *(Chevron Research Company)*

I've got three questions. First, you talked about customer octane requirements as opposed to the octane requirements determined by experts. I wondered, specifically, what adjustments you made?

Shannon

To give you a simple answer, about two numbers. It isn't constant across the entire band. The octane requirements tend to converge at the top end.

Spitler

It's been our observation that road octane doesn't necessarily correspond in a one-to-one relationship to lab octane, whether it's Motor octane as commonly used at the present time or some combination of Research and Motor. If you have an increase of two Motor octane numbers, you don't necessarily need an increase of two road octane numbers. Did you take that depreciation into account with higher octane non-leaded gasoline?

Shannon

Yes, within the range we looked at we considered depreciation.

Spitler

On the energy savings with diesel fuel, did that include any reduction in sulfur content of the diesel fuel?

Shannon

No, we didn't look at sulfur.

R. K. Pefley *(Santa Clara University)*

What kind of trade-offs are you using as far as emissions are concerned?

Shannon

We haven't made any yet. We're in the process of seeing whether our technique works. We haven't got the "smart" distributor hooked up to the accelerometers yet. First, we have to determine the emissions from this engine. It's a GM Chevy Nova 350 with a 1975 catalyst exhaust treatment system. We expect we can recalibrate the EGR for better NO_x control. We'll have to see how much more hydrocarbon comes out of the engine, and see whether the catalyst can handle it or not.

P. T. Vickers *(General Motors Research Laboratories)*

Could you tell us a little bit more about your knock sensor?

Shannon

The knock sensor itself is very simple, it's just an accelerometer. We talked to all of the automobile manufacturers about it, and Delco Remy was the only one who responded. They felt they could modify the current high energy ignition system. If we gave them a signal that indicated the engine was knocking, they could build in a time-delay following the pulse generated in the distributor. It wouldn't act as a proportional control, but by a step function. That is, it would go through a loop and retard the timing 3 to 4°. If the sensor still said it was knocking, it would go through another loop. We're not sure how many loops would be required. At some point in time this thing would get pretty unrealistic if you were really dramatically retarding ignition timing. It's a relatively simple thing. Delco Remy feels they're going to be able to incorporate it right into the circuitry of the current high-energy ignition system. The accelerometer really depends on whether our engineers can, in fact, sort out a frequency that really represents knock, and gate it. Historically, that hasn't been the case.

A. R. Sapre *(General Motors Research Laboratories)*

You said that the proportion of diesel fuel is likely to increase in the future. Would the relative costs of diesel fuel and gasoline change?

Shannon

We didn't really look at cost in this model, except insofar as these optimum grass roots refineries saved energy, saved investment, and so made it attractive economically as well as energetically to move in that direction. There were some small reductions in cost, maybe as much as 10%.

Bill Shapis *(Richard P. Mueller and Associates)*

I have two questions. First, what is the approximate thermal efficiency of these refineries and is there any possibility of improvement? Secondly, have you noticed any significant change in the quality or types of oils from the same areas over the last few years? Do you foresee any change in quality?

Shannon

For a high efficiency refinery you're talking about 10% of your crude disappearing into heat. A lot of work is being done on increasing refinery efficiency. We have a refinery on the west coast named Benicia. It is so efficient it can't start up by itself. It is so completely interlocked with inter-coolers, and so on, that they have to get an allocation of natural gas to start the beast. Every two years when they shut it down for turnaround they have a problem of getting natural gas to get the thing lit again. In terms of the quality of the oil, I don't know what you mean.

Shapis

I'm just wondering whether there is any change in the type of oil as you go to tertiary recovery?

Shannon

No. In our hypothetical model we looked at Arabian crude as one case and South Louisiana sweet crude as another. That would be a sour, high-sulfur crude and a low-sulfur sweet crude. While there are differences, the differences are small relative to all the other uncertainties you're dealing with. You break the molecules up so dramatically in the process of making gasoline that the characteristic of the crude gets lost in the shuffle. When you talk about heavier products, or lubes, then characteristics inherent in a crude do come through. Gasoline comes to the same molecules no matter where you start.

R. H. Perry, Jr. *(Mobil R&D)*

You showed when there's an increase in Motor octane number you have more miles of travel per barrel of crude. I have two questions. One, can you comment on the characteristics of the gasoline you were making, particularly volatility characteristics? And second, what would be the distribution of the other products from a barrel of crude when you increase the octane?

Shannon

The volatility requirements were an average of winter/summer volatility in terms of our specifications. The distribution of other products was kept constant.

R. A. Husted (*U.S. Department of Transportation*)

What kind of potential for fuel economy improvement do you see for the knock-sensitive spark advance device? Will it affect emission levels?

Shannon

Our first step is to raise compression ratio one unit. So we'd expect to see a 5% increase in fuel economy. Some limited data would seem to suggest that NO_x could go up by 20-25%. There should be a less dramatic change in hydrocarbons. We don't know if we can counter that or not. That's obviously going to be critical as to whether this is a rational approach to take or not.

AN OPPORTUNITY FOR MAXIMIZING TRANSPORTATION ENERGY CONSERVATION

E. M. JOHNSON, W. T. TIERNEY, and N. R. CRAWFORD

Texaco, Inc., Beacon, N. Y.

ABSTRACT

The best measure of energy conservation occurs when the vehicle, the fuel, and the refinery are considered as a system, and the impact of various combinations of these is evaluated in terms of miles of transportation attainable per barrel of crude. When this is done, potential savings in fuel consumption are greater than by any other approach, and warrant very serious consideration. In contrast, choice of the wrong option could seriously restrict the number of vehicles that could be produced and operated if crude oil availability should become limited.

Texaco Inc. has completed an evaluation of the optimization of vehicles, fuels and refinery processing, and its impact on energy conservation. In it, a number of engine-fuel combinations were considered. Among these were gasoline engines using leaded and unleaded fuels, diesel engines and future gas turbines. Additionally, an engine providing the fuel economy of the diesel, having no octane or cetane requirement, and operating on a 100-650°F broad boiling range fuel was included. Such an engine was represented by the direct injection stratified charge (DISC) engine.

The results clearly showed that the greatest energy penalty would occur if the manufacture of cars requiring unleaded gasoline is continued, while the DISC engine and its fuel offered the largest potential saving.

INTRODUCTION

The United States faces an increasing deficit between its demand for oil and gas and its ability to meet this demand domestically.

References pp. 90-91.

In the United States, approximately 77 percent of all energy used is derived from petroleum liquids and natural gas. Liquid petroleum accounts for 45.9 percent of U.S. energy consumption, and about 52 percent of these petroleum liquids are used for transportation. Within the transportation sector, roughly three-fourths of the fuel, or 39 percent of the total, is consumed in passenger cars and light trucks. Further, our dependence on oil as our major source of energy has grown during the past century, while our utilization of other energy resources has declined (1). This was an acceptable trend as long as we were able to produce enough crude oil domestically to satisfy our needs. Unfortunately, this capability has also changed. Not only has our domestic production ceased to increase, but it has actually turned around and is declining (1). With domestic production of about ten million barrels a day of petroleum liquids and consumption at approximately 16.5 million barrels, there is a deficit of over 6 million barrels daily which must be obtained from foreign sources. This deficit is not likely to decrease significantly in the future from the further development of our known reserves. Utilizing Alaskan North Slope oil, and emphasizing enhanced production recovery techniques will provide only a temporary respite.

There is, of course, considerable effort underway to develop alternative sources of energy. Unfortunately, it does not appear that these approaches will provide any significant degree of relief before the last part of this century. For the next two decades or more, we will be strongly dependent on oil as our major energy source. Therefore, we must not only continue to explore for additional petroleum reserves within our geographical boundaries, including the Outer Continental Shelf, but it is also vitally important that we conserve our presently known reserves.

Since passenger cars and light trucks consume such a large portion of our transportation fuel, they are prime candidates for improvement when seeking ways to conserve energy. Congress has recognized this and has mandated increased new vehicle fuel economy. Such a step may appear plausible at first glance. However, closer scrutiny discloses that it also incorporates certain pitfalls, particularly since other government regulations impose conflicting emissions, safety and damageability standards.

Many modifications are possible in automobiles to improve fuel economy. Possible contributors in the vehicle itself include weight reduction, lower driveline gear ratios, smaller engines, accessory efficiency improvement, aerodynamic drag reduction and use of radial tires. Within the engine, relaxation of unnecessarily severe emission control standards would permit modifications leading to higher power plant efficiencies. Additionally, certain alternate engine configurations offer a promise of fuel economy improvement over the current carbureted engine. Principal among these are the diesel and the direct injection stratified charge engines.

When optimally combined, the foregoing modifications can produce significant improvements in automobile fuel consumption characteristics. Unfortunately, the fuel economy improvements resulting from changes in engines and vehicles can also impose fuel characteristic requirements which generate fuel processing losses in the conversion of crude oil to the specified fuel. These losses can offset the gains made in the vehicle and produce a negative impact on overall energy conservation. It is, therefore, imperative to consider the vehicle, the fuel, and the refinery as a system in evaluating the net effect of any attempts at fuel conservation. Such an approach permits us to determine total miles of travel attainable from a barrel of crude. It is a much more valid measure of overall energy utilization than miles of travel per gallon of fuel burned.

To assess the magnitude of differences in miles of transportation per barrel of crude among various engine-fuel combinations, Texaco studied a number of engine options with significant differences in fuel consumption and impact on refinery operation. The study revealed potential savings in fuel consumption which are greater than those found by any other approach.

The Vehicle (engine)-Fuel-Refinery (VFR) systems optimization study considered representative engines, fuels and refinery cases. Each of these elements is discussed separately before interrelating to disclose the benefits to be gained from the systems approach.

ENGINES

The five engine types used in the study were:

a. Carbureted engines designed to operate on leaded gasoline (96 Research Octane Number—RON)

b. Carbureted engines designed to operate on unleaded gasoline (91 Research Octane Number)

c. Diesel engines designed to operate on 45 cetane number fuel.

d. Direct injection stratified charge or other engines capable of operation on 100-650°F boiling range fuel (without octane or cetane requirement) at a fuel economy approximately equal to the diesel.

e. A hypothetical engine with a fuel economy equal to a leaded gasoline (96 Research Octane Number) engine but capable of operation on a 100-650°F distillate fuel. A gas turbine engine of the future might, for example, typify this design.

Carbureted Engines — The basic engine configuration selected for the study was a carbureted engine typical of pre-1971 models operating on a leaded gasoline. A weighted average regular/premium octane number of 96 RON was considered satisfactory for 1971 and earlier cars based on Bureau of Mines 1972 annual data and *References pp. 90-91.*

volume production. The fuel economy of an engine using this fuel was taken as 1.0 and the fuel economies of all other engine configurations were stated relative to it.

Although it is recognized that future exhaust emission standards are uncertain, it was assumed for the study that all engine configurations would meet 1975 Federal interim standards of:

	Grams/Mile
Unburned Hydrocarbons (HC)	1.5
Carbon Monoxide (CO)	15
Oxides of Nitrogen (NO_x)	3.1

To our knowledge, there are no published data which provide direct comparisons of varying engine types which have been optimized to meet these standards. However, other investigators(2,3) have shown that compression ratio and the corresponding octane requirement can be varied within limits without seriously affecting HC or NO_x and with some change in CO. For study purposes, CO was also assumed to be negligibly affected.

Others(2) have shown the compression ratio - octane requirement noted in Fig. 1. Further, the relative fuel economies for this relationship are as indicated in Fig. 2. Although other investigators(4,5,6) have additional data on these interrelationships, it is believed that when considered on a similar relative fuel economy basis, they will not alter the significance of the findings of this paper.

The relative fuel economies established from the foregoing and used in the study are noted below for carbureted engines satisfied with the two fuels selected.

Engine Configuration	Relative Fuel Economy
96 RON Leaded	1.0
91 RON Unleaded	0.93

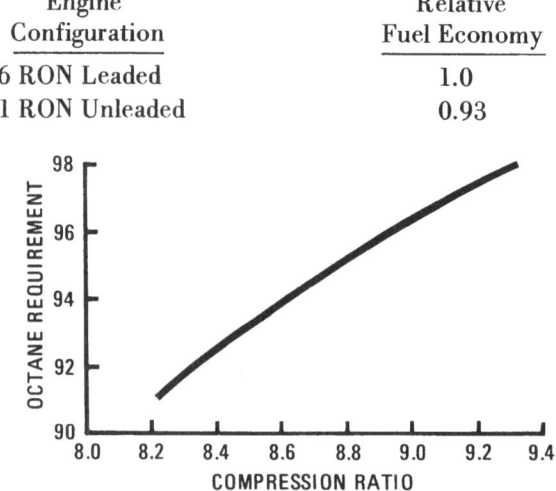

Fig. 1. Compression ratio versus octane requirement.

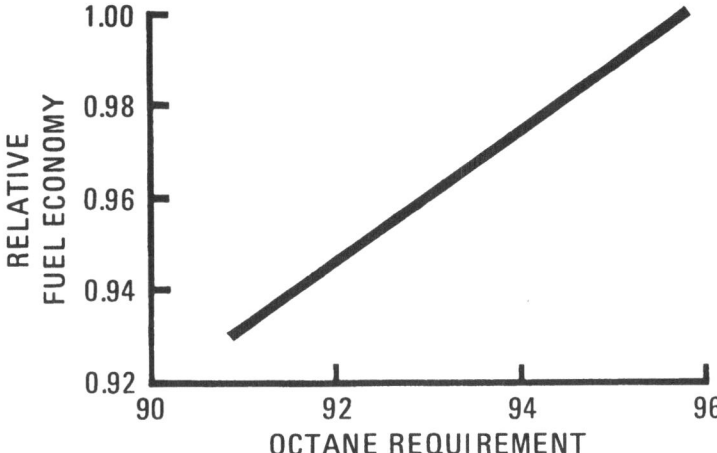

Fig. 2. Fuel economy versus octane requirement.

Diesel Engines — Readily discernible problems arise when attempting a comparison among various engine types for use as passenger car power plants. This is particularly true for comparisons between diesel and gasoline engines. Past practice of comparing these engines on an equal displacement basis would have been improper for the present study, since such a diesel would have exhibited significantly poorer performance characteristics. It was most desirable that relative fuel economies be based on engines of equal performance. Diesel-gasoline engine comparisons have been made somewhat easier by the recent trend to derated outputs for gasoline engines of a given size due to the effect of emission controls(7). This has tended to narrow the performance/weight gap historically existing between gasoline and diesel engines. Additionally, relative fuel economies for the two engine types may vary from essentially no difference at maximum output to substantial advantage for the diesel under part load, lower speed conditions. On an overall driving cycle basis, the diesel will have superior fuel economy and the question is — "By how much?"

Work published by Barnes-Moss and Scott(8) permits estimating a 30 per cent fuel economy advantage for the future passenger car diesel engine relative to a gasoline engine of essentially comparable performance and weight on the basis of a reasonable driving cycle. They, and others(9), indicate that the diesel would also be capable of meeting the exhaust emissions constraints previously noted.

Therefore, for this study, a relative fuel economy factor of 1.3 was assigned to the diesel engine.

Stratified Charge Engines — Direct injection stratified charge (DISC) engines can be designed to operate with no octane requirement and to exhibit decidedly superior fuel economy characteristics when compared to present gasoline engines. In addition, their fuel requirements are simplified by a tolerance for a wide variety of fuels

References pp. 90-91.

without serious performance degradation. For the long term view, however, an entirely satisfactory broad distillation fuel, 100-650°F, can be specified to increase refinery output of transportation fuel.

DISC engines are of similar size and weight when compared to current gasoline engines, and are of essentially the same basic structural design. Thus, their adoption would probably not require a major retooling of automotive engine manufacturing facilities, except for that required for increased production capacity of fuel injection systems.

Generally, DISC engines demonstrate a slightly lower output at maximum power due to limitations on air utilization when compared to equivalent size gasoline engines. However, with the trend to derating of the latter to control emissions and with improvement in the development of the former, this difference has diminished.

There are now several versions of the DISC engine(10,11). Typical of these is the Texaco Controlled-Combustion System (TCCS) which has been described(12). Because of our familiarity with the TCCS, it has been used as the example of a DISC engine in the study.

The performance of the DISC engine has been well documented including examples of its potential to achieve equivalent installed performance to that of the same engine operated on a carbureted basis(13).In addition, the marked fuel economy superiority of the TCCS has been confirmed on numerous occasions in both normally aspirated and turbocharged designs. This superiority has been consistently substantial, running at times to over 70 per cent compared to equivalent gasoline engines while operating under various driving cycles(13). Further, the fuel economies of the TCCS and current passenger car diesel engines are comparable when measured during the 1975 Federal Test Procedure (FTP) (14).

Based on this background, a 30 per cent average improvement over current gasoline engines was considered reasonable, and a relative fuel economy of 1.3 was assigned to DISC engines.

In addition, dynamometer stand studies have indicated the TCCS engine is capable of meeting exhaust emission standards of 1.5 HC, 15 CO and 3.1 NO_x in grams/mile without a catalytic converter or other hang-on emission control devices. Confirmation of this will require vehicle testing under the FTP conditions.

Other Engine Concepts — The impact of still another engine which might be operable on a 100-650°F broad boiling range fuel at the same fuel economy as that of leaded gasoline reciprocating engines was included in the study. An engine of this type might, for example, be represented by a future gas turbine. Its relative fuel economy was therefore assigned as 1.0, and it was assumed that compliance with emissions constraints would be achievable.

The foregoing relative fuel economies of the five engine options studies are summarized below.

Engine Option	Relative Fuel Economy
96 RON Leaded	1.0
91 RON Unleaded	0.93
Diesel	1.3
DISC	1.3
Future Concept	1.0

FUEL-REFINERY-MODELING CONSIDERATIONS

Fuel Specifications — To develop a refinery's capability for maximizing different types of fuels, it was necessary to establish modeling specifications for fuels which would satisfy the engine configurations previously described. Fuel specifications used in the various case studies are shown in Table 1. The gasoline and diesel specifications are based on U.S. Bureau of Mines data (15) for 1972. For gasoline specifications, the average of winter and summer qualities was used. For the wide boiling range fuel it may be noted that only sulfur and vapor pressure specifications had to be met.

Having selected the engine configurations and fuels to be used for the study, it was then necessary to generate refinery simulation models which would maximize the production of each fuel type and to designate a datum or base case to which the other modeling runs could be compared.

Refinery Modeling System — The year 1972 was selected as the base for model development since the statistics on refinery input and output(15,16) were readily available, the capacities(17) of each unit in each United States refinery had been published, and crude values(16) and average product prices(18,19,20) were a matter of record. Also, 1972 was the latest year for which leaded gasoline production figures were not significantly affected by low lead and unleaded gasoline manufacture and it, therefore, provided a base case for measuring differences.

A linear programming modeling system(21) was used. In addition to the computer program, the system contains an extensive library representing many crudes, models of refinery processes, and blending data for many of the physical properties of each refinery stream. Each process model is a linear representation of a refinery unit. The complete refinery model is constructed by the program in a modular fashion.

From user-supplied definitions and the library, the program will define the set of linear equations (model) representing the possible refinery input, output and refining sequences. The solution of this system of simultaneous equations is the one which maximizes profit.

References pp. 90-91.

TABLE 1

Fuel Specifications

	Leaded Premium Gasoline		Leaded Regular Gasoline		Lead Free Gasoline	
	Min.	Max.	Min.	Max.	Min.	Max.
Research Octane Number (RON)	99.8	- -	94.0	- -	91.0	- -
Motor Octane Number (MON)	92.2	- -	86.4	- -	83.0	- -
Lead, gm/gal.	- -	2.48	- -	1.96	- -	- -
Specific Gravity	--------------------------------- 0.72		0.76 ---------------------------------			
Reid Vapor Pressure, psia	--------------------------------- - -		10.7 ---------------------------------			
Percent Evap. at 160°F	--------------------------------- 24.0		40.0 ---------------------------------			
Percent Evap. at 230°F	--------------------------------- 55.0		64.0 ---------------------------------			
Percent Evap. at 360°F	--------------------------------- 90.0		100.0 ---------------------------------			

	Diesel (No. 2 Distillate) Fuel		100-650°F Fuel	
	Min.	Max.	Min.	Max.
Sulfur, wt. percent	- -	0.222	- -	0.10
Flash, °F	140	194	- -	- -
Pour Point, °F	- -	20	- -	- -
Cetane Number	45.0	- -	- -	- -
Specific Gravity	0.835	0.860	- -	- -
Reid Vapor Pressure, psia	- -	- -	- -	12.0
Percent Evap. at 400°F	2.0	10.0	- -	- -

Base Case Modeling — The refinery configuration and unit capacities used in the Base Case reflect the 1972 U.S. refinery process unit configurations and capacities as reported in the Oil and Gas Journal on April 2, 1973. The refinery charge mix and product slate reflect U.S. Bureau of Mines data(16). In essence, the Base Case refinery is a 100,000 BPCD* representation of the total U. S. refining capacity in 1972. The refining process units and sizes for the Base Case are presented in Table 2.

A considerable effort was made to have the Base Case model represent total 1972 U.S. refining capacity and configuration. When the 1972 refinery model is supplied with the average U.S. 1972 charge mix, it will produce, within close tolerances, a product slate that represents the average refined products made by all 1972 U.S. refineries and meeting 1972 product specifications defined in Table 1. The foregoing capability of the refinery model permitted it to be used as a means for establishing transportation type fuel availability for a series of alternative engines and fuels. Transportation fuels as discussed herein refer entirely to surface transportation.

Barrels per calendar day

TABLE 2

Refinery Unit Process Capacities

Unit	Capacity, BPCD
Crude Distillation	
Atmospheric	100,000
Vacuum	45,000
Gasoline Hydrotreater	19,149
Catalytic Reformer	19,007
Fluid Catalytic Cracker	31,636
Alkylation	5,200
Coker	2,250
Hydrocracker	5,700
Distillate Hydrotreater	3,993
Cryogenic Hydrogen Purifier (MSCF/D)	11,700
Sulfur Recovery (T/D)	24

CASE STUDIES AND RESULTS

Four case studies were made to determine the maximum quantity of various transportation fuels that could be produced. These included:

A. Production of fuels meeting 1972 specifications.

B. Production of 91 RON lead free gasoline.

C. Production of diesel fuel oil.

D. Production of a 100-650°F EP fuel with no octane or cetane requirement.

Certain freedoms and constraints were imposed on the modeling runs. Crude input was maintained constant at 100,000 BPCD for all cases. Expansion of catalytic reforming capacity was permitted in order to maximize lead free fuel availability. Further, it was recognized that certain other product markets had to be maintained. Therefore, petrochemical feed stocks, special naphthas, residential heating oil, aviation jet fuel, diesel fuel, lube oil, wax, and asphalt yields were protected to the same degree in each case.

Table 3 is a summary of the modeling results for each case studied showing the yield distributions for the Base Case and other alternatives for a refinery processing a nominal 100,000 BPCD of 1972 average quality crude oil.

References pp. 90-91.

TABLE 3

Refinery Charging 100,000 BPCD of Crude Oil
Maximum Transportation Fuel Production

Case	A	B	C	D
Identification	Base Case	91 RON (83 MON)	Maximum Diesel	Maximum 100-650°F
Yields - BPCD				
A. Refinery Fuel:				
1. Fuel Gas	1,461	1,649	1,172	975
2. Propane - Propylene	191	2,420	88	- -
3. No. 6 Fuel Oil	4,526	3,568	4,150	4,005
4. FC Coke (400 lbs./Bbl.)	1,575	1,652	1,094	1,041
Sub-Total	7,753	9,289	6,504	6,021
B. Transport Fuels:				
5. Premium Gasoline	20,910	- -	18,389	- -
6. Regular Gasoline	33,944	- -	29,853	- -
7. Lead Free Gasoline	- -	52,759	- -	- -
8. Diesel Fuel	10,131	10,131	17,349	10,131
9. 100-650°F EP Fuel	- -	- -	- -	56,609
Sub-Total	64,985	62,890	65,591	66,740
C. Protected Products:				
10. Avjet	9,100	9,100	9,100	9,100
11. Home Heating Oil	13,651	13,651	13,651	13,651
12. Petrochemicals & Special Naphthas	3,400	3,400	3,400	3,400
13. Lube Oils & Wax	1,800	1,800	1,800	1,800
14. Asphalt & Road Oils	3,300	3,300	3,300	3,300
Sub Total	31,251	31,251	31,251	31,251
D. Other Fuels (Coke, LPG, #4, 5 & 6 Fuel Oils)	8,392	8,561	8,378	8,388
TOTAL	112,381	111,991	111,724	112,400

Base Case (Case A) — The Base Case represents a 100,000 barrel "slice" out of the total U.S. refining industry charges, yields and throughputs for the year 1972. Leaded gasolines and diesel fuel oil were produced for transportation uses. Gasoline production was 54,854 BPCD, and diesel fuel production was 10,131 BPCD, which is a total of 64,985 BPCD of fuels for transportation. In addition, production was protected for a number of products throughout the study as noted in Table 3. The Base Case will be used to make yield comparisons for the other three runs representing the different fuel types.

The energy required to run the refinery in the Base Case was 50.2 billion BTU/day or 502,000 BTU per barrel of crude. This represents the requirement for operating the Base Case units only. In contrast, Bureau of Mines data indicate a total refinery fuel requirement for all purposes in 1972 of 679,800 BTU per barrel of crude. The composition of the refinery fuel that was used to supply this energy is shown in Table

3. Similar information is also given in Table 3 for the other cases considered.

91 RON Unleaded Gasoline (Case B) — The 91 RON unleaded gasoline run was made to produce the maximum possible amount of this type of fuel.

To accomplish this objective, it was necessary to install 20.8 percent more catalytic reforming capacity capable of sustained operation at a 100 RON severity level than used in the Base Case, and to provide the associated hydrotreating capacity. The additional feed stock required for the increased catalytic reforming capacity was a 145-200°F fraction of the straight run naphthas, and the hexanes and heptanes obtained from natural gasolines. In addition, the charge rates to the Coking Unit and the Fluid Catalytic Cracking Unit were increased. Some additional distillate hydrotreating was needed to meet sulfur specifications due to a shift in stocks used for cracking and middle distillate blending in order to maximize unleaded gasoline production.

The net result of these changes was the production of 52,759 BPCD of 91 RON unleaded gasoline, which is 2,095 BPCD less than than gasoline production in the Base Case. Diesel fuel production was held constant in both cases. This comparison clearly demonstrates the production penalties involved in producing unleaded gasoline. In addition, the engine operates more inefficiently on the lower octane number unleaded gasoline, as previously noted.

The energy required to run the refinery to produce 91 RON unleaded gasoline (Case B) was 57.5 billion BTU/day or 575,000 BTU per barrel of crude.

Maximum Diesel Fuel (Case C) — Maximum diesel fuel was produced by routing distillates from the crude still directly to middle distillate blending, rather than to the fluid catalytic cracking unit for gasoline manufacture. The conversion on the fluid catalytic cracking unit was reduced from 67.1 percent in the Base Case to 58.7 percent in this case, in order to increase the production of Light Fluid Cracked Cycle Gas Oil, which is suitable for middle distillate manufacture.

The computer program used to make these studies was based on perfect fractionation, and it would have produced 50.8 percent of total middle distillates unless some restraint was used. To allow for the less than perfect fractionation encountered in a typical refinery, the total middle distillate production was more realistically limited to 40 percent of the crude oil.

From Table 3, it may be noted that it is not possible to manufacture diesel fuel to the complete exclusion of gasoline. In this case, leaded gasolines were made to the same specifications as those of the Base Case.

The change in gasoline and diesel fuel production compared with the Base Case is shown on the next page.

References pp. 90-91.

Case	Gasoline - BPCD	Diesel Fuel - BPCD
Base Case	54,854	10,131
Case C	48,242	17,349
Difference	-6,612	+7,218

Diesel fuel production is increased appreciably in this case compared with the Base Case. Considering the increases in the miles of transportation that can be obtained with a diesel engine, this case will generate more miles of transportation than either of the gasoline cases considered up to this point.

Less energy is required to operate the refinery than in the Base Case, as shown below.

	BTU/Bbl. Crude
Base Case	502,000
Case C	433,000

Maximum 100-650°F Fuel (Case D) — The 100-650°F boiling range fuel that can be used to operate a DISC engine does not require octane or cetane specifications for satisfactory engine performance. To maximize the production of this type of fuel, the charge rate to the conventional catalytic reforming units was reduced to that required to produce the hydrogen needed for hydrotreating and hydrocracking operations. (11,962 BPCD versus 19,007 BPCD in the Base Case). In addition, severity was lowered to 86.2 RON in the catalytic reformer unit.

It may not be desirable, or economical, in the future to charge DISC engine fuel components to the catalytic reforming units to produce hydrogen. Alternate methods for hydrogen production could be used. Future refinery construction studies would have to compare the cost of hydrogen manufacture by a number of alternative processes to obtain the most economical source.

To further maximize the production of this fuel, the charge rate to the fluid catalytic cracking unit was reduced from 31,636 BPCD in the Base Case to 24,544 BPCD. This reduction was accomplished by routing distillate streams from the crude still directly to the blend of 100-650°F boiling range fuel.

The total production of transportation fuel in this case, compared with the Base Case is shown below.

	Base Case	Case D
Gasoline, BPCD	54,854	—
Diesel Fuel, BPCD	10,131	10,131
100-650°F Fuel, BPCD	—	56,609
Total	64,985	66,740

In this case 66,740 BPCD of transportation fuels are produced, which when used in a DISC engine will yield the maximum miles of transportation per barrel of crude oil, as will be shown in the following section of this paper.

Less energy is required to operate the refinery to produce a 100-650°F fuel because the charge rate and severity of the catalytic reforming and fluid catalytic cracking units have been reduced. The resulting savings in energy are appreciable, as noted in the following comparison with the Base Case.

	Refinery Fuel BTU/Bbl. Crude
Base Case	502,000
Case D	376,000

VFR SYSTEM FUEL CONSERVATION IMPACT

Fuel supplied for cars of the 1980's will likely continue to be based almost entirely on petroleum. Conservation of this resource is, therefore, an absolute requirement. The choice of engine type, its fuel, and the refining required to process the fuel have a dominant influence on petroleum resources and on the imports required to supplement them. The extent of the differences that can be encountered in these interrelationships is presented in Table 4. It shows the total amount of crude oil* that must be refined to achieve a fixed amount of vehicular transportation. Deficiencies noted are amounts which would have to be made up from imports, while a surplus indicates a decreased dependence on imports.

Case A in the table is the Base Case representing historical data for the year 1972. A total of 12.56 million barrels per day of crude oil were actually processed(16) to produce enough leaded gasoline and diesel fuel to obtain an estimated 4.33 billion miles of transportation per day, based upon an average fuel economy of 12.07 miles per gallon(22) for all gasoline powered vehicles on the road in 1972. Calculations used to develop the transportation mileage are presented in Table 5. This 4.33 billion miles was maintained as a standard in subsequent cases against which the crude oil requirements for various engine-fuel combinations could be compared. It is obvious that the crude requirements differed substantially as the engines and type of fuel were varied.

The maximum amount of crude oil was required to permit 4.33 billion miles of transportation a day when unleaded gasoline (Case B) production was maximized. This is the course being followed in the United States today. Automobiles are equipped with catalytic converters to meet emission standards and the petroleum industry is required to produce 91 RON unleaded gasoline to protect the catalyst. Compared to the Base Case, not only was there a loss from 345 to 316 in miles of transportation attainable per barrel of crude, but crude oil requirement also increased to 13.70 from 12.56 million barrels per day.

*Crude oil as used here refers to crude oil plus natural gas liquids.

References pp. 90-91.

TABLE 4

Interrelationships of the VFR System

Case	Billion Miles of Transportation per day	Miles of Transportation per barrel of crude oil	Crude Oil - Million BPCD		
			Required	Produced in U. S.	Deficiency or (Surplus)
A (Leaded)	4.33	345	12.56	11.22	1.34
B (91 RON)	4.33	316	13.70	11.22	2.48
C (Diesel)	4.33	358	12.09	11.22	0.87
D (DISC)	4.33	440	9.84	11.22	(1.38)
E (Turbine)	4.33	357	12.13	11.22	0.91

TABLE 5

Calculations of
Transportation Mileage and Crude Requirement

1972 Leaded Gasoline Base Case

1. 54,854 BPD gasol.   x 42 gal./bbl x 12.07 mpg x 1.0 rel.   FE = 27.808 x 10^6 mi/day

    10,131 BPD diesel   x 42 gal./bbl x 12.07 mpg x 1.3 rel.   FE =   6.677 x 10^6 mi/day

    Σ = 34.484 x 10^6 mi/day

2. 34.484 x 10^6 mi/day $\div$ 100 M BPD crude = 344.8 mi/bbl. transportation

    $\approx$ 345 mi/bbl.

3. 345 mi/bbl x 12.56 MM BPD crude processed in U.S. in 1972 =

    = 4.33 x 10^9 mi transportation/day in U.S.

Other Cases

    Use 91 RON Unleaded, for example:

1. 52,759 BPD$_{91\ RON}$ x 42 gal./bbl x 12.07 mpg x 0.93 rel.   FE = 24.873 x 10^6 mi/day

    10,131 BPD$_{diesel}$ x 42 gal./bbl x 12.07 mpg x 1.3   rel.   FE = 6.676 x 10^6 mi/day

    Σ = 31.55 x 10^6 mi/day

2. 31.55 x 10^6 mi/day $\div$ 100 MBPD crude = 315.5 mi/bbl. transportation

    $\approx$ 316 mi/bbl.

3. To obtain same total miles transportation/day in U.S. as for base case (4.33 x 10^9), requires:

    4.33 x 10^9 mi/day $\div$ 316 mi/bbl. = 13.70 x 10^6 BPD crude

Additionally, this study permitted using only processes available in the Base Case. It is recognized that by modifying the existing refinery through other process approaches, the gasoline yield will be affected, and some effort was devoted to reviewing this possibility. One such case considered the potential for 94 RON unleaded gasoline. In it, a simplified approach was taken whereby all straight run components were reformed at a very high severity level to gain some insight into the miles of transportation per barrel of crude attainable from engine modifications requiring this type fuel. Even by using marginal charge stocks and running reformers to an unrealistic 103 RON constant severity level, or by using alternate refining processing, only a very small improvement over the 91 RON lead free case resulted. The efficiency improvements realized from engine modifications were largely negated by increased processing energy requirements. We are confident, therefore, that the route of increasing unleaded octane numbers does not offer any potential for improving transportation efficiency. Thus, the conclusions drawn in this paper with respect to other engine types such as the diesel and DISC engines remain valid.

An increase in transportation capability per barrel of crude oil can be obtained by producing maximum diesel fuel with the gasoline fraction supplied as leaded instead of unleaded gasolines. This is demonstrated in Case C which shows a potential of 358 miles per barrel of crude oil and a crude oil requirement of 12.09 million barrels per day to maintain a total transportation capability of 4.33×10^9 miles per day.

It is interesting to note that the assumed gas turbine engine of the future (Case E) provides an improvement in transportation miles essentially equal to that attainable from the diesel (Case C). However, the fuel for Case E is the same 100-650°F broad range product used in the DISC engine.

The best transportation capability, occurs in Case D, which uses a DISC engine and a fuel with a boiling range between 100°F and 650°F. It produces 440 miles of transportation per barrel of crude oil and requires less crude oil per day than any of the other cases.

The extent of the foregoing differences is more vividly apparent when displayed graphically as in Fig. 3. It illustrates that an engine such as the DISC, with its superior fuel economy and ability to operate on a broad boiling range fuel, has the potential for achieving 35 percent more miles of transportation per barrel or crude than engines requiring 91 RON unleaded gasoline. Further, the DISC type approach was the only one in the study which alleviated the problem of transportation dependence on foreign crude. In contrast, maximizing the diesel and future gas turbine approaches provided intermediate levels of improvement in miles of transportation attainable per barrel of crude, but a deficiency in crude requirements still remained.

In addition to the preceding factors, there are certain other important considerations associated with the VFR systems approach which should not be overlooked or dismissed lightly.

References pp. 90-91.

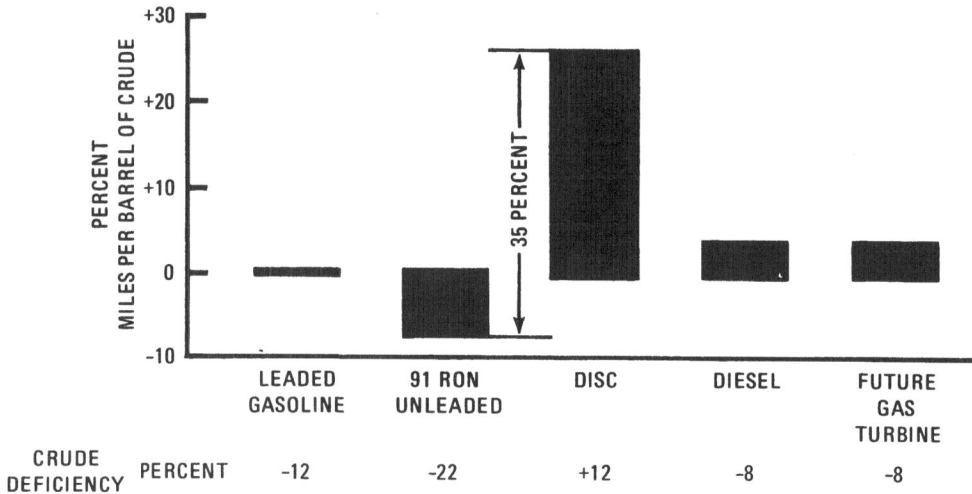

Fig. 3 VFR effects.

National preparedness for possible future limitations on crude oil supplies and energy conservation can be associated. Assume, for example, that if future crude availability were to be limited to crude produced in the United States in 1972 (11.22 million BPCD), transportation mileage would have to be reduced about 18 percent under Case B (91 RON unleaded gasoline), wheras under Case D (the DISC engine), attainable transportation mileage could increase about 14 percent. The net difference between these two, 32 percent, represents the achievable improvement in transportation mileage over current engine-fuel practices. These gains have been conservatively estimated from 1972 data. Added to them would be other fuel saving benefits to be derived from non-engine design changes which would increase the total even further. Stated simply, another advantage for DISC type engines is that more automobiles could be fueled and more miles driven, if crude supplies should be limited in the future.

Consideration of the nation's economy, too, should be noted. The maximum deficiency among all the cases studied occurs in Case B, 91 RON unleaded gasoline, and amounts to 2.48 million barrels a day or almost one half of total imports in 1975. If the imported crude costs $10.00 per barrel, this 2.48 million barrels per day represents about $9.0 billion a year, which is an adverse balance of payments that could be avoided by using a DISC engine and its wide boiling range fuel.

Attention should also be called to another nationwide advantage of the DISC engine approach. Over half the refineries in the United States are small (less than 50,000 BPCD) in terms of their throughput capacity. For these refineries, radical fuel requirement changes could be economically intolerable and could force a decrease in their output or even a closure in some cases. This would be a situation the nation could ill afford. Future adoption of DISC engines and a broad boiling range fuel would permit even the small refiners to move more rapidly in this direction, and at less expense than toward other more demanding fuel process requirements.

In summary, satisfaction of our future transportation energy requirements by an integrated Vehicle, Fuel, Refinery approach would be beneficial to energy producers, vehicle manufacturers, consumers and the government, and would provide for a greater degree of national energy self-sufficiency. In view of this, and the magnitude of the potential savings involved, we believe the VFR approach merits very serious consideration.

FUTURE REFINERIES

One of the constraints imposed upon this study required that refineries be representative of current configurations. Construction of new refineries could be quite different from these if a 100-650°F fraction were the fuel of the future. Processes such as Catalytic Reforming, Fluid Catalytic Cracking and Alkylation conceivably could be eliminated since there would be no octane or cetane specifications. Hydrogen, though, would likely be required for desulfurization. If by-product hydrogen were not available from a Catalytic Reformer, it would have to be produced by an alternate process. The economics of destroying transportation fuel on Catalytic Reforming Units to produce hydrogen, versus alternate methods, would be determined by the future course of prices.

In view of the many uncertainties in projecting future refinery construction, no attempt was made in this study to model one designed specifically to maximize the production of 100-650°F fuel. It is recognized, however, that this might be a viable subject for a future study.

CONCLUSIONS

Based upon the study discussed herein, the following conclusions are drawn:

1. Maximum transportation per barrel of crude oil is required to support the U.S. economy under conditions of declining crude supplies and increasing costs of crude oil.

2. A direct injection stratified charge engine, and its 100-650°F boiling range fuel with no octane or cetane requirement, provide the maximum transportation in terms of miles per barrel of crude oil and minimum refinery energy losses for the cases studied.

3. An internal combustion engine that requires unleaded gasoline provides the least transportation per barrel of crude oil.

4. Imported crude oil requirements can be reduced appreciably by adoption of direct injection stratified charge engines as motor vehicle power plants.

References pp. 90-91.

ACKNOWLEDGEMENT

The authors express their appreciation to Messrs. J. F. Hertweck, R. F. Huhndorff, W. D. Leary, R. J. Pecora and R. M. Reuter, and to Dr. W. F. Brown for their valuable assistance in the analyses on which this paper was based.

Acknowledgement is also made of the contribution of the late Henry D. Moorer who provided considerable early guidance in the study.

REFERENCES

1. *"A National Plan for Energy Research, Development and Demonstration: Creating Energy Choices for the Future,"* United States Energy Research and Development Administration, Volume 1, ERDA-48, June 28, 1975.

3. P. E. Oberdorfer, *"Compression Ratio, Emissions, Octanes and Fuel Economy – Experimental Study,"* Paper No. 60-72, API Division of Refining, 37th Midyear Meeting, May 11, 1972.

3. A. E. Felt and S. R. Krause, Ethyl Corporation, *"Effect of Compression Ratio Changes on Exhaust Emissions,"* Society of Automotive Engineers National Combined Fuels and Lubricants, Powerplant and Truck Meeting, October 26-29, 1971.

4. Clayton LaPointe, Ford Motor Company, *"Factors Affecting Vehicle Fuel Economy,"* Paper No. 730791, SAE Combined National Farm, Construction, Industrial Machinery and Fuels and Lubricants Meeting and Manufacturing Forum, Milwaukee, Wisconsin, September 11, 1973.

5. L. E. Furlong, E. L. Holt and L. S. Bernstein, Esso Research and Engineering, *"Emission Control and Fuel Economy,"* American Chemical Society, Los Angeles, California, April 1, 1974.

6. *"Gasoline Saved Equals Crude Conserved,"* Air Conservation Notes Number 75-1, Ethyl Corporation, January 1975.

7. *"Statement of Rudolf Uhlenhaut, Director, Passenger Car Development-Daimler-Benz Aktiengesellschaft,"* Before the Panel on Environmental Science and Technology Subcommittee on Air and Water Pollution – U. S. Senate Committee on Public Works, March 14, 1972.

8. H. W. Barnes-Moss and W. M. Scott, Ricardo & Co., Engineers, Inc., *"The High Speed Diesel Engine for Passenger Cars,"* The Institution of Mechanical Engineers Conference on Power Plants and Future Fuels, London, England, January 21-22, 1975.

9. *"Emissions from Diesel and Stratified Charge Power Cars,"* EPA-460/3-75-001-a, United States Environmental Protection Agency, December 1974.

10. Gerhard Finsterwalder, *"The New Deutz Multi-Fuel System,"* Paper No. 720103, Automotive Engineering Congress, Detroit, Michigan, January 10-14, 1972.

11. H. Hagan, P. Kuhlmann and A. Urlaub, *"M.A.N. Activities in Power Plants for the Future,"* Paper No. c36/75, Conference on Power Plants and Future Fuels, The Institute of Mechanical Engineers, London, England, January 21-22, 1975.

12. W. T. Tierney, E. Mitchell and M. Alperstein, Texaco, Inc., *"The Texaco Controlled-Combustion System – A Stratified Charge Engine Concept – Review and Current Status,"* Paper No. C1-75, The Institution of Mechanical Engineers Power Plants and Future Fuels Conference, London, England, January 21-22, 1975.

13. W. T. Tierney, E. M. Johnson and N. R. Crawford, Texaco, Inc., *"Energy Conservation – Optimization of the Vehicle-Fuel-Refinery System,"* SAE Paper No. 750673, SAE Fuels and Lubricants Meeting, Houston, Texas, June 3-5, 1975.

14. Karl J. Springer, Southwest Research Institute, and Ralph C. Stahman, Environmental Protection Agency, "Emissions and Economy of Four Diesel Cars," SAE Paper No. 750332, 1975 Automotive Engineering Congress & Exposition, Detroit, Michigan, February 24-28, 1975.

15. Mineral Industry Surveys, United States Department of Interior, Bureau of Mines, Petroleum Product Surveys, No. 75, No. 76, No. 78 and No. 79.

16. Mineral Industry Surveys, United States Department of Interior, Bureau of Mines, Petroleum Statement, Annual Crude Petroleum, Petroleum Products and Natural Gas Liquids: 1972.

17. Oil and Gas Journal, Annual Refining Issue, April 2, 1973.

18. Oil and Gas Journal, December 25, 1972.

19. National Petroleum News Factbook Issue, Mid-May, 1973.

20. Minerals Yearbook, Vol. 1, United States Department of Interior, Bureau of Mines, 1972.

21. "Refinery and Petrochemical Modeling System," Software Systems, Bonner and Moore, Houston, Texas.

22. 1973-74 Automobile Facts and Figures — Motor Vehicle Manufacturers Association of the United States, Inc., Detroit, Michigan.

DISCUSSION

R. M. Campau (Ford Motor Company)

I have a couple of questions with regard to the direct injection stratified charge engine. Could you comment on the feed gas to a catalytic converter or whatever you put downstream of the engine? Will it control emissions to the 1.5 HC, 15 CO, and 3.1 NO_x g/mile standards? The second question is, have you made any projections in this study regarding the investment required for production of such an engine?

Johnson

The first one is easiest to answer. We have not optimized an engine in a vehicle for the 1.5, 15, 3.1 standards. But based on our dynamometer tests we believe that 1.5, 15, 3.1 is achievable without any hang-on catalyst. This is probably about the maximum you could go. We haven't estimated tooling costs for such an engine configuration. We've made an assumption that this could be done reasonably because, with the DISC engine, all that is involved is a change in the cylinder head and the piston, as well as the adoption of a diesel type fuel injection system.

J. B. Bidwell (General Motors Research Laboratories)

It seems to me when you get all done, what you're really pleading for is improved engine efficiency, regardless of what fuel is used. In other words, if I could build an engine that used unleaded fuel that had the same efficiency as the DISC engine, and I used the same fuel product mix you've got now, the improvement could be almost as great. There might be some slight differences because of losses in refinery energy. But by far the biggest improvements are simply due to the increased engine efficiency. Is that correct?

Johnson

Yes, I'm glad you brought that up because I did forget the point I wanted to make there. The 3800 barrels per day difference in production of transportation fuels for 100,000 barrels of crude charged is less than 4%, but if you convert that to equivalent mileage it turns out to be over 300,000,000 miles per day of transportation. When you start thinking about it in those terms, it sounds a lot bigger.

Bidwell

The other point is that your production of diesel fuel in your maximum diesel case is considerably less than Hugh Shannon's. I presume that's due to the refining constraints you had, and that had you assumed a new refinery, you'd have ended up with a considerably larger improvement for the diesel fuel case. Is that right?

Johnson

We could have increased it if we took off some of the constraints. We felt we might have been able to move up to another 7,000 barrels per day under certain circumstances. So, I'd say that, probably, our differences are due to different approaches. One thing we did was to put a limitation on our atmospheric stills at 40% fractionation, where theoretically you can go up to, I think, 52%. But as far as I know, nobody operates their refinery up at the theoretically ideal fractionation limit.

J. B. Heywood *(Massachussetts Institute of Technology)*

I have a comment about the diesel case along the lines of Mr. Bidwell's question. Isn't there a question as to the amount of diesel fuel you have produced? If you look at the long-term time-frame that we're talking about, is that really a realistic estimate? Let's assume we've gone to 100% production of wide-range distillate, 1/3 diesel, 2/3 gasoline. We're talking about a 10 to 20 year time-frame. So I think your comparison of the maximum diesel and wide-range distillate cases, in terms of vehicle miles traveled, is not realistic.

Johnson

Are you saying you think we could increase the diesel production still further to come up to the equivalent to the 100-650 production? Is that basically what you're saying?

Heywood

I'm suggesting that in the long-term it will come a lot closer than your numbers indicate.

Johnson

I'm not sure that you can because, remember, your diesel fuel distillation range is quite a bit different than your 100-650. Your 100-650 includes most of the gasoline fraction. Whenever you try to make diesel fuel, you're also going to make a certain amount of gasoline, and if you can't burn that gasoline in diesel fuel, you've got to use it somewhere else.

L. A. McReynolds *(Phillips Petroleum Company)*

You show 440 miles per barrel of crude oil for the DISC engine, and 357 for the turbine engine. You used the assumption that the turbine engine could use a broad range fuel. Therefore the difference must be in the efficiency of the engines. Do you think that the assumed differences in efficiency are realistic?

Johnson

This case which you're mentioning is a hypothetical engine which might be representative of an engine of the future. As we were working on this study we mentioned it to different people and a major automotive company said it would be very nice if you would include in your study a case for a hypothetical engine whose relative fuel economy is the same as the base case, and see what would happen to miles traveled per barrel of crude. So we included it as a possibility.

W. E. MacDonald *(Marathon Oil Company)*

When you were comparing leaded gas and unleaded, did you credit the unleaded with the use of a catalytic converter and better tuning in the engine to get better fuel economy?

Johnson

No. We didn't take into consideration the tuning that you mentioned with the catalytic converter. We're talking about the basic engine itself, which should be able to meet the 1.5, 15, 3.1 g/mile emission standards.

J. P. Longwell *(MIT and Exxon Company)*

I judge that you didn't include aviation jet fuel as a transportation fuel, and that's just a matter of bookkeeping. But the point I want to make is that most projections I've seen show that aviation fuel requirements are growing more rapidly than ground transportation fuel requirements in terms of mid-distillate. They require a lot of paraffins. If we load up the mid-distillate area by a lot more diesel, or fuels for stratified charge engines, you're going to be stressing the availability of paraffinic fuels for aviation. That would impose a constraint which would lead to hydrogenation-produced, low-aromatic fuels which cut down the efficiency of refineries and tend to put us back where we were.

Johnson

The jet fuels were protected, that was one of the constraints. If your markets grow unevenly, or out of the proportions that we used in the study, then you would have to rebalance and see how much of each fuel could be made. I'm sorry I didn't make it clear, but what we refer to as transportation fuels are only ground transportation fuels.

R. W. Burgett *(AC Division, General Motors Corporation)*

I have a question regarding your relative fuel economy. You listed the relative fuel economy as 1.3 for your DISC engine. Being from a spark plug company, we were very interested in the spark plugs it requires. We duplicated the spark plugs and the repetitive-spark ignition system in a conventional engine and got the same improvement in fuel economy. I wonder if the improvement in fuel economy with the direct injection, stratified charge engine is due to the stratified charge or to the two-inch long spark plugs sparking for a prolonged period?

Johnson

Do I understand correctly that by modifying the spark plugs and putting them into a standard auto cycle engine you got the same improvement in fuel economy?

Burgett

No. What it amounts to is, we took an engine off the road and put it into a dynamometer lab. We located the spark gap geometrically in the center of the combustion chamber and with repetitive sparks we can attain this same improvement in fuel economy with a conventional cycle engine that we can with direct injection stratified charge.

Johnson

O.K. The next question then I guess would be, would that engine also use a broad-boiling range fuel?

J. L. Beckham *(Cadillac Motor Division, General Motors Corporation)*

I'm going to go back to your 0.93 number for the 91 octane which you said was derived from a non-catalyst vehicle. I assume that this study was done in 1974. I wonder if you had any feel for what that number would be now that we get much better fuel economy with a catalytic converter vehicle.

Johnson

At the time we started this study back in 1974, the relative fuel economies selected appeared to be reasonable. I don't think we're really trying to sell anything, except one thing. We'd like to get people started thinking in terms of maximizing miles of transportation per barrel of crude. I think anybody who wants to should take their own relative fuel economies, set up a similar program, and see where they come out. I do think that the numbers may not be exactly the same. I'm sure they won't be. But I think, on a relative basis, you will generally find that as you move toward a broader boiling-range fuel, you will tend to move in the direction of more miles per barrel of crude.

J. M. Colucci *(General Motors Research Laboratories)*

Now that we know that the emissions standards which you have used in your study are not going to be permitted in the future, do you have any plans to repeat the study with emission standards that we will be expected to meet in the future? What do you think the results will be?

Johnson

In setting this study up, there were a number of constraints that we had to establish and one of them was the emission standards. Perhaps we should go more severe, and this may change the fuel economy relationships. If you go severe enough you're going to wipe out some of the alternatives. We have no plans to look at this, but we have thought that we might want to look at the possibilities of going to a grass-roots refinery instead of assuming existing refinery configurations. In answer to your question, I have to say no, we don't have any plans to study tighter emission standards.

MATCHING FUTURE AUTOMOTIVE FUELS AND ENGINES FOR OPTIMUM ENERGY EFFICIENCY

R. F. STEBAR, W. A. DANIEL, A. R. SAPRE and B. D. PETERS

General Motors Research Laboratories, Warren, Michigan

ABSTRACT

In the future, if fuels from different sources (oil shale, coal, nuclear, biomass) are to be used in automobiles powered by current and alternative engines, we must be concerned with minimizing the overall energy expenditure for converting primary energy sources to power at the wheels. It is recognized, however, that many other factors, including environmental concerns, economics, and politics will have strong, and perhaps overriding influences on the final decision about which fuel-engine combinations should be and will be utilized.

This paper is concerned with only those issues which are involved in optimizing the overall energy utilization efficiency of automobiles. In turn, this efficiency depends on four intermediate efficiencies:

1. Resource acquisition efficiency,

2. Resource conversion efficiency (into various fuels),

3. Fuel distribution efficiency (from producer to consumer), and

4. Fuel-utilization efficiency (by the automobile).

Only the second and fourth factors were investigated because they play the major roles in determining the overall energy utilization efficiency. The efficiencies associated with resource acquisition and fuel distribution are generally high compared with the other efficiencies.

The conversion efficiencies for obtaining potential automotive fuels (mainly various liquid hydrocarbons and alcohols) from the primary energy sources were estimated from a critical review of the literature. The compatibility of each of these fuels with current and alternative automotive powerplants was examined with emphasis on fuel combustion quality (octane or cetane number), volatility, and impurities (sulfur, nitrogen, and ash). Powerplants considered include: conventional spark-ignition and diesel engines; stratified-charge engines; and steam, gas turbine, and Stirling engines.

Estimates of the fuel-utilization efficiency of these engines should be based on vehicle data for comparably sized vehicles having the same exhaust emission levels and having similar performance. These data are not available for most of the alternative automotive powerplants. Therefore the concept of optimum energy utilization is illustrated with an example using data from conventional spark-ignition and diesel engines. Based on this comparison, the overall energy utilization efficiency appears highest with fuels from petroleum, followed by fuels from oil shale, coal, biomass, and nuclear.

To apply this approach to all alternative fuels and engines in order to firmly conclude which combinations are the most energy efficient, improvements in the data bases are required. The efficiencies of converting energy resources to fuels must be established with greater certainty than heretofore possible, the properties of the fuels obtained must be determined, and the fuel-utilization efficiency of automobiles powered with various fuel-engine combinations must be measured. These last measurements must be made with the fuels and engines in automobiles of comparable size, performance, and exhaust emissions.

In addition, future studies are justified for those fuel-engine combinations currently deemed incompatible, but with potential for greater energy utilization efficiency than compatible combinations.

INTRODUCTION

The desire for domestic energy self-sufficiency could, at some time in the future, dictate that fuels from nonpetroleum sources also be used to power automotive engines. If energy is to be conserved in that situation, the following questions arise. What fuels can be produced efficiently? From what resources? What types of engines should be used?

To obtain meaningful answers to such questions requires assessment of the energy efficiency of the overall fuel production and utilization system, from the raw material in the ground to conversion of fuel energy into mechanical work. Such a treatment should include all potential fuels, all major sources of those fuels, and actual testing of potential fuels in viable automotive engines. Fuel and engine combinations which provide high energy efficiency in fuel production and utilization should be singled out for more concentrated development efforts, to achieve the most efficient use of energy resources by automobiles.

Of course energy efficiency is not the only criterion for selecting future fuel-engine combinations. Environmental constraints, political factors, and economic considerations all play a part in the selection. The environmental and economic considerations

References pp. 114-116.

are extremely complex, and the political factors are unpredictable. Yet these issues could well be dominant factors in the final selection of future fuel-engine combinations.

This paper concentrates on the rationale for assessing the energy efficiency of potential fuel and engine combinations, and does not address the environmental, political, and economic aspects.

METHODOLOGY OF ASSIGNING ENERGY EFFICIENCIES

Regardless of what the energy source might be, several technical steps are involved in utilizing an automotive fuel from a natural resource. The steps include:

1. Acquisition of the raw material from the source (by mining, drilling, harvesting, etc.);

2. Conversion of the raw material into fuels;

3. Distribution of fuels to the customer; and

4. The utilization of the fuel in the private passenger car.

Some of these steps are more energy efficient than others and the overall energy efficiency, which is the product of the efficiencies of all of the steps, could vary depending on the source of the fuel, the specific fuel produced, and the type of automotive engine involved. To minimize energy use by automobiles, the efficiency of the overall process, from acquisition to end use, must be maximized.

To simplify our assessment of potential systems, two of the four steps in the overall fuel production and utilization process were ignored, namely, resource acquisition and fuel distribution. These steps involve mostly mechanical operations rather than chemical upgrading and are generally considered to have high efficiencies compared with the others (1,2). The efficiencies of the two remaining steps, fuel conversion and fuel utilization, were multiplied to yield "combined energy efficiency," as shown by the following expression:

$$\text{Combined Energy Efficiency} = \text{Fuel-Conversion Efficiency}$$
$$\text{x Fuel-Utilization Efficiency}$$

Obviously, combined energy efficiency is not as comprehensive a term as overall efficiency. Nevertheless, it provides a means of identifying noteworthy future fuel-engine combinations.

The remainder of this paper describes our assessment of potential fuel and engine systems. Attention is directed to the following topics:

1. Energy resources and potential fuels;

2. Conversion efficiencies for obtaining various fuels from the different primary

energy sources;

3. Fuel properties and fuel-engine compatibility;

4. The efficiency with which various fuels can be used by cars (fuel-utilization efficiency) equipped with different types of engines and operated in a meaningful manner; and

5. Combined energy efficiency (which provides the basis for identifying energy efficient fuel-engine combinations for the future).

ENERGY RESOURCES AND POTENTIAL FUELS

The primary energy resources believed to be sufficiently abundant to serve as future sources of automotive fuels are listed in Table 1. Of these resources, the petroleum, oil shale, coal, biomass, and nuclear sources are considered the most likely for providing liquid fuels for automobiles (3,4). Natural gas is excluded because of limited supply. Geothermal, hydro, solar, tidal, and wind sources would most likely provide electricity and therefore could be treated in much the same way as a nuclear source.

TABLE I

Energy Resources

Fossil	Nonfossil
Petroleum	Biomass
Natural Gas	Nuclear
Coal	Geothermal
Oil Shale	Hydro
	Solar
	Tidal
	Wind

A variety of fuels can be obtained from each of the different energy resources via adequate processing. For example, some of the combustibles that can be derived from petroleum, oil shale, and coal sources are shown in Fig. 1. However, the automotive fuels most likely to be produced from primary energy resources are those listed in Table 2 (3,4). In considering automotive fuels from potential sources, all gaseous fuels such as hydrogen, light hydrocarbons, and ammonia were excluded due to their low volumetric energy density and the consequent problem of on-board storage (3,4). Similarly, hydronitrogen fuels were omitted from serious contention since they present additional problems related to toxicity and safety (3,4).

References pp. 114-116.

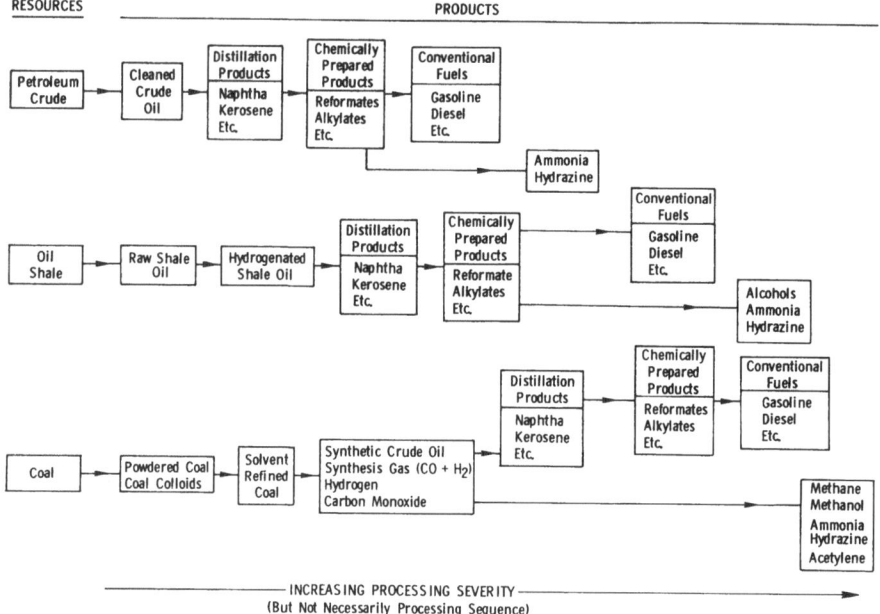

Fig. 1. Combustibles obtainable from fossil energy resources.

TABLE 2

Likely Automotive Fuels from Various
Primary Energy Resources

Energy Resource	Fuels
Petroleum	Cleaned crude oil,* distillate fuel,** gasoline, and diesel fuel
Oil Shale	Hydrogenated shale oil, distillate fuel, gasoline, and diesel fuel
Coal	Powdered coal, pulverized solvent-refined coal, distillate fuel, gasoline, diesel fuel, methanol, Fischer-Tropsch liquid hydrocarbons
Biomass	Ethanol, liquid hydrocarbons
Nuclear Sources	Methanol, Fischer-Tropsch liquid hydrocarbons
All Sources	Electricity

* *Water, solid materials, and gases removed.*

** *In this paper, the term distillate fuel will be used to describe a fuel obtained via distillation but with no specific octane or cetane rating.*

As indicated in Fig. 1 and Table 2, gasoline and diesel fuels are obtainable from several energy sources. However, more extensive processing is required to produce these specialized fuels from oil shale or coal than from petroleum, and therefore greater energy expenditure is required. Consequently, fuel-conversion efficiency is a critical parameter in identifying energy efficient fuel-engine combinations.

FUEL-CONVERSION EFFICIENCY

Definition of Conversion Efficiency — The amount of energy consumed in producing a particular fuel varies depending on the specific fuel and the source. Fuel-conversion efficiency indicates the magnitude of that energy loss. Throughout this paper, fuel-conversion efficiency is defined as:

$$e = \frac{H_p}{H_m} \times 100$$

where:

e = conversion efficiency, in percent;

H_p = total heating value of net "useful"* fuel products;

H_m = total heating value of primary source material used in producing the fuel products, plus the energy used in processing and in maintaining auxiliary facilities required to generate hydrogen, oxygen, electricity, etc., as needed.

In this definition, the fuel producing facility is assumed to be self-contained, i.e., all energy input to the facility is in the form of a primary source used for fuel production. Thus, a fuel that can be obtained with a minimal energy expenditure has a high energy efficiency, and the use of such a fuel would offer an opportunity to increase the combined energy efficiency of the fuel-vehicle system.

Qualifications — Estimating conversion efficiencies for different fuels would be relatively straightforward for cases where process technology is well established and where the following information is available for a large scale fuel producing facility:

1. The amount of source material required to make fuel products.

2. The quantities of useful fuel products manufactured.

3. The amount of source material required to provide energy for processing and auxiliary functions.

However, in the United States no large scale facilities exist for making automotive fuels from nonpetroleum sources. Even the petroleum facilities are not self-contained,

---

* For example, hydrogen sulfide and ammonia are not considered as "useful" fuel products even though they possess heating value.

References pp. 114-116.

i.e., crude oil is not the only energy input to the facility. Thus, even in the case of established technology, available information cannot be used directly for calculating the conversion efficiencies of petroleum-derived fuels.

The situation is even more complex in the case of newly-developing fuel conversion technologies. The available information usually applies to laboratory scale experiments or, at best, to small scale pilot plant operations. Data pertaining to the amounts of primary source material used and the quantities of useful fuel products manufactured are available only for a few generally less-refined fuels such as hydrogenated shale oil from oil shale or for synthetic oils from coal. The data do not contain energy requirements for processing and for maintaining auxiliary functions. In addition, the data are not based on self-contained facilities and are generally applicable to only a specific set of process conditions which haven't been optimized. Often, several alternative schemes for producing a given fuel from a given source are still undergoing development.

Thus, the fuel-conversion efficiencies cited in this paper are estimates based both on available information and our engineering judgment. The judgment is required when considering source-related, process-related, and product-related effects on fuel-conversion efficiency. Among the source-related factors, hydrogen and impurity (S, N, O, and H_2O) contents can have large influences. High hydrogen content and low impurity content favor high conversion efficiency. Among the process-related factors, processing severity is the single most important item. Increasing processing severity tends to decrease conversion efficiency. In turn, increased processing severity can yield more refined fuels. Of the product-related factors, stringent specifications for fuel properties and impurities content can reduce conversion efficiency.

In arriving at the conversion efficiencies for the various fuels from a potential resource, it was assumed that only that resource was available for fuel production, and that only one fuel was produced. It is likely, however, that production of more than one fuel from a given resource using a given technology could change the fuel-conversion efficiency.

Results — Conversion efficiencies estimated for the potential automotive fuels most likely to be produced from petroleum, oil shale, coal, biomass, and nuclear sources are shown in Fig. 2. The average conversion efficiency for each fuel is listed on the figure beside the shaded bar for that fuel. The shaded bar indicates the range or degree of uncertainty associated with the average value. Future developments could narrow the ranges shown, or even place some fuels in higher relative positions on this chart. Each of the potential fuels in this figure, except methanol and ethanol, represents a general fuel type rather than a specific fuel with rigid properties.

Fuels from Petroleum — Four types of fuels from petroleum were considered. As can be seen from Fig. 2, the most efficient fuel would be cleaned crude oil; i.e., crude oil with water, solid materials, and gases removed. The conversion efficiency for cleaned crude oil could be 98 percent (5) depending on the amount of cleaning

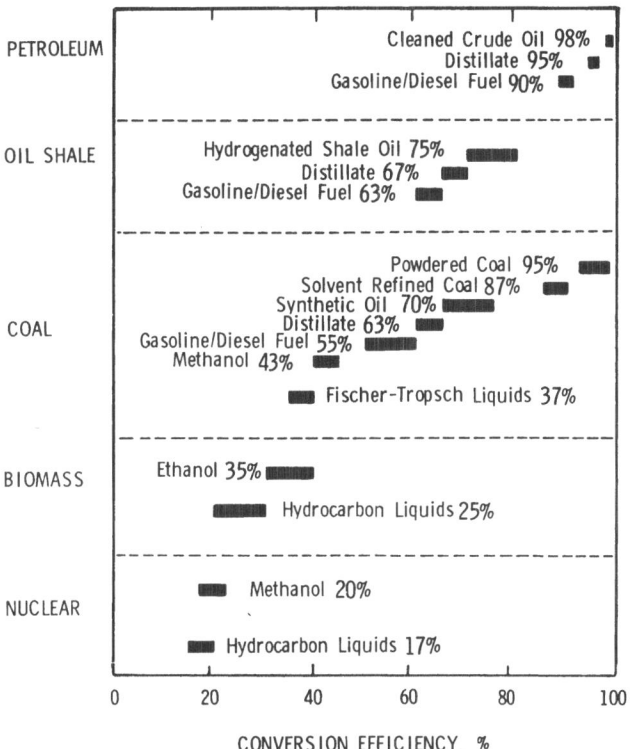

Fig. 2. Fuel conversion efficiencies.

required. The next most efficient fuel would be distillate, which we have defined as a fuel with no specific octane or cetane rating but derived via primary distillation. The conversion efficiency for obtaining this type of fuel from petroleum is estimated to be about 95 percent (5,6). For diesel fuel and/or gasoline, the estimated conversion efficiency is approximately 90 percent (5,6,7). We recognize that the efficiencies for producing diesel fuel and gasoline are not identical. However, in the context of the wide range of fuel-conversion efficiencies shown in Fig. 2, this difference is not significant.

The exact conversion efficiencies for distillate, diesel fuel, and gasoline would depend on a number of factors such as the quality of the petroleum crude, the proportion of automotive fuels in the total output of the refinery, and the specific fuel properties desired (8).

Fuels from Oil Shale — As previously mentioned, oil shale, like petroleum, may yield primarily four types of fuels: hydrogenated shale oil, distillate, diesel fuel, and gasoline. The conversion efficiency is highest, about 75 percent, for hydrogenated shale oil (9,10) while it is the lowest for diesel fuel and gasoline, approximately 63 percent (9-15). A distillate fuel can be obtained from oil shale with an efficiency of

References pp. 114-116.

about 67 percent (9-12). More energy is required to obtain highly refined fuels like gasoline and diesel fuel from the shale oil than from petroleum because shale oil generally contains more high molecular weight paraffinic compounds and substantially more nitrogen than does petroleum. These characteristics of shale oil could require more extensive and complex refining to obtain fuels with stringent property specifications.

Conversion efficiencies for the potential fuels from oil shale will vary somewhat depending on the process used in extracting shale oil from oil shale and also on the grade of oil shale.

Fuels from Coal — Powdered coal would have a high conversion efficiency, approximately 95 percent (16,17), depending on the particle size desired, and the final moisture and ash contents. Pulverized solvent-refined coal is obtained with a slightly lower conversion efficiency, around 87 percent (18,19). However, it contains significantly less ash and sulfur than powdered coal (19). Liquid fuels may be obtained via either hydrogenation or gasification.

Among the liquid fuels obtained via hydrogenation, synthetic oils (such as SYNTHOIL) appear to have the highest conversion efficiency. This type of low-sulfur, liquid fuel with practically no ash may be obtained with an approximate efficiency of 70 percent (9,20-29). Additional processing of these synthetic oils, with a consequent loss in conversion efficiency, would yield a distillate fuel with low levels of sulfur and nitrogen. Finally, fuels similar to diesel fuel or gasoline can be obtained with a conversion efficiency of about 55 percent (20-29).

Compared to the hydrogenation route, production of liquid fuels from coal-derived synthesis gas requires more process energy (30-32). Two kinds of liquid fuels can be obtained: methanol, and paraffinic and olefinic liquid hydrocarbons (30-32) (via the Fischer-Tropsch process). For methanol, the efficiency estimate is 43 percent (9,20); whereas, for Fischer-Tropsch liquids, the estimate is 37 percent (9,20,30-32).

More energy is required to obtain the same types of fuels from coal than from petroleum or oil shale. This larger energy requirement occurs because coal contains significantly less hydrogen than either petroleum or kerogen (the oil yielding component of oil shale) (19), and it is chemically more complex. Additionally, coal generally has much higher ash, moisture, sulfur, and nitrogen contents than petroleum (19,33,34).

Conversion efficiencies for fuels from coal will vary significantly depending on the rank,* moisture content and composition of coal. They would also depend on the conversion process used and the specific fuel properties required. In most cases, technology for obtaining the various fuels from coal, as from oil shale, is still in the development stage. Hence it will be some time before these fuels could be made available on a large scale.

---

* *Rank of coal is related to the percentage of fixed carbon and heat content. There are four main ranks of coal: 1) anthracite, 2) bituminous, 3) subbituminous, and 4) lignite.*

Fuels from Biomass — Primarily two types of fuels can be obtained from biomass: alcohols (mainly ethanol) and hydrocarbon liquids (similar to distillate fuel). The conversion efficiency is estimated to be about 35 percent for ethanol (33,35) and 25 percent for the hydrocarbon liquids (36-38).

Liquid Fuels from Nuclear Energy — With a water-cooled, nuclear-fission reactor, either heat or electricity could be used to obtain hydrogen from water. The hydrogen can be combined with oxides of carbon, obtained via decomposition of limestone, to form either methanol or the Fischer-Tropsch liquid hydrocarbons. The conversion efficiency of obtaining nuclearly generated hydrogen is 25 to 30 percent (39,40). We estimated that the conversion efficiencies for the overall processes in obtaining methanol and the Fischer-Tropsch liquids, starting with a nuclear-fission reactor, would be around 20 and 17 percent, respectively (39-41). We did not consider either the breeder or the fusion reactor because of their uncertain future.

Electricity from Primary Sources — An alternative way of using primary sources of energy to power automobiles is to generate electricity for use in battery-electric vehicles. Energy resources, like petroleum, oil shale, coal and biomass require some processing before they can be used for electricity generation. Hence, the efficiency of generating electricity from energy resources is the product of fuel-conversion efficiency and powerplant efficiency. We assumed this latter efficiency to be 40 percent for nonnuclear powerplants (42,43). The estimated generation efficiencies for producing electricity from the five potential sources of automotive fuels are shown in Fig. 3.

Cleaned crude oil, hydrogenated shale oil, and powdered coal would be satisfactory utility powerplant fuels from petroleum, oil shale, and coal, respectively. Accordingly, multiplying the conversion efficiencies of these fuels by 40 percent yields

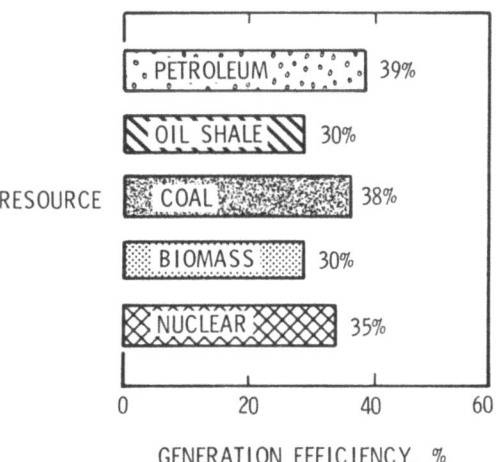

Fig. 3. Estimated efficiencies for generating electricity from various energy resources.

References pp. 114-116.

electricity-generation efficiencies of 39, 30, and 38 percent, respectively.

Biomass would have to be dried, cleaned and pulverized before use. We assumed that this usable form of biomass may be obtained with 75 percent conversion efficiency (44); thus, the efficiency for generating electricity from biomass would be 30 percent (44,45).

Utilizing a nuclear reactor as an energy source, electricity-generation efficiency and powerplant efficiency are one and the same. In general, a nuclear powerplant has a lower efficiency than a modern fossil-fueled steam plant. Consequently, we assumed that electricity from a water-cooled nuclear reactor would be obtained at 35 percent efficiency (42,43).

FUEL PROPERTIES AND FUEL-ENGINE COMPATIBILITY

While fuel-conversion efficiency is important, it is only one of the many criteria for selecting potential future fuels. Physical and chemical properties of fuels are also critical because they control compatibility with the engines that will be available for use. The engines of tomorrow will likely evolve from those undergoing substantial research and development today. These include the diesel; the conventional spark ignition engine; various stratified-charge, spark ignition engines; the gas turbine; the steam engine; the Stirling engine; and battery-electric systems. Some indication of the fuel properties requirements of these alternative powerplants may be obtained from developmental experience with petroleum-derived fuels.

The fuel properties important for satisfactory operation of engines are: combustion characteristics, volatility, impurity concentrations, viscosity, volumetric energy density, and safety of handling. The combustion characteristics of a fuel are especially important in intermittent combustion engines, since they determine a fuel's resistance to knock (octane quality) or tendency to ignite quickly after being compressed (cetane quality). The volatility of a fuel must be matched with the combustion system employed. Impurities such as sulfur, nitrogen, and ash are undesirable. High sulfur and nitrogen contents can lead to sulfur dioxide, sulfate, and nitrogen oxide exhaust emission problems. Also, high sulfur and ash content could cause engine durability problems. Removing these impurities generally requires additional energy expenditures, and may force alternative fuels to be much like current fuels derived from petroleum. Viscosity and volumetric energy density may be less critical from the standpoint of achieving satisfactory engine operation, but could alter the design of the hardware needed for storing, pumping, and metering the fuel.

Establishing the properties of potential alternative fuels is considerably more difficult than defining fuel properties requirements for future engines. Most alternative fuels have not been available in sufficient quantities for laboratory and engine testing. Consequently, a precise appraisal of the physical and chemical properties of alternative fuels has not been possible. Nevertheless, meaningful predictions of fuel properties can be made in advance via information in the literature (3,4,19,23,32,46-51) and by examining both the nature of the primary energy resources (15,19,25,34,52) and the characteristics of the fuel conversion processes involved (3,9,12,15,19,30,53).

Fuel properties predictions for the alternative fuels in Fig. 2 are presented in Table 3. In this table, "high" and "low," as applied to the individual fuel properties, are relative terms consistent with current practice when referring to petroleum-derived fuels. The yes or no classifications listed represent gross oversimplifications of relative fuel properties. It is expected that, in the near future the different fuels will be available in quantities sufficient for definitive laboratory testing leading to more precise characterization.

The octane and cetane quality designations for the oil shale and coal-derived distillates in particular may require a brief explanation. If distillate from coal is highly aromatic in nature (19,23,34), it would be expected to have high octane quality and low cetane quality. Similarly, if oil-shale-derived distillate is highly paraffinic in nature (12,47,49,50), it would likely be high in cetane quality, and low in octane quality. Consequently the designations are only approximations of actual octane and cetane values that fuels of these types will have. The specific chemical composition of the final individual fuels will determine the actual octane and cetane qualities.

The impurities designations also deserve comment. The less-refined fuels from coal and oil shale are likely to have high nitrogen and/or sulfur contents (12,19,49,50,53,54). The coal-derived fuels, such as solvent-refined coal and synthetic oils, may also contain appreciable amounts of ash (18,19,53).

Matching fuel properties and engine requirements is obviously necessary in a meaningful assessment of future fuel-engine combinations. Our view of the compatibility of petroleum-derived fuels with the various alternative engines is illustrated in Table 4. An "X" beside a given engine type designates possible compatibility with the fuel listed above. One must recognize, of course, that compatibility is not necessarily a yes/no condition. Often, there are trade-offs in important areas such as emissions, durability, reliability, and customer acceptability. More data are needed to accurately define these trade-offs.

As indicated in Table 4, the diesel engine requires a fuel with adequate cetane quality. Likewise, the conventional spark ignition engine requires gasoline, because of octane and volatility considerations. Some stratified-charge engines have a fairly wide fuel tolerance with regard to octane quality, volatility, and viscosity. The gas turbine, the steam engine, and the Stirling are probably even more fuel tolerant, because combustion quality is somewhat less critical than with piston engines. With battery-electric systems, almost any fuel could be used at the powerplant. Although not shown in this table, alcohol, obtainable primarily from coal or biomass, is probably compatible with all the engines listed except the diesel.

References pp. 114-116.

TABLE 3

Properties of Potential Automotive Fuels

Fuel Sources and Potential Fuels	High* Octane Quality	High Cetane Quality	High Volatility	S	Low* N	Ash	Low Viscosity	High Volumetric Heating Value
Petroleum								
Crude Oil	No	No	No	No	Yes	Yes	No	Yes
Distillate	No	No	No	Yes	Yes	Yes	Yes	Yes
Coal								
Powdered Coal	?	?	No	No	No	No	(Solid)	No
Pulverized Solvent-Refined Coal	?	?	No	No	No	No	(Solid)	No
Synthetic Oil	No	No	No	No	No	Yes	No	Yes
Distillate	Yes	No	No	No	No	Yes	Yes	Yes
Diesel Fuel	No	Yes	No	Yes	Yes	Yes	Yes	Yes
Gasoline	Yes	No	Yes	Yes	Yes	Yes	Yes	Yes
Methanol	Yes	No	Yes	Yes	Yes	Yes	Yes	No
Fischer-Tropsch Liquids	No	Yes	Yes	Yes	Yes	Yes	Yes	Yes
Oil Shale								
Hydrogenated Shale Oil	No	No	No	Yes	No	Yes	No	Yes
Distillate	No	Yes	No	Yes	Yes	Yes	Yes	Yes
Diesel Fuel	No	Yes	No	Yes	Yes	Yes	Yes	Yes
Gasoline	Yes	No	Yes	Yes	Yes	Yes	Yes	Yes
Biomass								
Ethanol	Yes	No	Yes	Yes	Yes	Yes	Yes	No
Hydrocarbon Liquids	No	Yes	Yes	Yes	Yes	Yes	Yes	Yes
Nuclear Reactor								
Methanol	Yes	No	Yes	Yes	Yes	Yes	Yes	No
Hydrocarbon Liquids	No	Yes	Yes	Yes	Yes	Yes	Yes	Yes

* The terms *High* and *Low* are relative to and consistent with current practice when referring to petroleum-derived gasoline and diesel fuels.

TABLE 4

Fuel-Engine Compatibility
Petroleum Resource

| Engine | Distillate | Fuel | | Electricity |
		Diesel	Gasoline	
Diesel		X		
Conventional Spark Ignition			X	
Stratified-Charge Spark Ignition	X	X	X	
Gas Turbine	X	X	X	
Steam	X	X	X	
Stirling	X	X	X	
Battery-Electric				X

Table 4 deals only with fuel-engine compatibility of petroleum-derived fuels. A similar assessment is also required for fuels from oil shale, coal, biomass and nuclear sources. But whereas not enough is known about fuel requirements of alternative engines, even less is known about the properties of the potential fuels from the different resources. Consequently, a complete analysis of fuel-engine compatibility cannot be made until the nonpetroleum-based alternative fuels become available for testing. Needless to say, fuel-engine compatibility is essential to determinations of fuel-utilization efficiency and combined energy efficiency of potential fuel-engine combinations.

FUEL-UTILIZATION EFFICIENCY IN AUTOMOBILES

In our approach to examining the efficiency of energy use by automobiles, we defined fuel-utilization efficiency as the energy required to overcome the inertia forces and the dynamic drag of a vehicle performing representative driving maneuvers, divided by the fuel energy consumed by the vehicle while performing those maneuvers. (Including vehicle inertia in this calculation factors out the effect of differences in vehicle weight.)

Fuel-utilization efficiency is affected by several important factors listed in Table 5. Vehicle performance, exhaust emissions levels, and vehicle driving schedule are considered primary constraints. Some engines providing high performance will suffer a loss in efficiency when operated at light load. Emission constraints not only affect the efficiency of many engines, but very strict standards (such as the 0.4 gram per mile NO_x requirement) may eliminate some engines, such as the diesel. Vehicle driving schedule comes into play because both the steady-state fuel consumption of a given engine-vehicle combination and the relative fuel consumption for different types of engines can vary appreciably with the type of driving (55,56,57).

References pp. 114-116.

Consequently, fuel-utilization efficiency must be determined, not for ideal steady-state conditions, but for the way engines will be operated in representative driving.

TABLE 5

Factors Affecting
Fuel-Utilization Efficiency

Primary Constraints	Secondary Constraints
Vehicle Performance	Durability
Exhaust Emission Levels	Reliability
Vehicle Driving Schedule	Cost
	Customer Acceptability

Durability, reliability, cost, and customer acceptability are regarded as secondary constraints which also may involve trade-offs with efficiency. These three, in addition to the primary constraints, should be considered to obtain a valid comparison of different fuel-engine combinations.

Fuel-utilization efficiencies for two familiar engines, the conventional spark ignition engine and the diesel, are shown in Table 6. Details of the complete fuel-utilization efficiency calculation procedure are given in Appendix A. Both engines were operated on their respective petroleum-based fuels, and both were reasonably well matched with regard to the primary constraints. As can be seen from Table 6: the 0-96.6 km/hr acceleration times were similar; exhaust emissions were not identical, but were reasonably close. Both types of engines were tested on the EPA composite urban-highway driving schedules. These matched vehicles resulted in fuel-utilization efficiencies of 17 percent for the diesel and 15 percent for the gasoline engine. This constitutes a 13 percent advantage for the diesel relative to the conventional spark ignition engine. (The above efficiency values were determined on a BTU/km basis, taking into account differences in fuel heating value.)

Obviously, automobile fuel-utilization efficiencies should be determined for all the potential fuel and engine combinations, including each of the alternative engines. Unfortunately this is not possible at present, because as stated previously most of the alternative fuels of interest are not available, and because vehicle data satisfying the constraints listed in Table 5 are not available for all the different types of engines. But such information will ultimately have to be obtained in order to arrive at a meaningful appraisal of future fuels and engines.

COMBINED ENERGY EFFICIENCY

As defined earlier, combined energy efficiency is the product of fuel-conversion

TABLE 6

Characteristics of Vehicles Used*

Performance	Conventional Spark Ignition Engine	Diesel Engine
0-96.6 km/hr, sec	18.7	19.4
Exhaust Emissions		
HC, g/km	0.25	0.43
CO, g/km	2.4	1.4
NO_x, g/km	1.2	1.4
Fuel-Utilization Efficiency, %		
EPA Composite Urban-Highway Driving Cycles	15	17

Values in this table were obtained by averaging data from two cars.

efficiency and fuel-utilization efficiency, and provides a meaningful basis for comparing potential fuel-engine combinations from an energy conservation standpoint. Fuel-engine combinations that exhibit a high combined energy efficiency represent strong candidates for automotive application in the future, provided they are acceptable environmentally, politically, and economically.

Combined energy efficiency values for the conventional spark ignition and diesel engines operated with their respective petroleum-based fuels are shown in Fig. 4. Applying the fuel-conversion efficiency of 90 percent (from Fig. 2) for diesel fuels to the fuel-utilization efficiency of 17 percent for the diesel engine (from Table 6) resulted in a combined energy efficiency of 15 percent. Similar treatment of the spark ignition engine operating with gasoline gave a combined energy efficiency of 13 percent. Since a 90 percent fuel-conversion efficiency was used in both cases, the diesel retains its advantage relative to the gasoline engine on the combined energy efficiency basis.

Assuming that the ratio of the thermal efficiencies of diesel and spark ignition engines operated with conventional fuels from petroleum would be the same when they are operated with fuels from alternative sources, the combined energy efficiencies would decrease for the alternative fuels in the manner shown in Fig. 5. The combined energy efficiency would be less for oil shale-derived fuels than for petroleum fuels and would be even lower for coal-derived fuels. This trend results directly from changes in fuel-conversion efficiencies for the different fuel sources. However, since fuel-conversion efficiencies are about the same for gasoline and diesel fuels from a given source, the relative ranking of the two engine types remains about the same for all sources.

References pp. 114-116.

ENGINE TYPE AND FUEL-UTILIZATION EFFICIENCY	FUEL TYPE AND FUEL-CONVERSION EFFICIENCY			
	DISTILLATE 95%	DIESEL 90%	GASOLINE 90%	ELECTRICITY 39%
Diesel (17%)		**15**		
Conventional Spark-Ignition (15%)			**13**	
Stratified Charge Spark-Ignition				

Fig. 4. Combined energy efficiencies for petroleum-derived fuels.

ENGINE TYPE AND FUEL-UTILIZATION EFFICIENCY	PETROLEUM				OIL SHALE				COAL							BIOMASS		NUCLEAR		ALL
	Crude 98%	Distillate 95%	Diesel 90%	Gasoline 90%	Hydrogenated 75%	Distillate 67%	Diesel 63%	Gasoline 63%	Powdered 95%	Solvent Refined 87%	Synthetic Oils 70%	Distillate 63%	Diesel 55%	Gasoline 55%	Alcohol 43%	Alcohol 35%	Liquid HC 25%	Alcohol 20%	Liquid HC 17%	Electricity 30-39%
Diesel 17%			**15**				**11**						**9**							
Conventional Spark-Ignition 15%				**13**				**10**						**8**						
Stratified-Charge Spark-Ignition ?																				
Gas Turbine ?																				
Steam ?																				
Stirling ?																				
Battery-Electric ?																				

*FUEL - CONVERSION EFFICIENCY

Fig. 5. Combined energy efficiencies.

The many voids in Fig. 5 illustrate the vast amount of work still to be done. Combined energy efficiency values are needed for all the potential engines operated on all likely (and compatible) fuels from the different sources. Once that missing information has been provided, one will be able to identify the most efficient fuel and fuel source for each engine type, and to identify the most efficient engine for any given fuel or fuel source.

SUMMARY

In the interest of energy self-sufficiency and conservation of energy resources, a rationale was developed to aid in identifying fuel-engine combinations that might minimize energy consumption by automobiles in the future. Combined energy efficiency was chosen as the basis for appraising potential fuel-engine combinations. Combined energy efficiency is a composite of the efficiency of producing an alternative fuel from an energy source and the efficiency with which that fuel can be used in automobiles powered by alternative engines.

The fuels likely to be produced from available energy resources were identified as follows: from petroleum - cleaned crude oil, distillate fuel, and gasoline and diesel fuels; from oil shale - hydrogenated shale oil, distillate fuel, and gasoline and diesel fuels; from coal - powdered coal, pulverized solvent-refined coal, synthetic oil, distillate fuel, gasoline and diesel fuels, methanol, and Fischer-Tropsch liquids (paraffinic and olefinic hydrocarbons); from biomass - ethanol and hydrocarbon liquids; from nuclear sources - methanol and hydrocarbon liquids. Electricity can also be produced from each of the five primary energy sources.

Efficiencies of producing the above fuels were estimated utilizing information in the literature and the authors' engineering judgment. Depending on the specific fuel produced, the fuel-conversion efficiency estimates ranged from 90-98 percent for petroleum-derived fuels, 63-75 percent for oil shale-derived fuels, 37-95 percent for coal-derived fuels, 25-35 percent for fuels from biomass, and 17-20 percent for liquid fuels from nuclear sources. Electricity, depending on the source, can probably be produced at 30-39 percent efficiency.

Compatibility of alternative fuels and engines was acknowledged as an important criterion for assessing candidate fuel. Since most alternative fuels have not been available for testing, their general properties were forecast from knowledge of the nature of the fuel source and the characteristics of the fuel-conversion processes used. The need for comparing alternative fuel-engine combinations in actual vehicles was also acknowledged. It is imperative that the test vehicles be driven in a realistic manner and have similar performance and emissions characteristics.

Because of unavailability of alternative fuels and the lack of appropriate data for vehicles with alternative engines, meaningful fuel-utilization and combined energy efficiency comparisons were restricted to conventional fuels and diesel and spark ignition engines. The combined energy efficiencies for those engines operated on their petroleum-based fuels were 15 percent and 13 percent, respectively. Decreases in combined energy efficiency were predicted for the two engines in cases where gasoline and diesel fuels were obtained from sources other than petroleum, such as oil shale and coal. However, the relative ranking of the two engines remained unchanged, regardless of the fuel source.

Much work remains to be done by both the energy and the automotive industries in providing the various alternative fuels from the different resources, in defining fuel properties of alternative fuels, in determining compatibility with the various potential engines, and in obtaining valid vehicle evaluations of the various compatible fuels and engines. Such information is essential if the most energy-efficient fuel-engine

References pp. 114-116.

combination for the future is to be identified. Of course, the ultimate selection of fuel-engine combinations for the future will be strongly influenced by environmental, political, and economic factors in addition to those regarding overall energy efficiency.

ACKNOWLEDGMENT

We wish to thank M. B. Young for calculating driving cycle energy requirements, J. M. Colucci and N. E. Gallopoulos for their direction and critical review, J. C. Christensen and L. J. Lamar for their perseverance in typing, G. S. Steadman for preparing the artwork, and other Research Laboratories personnel for providing reference material and technical comments.

REFERENCES

1. P. F. Chapman, "Energy Costs: A Review of Methods," Energy Policy, June 1974.
2. R. H. Williams, Editor, "The Energy Conservation Papers," Ballinger Publishing Company, Cambridge, Massachusetts, 1975.
3. F. H. Kant, et al., "Feasibility Study of Alternative Fuels for Automotive Transportation," Vol. I, II and II, EPA-460/3-74-009-a, b, c, U. S. Environmental Protection Agency, Ann Arbor, Michigan, June 1974.
4. J. Pangborn and J. Gillis, "Alternative Fuels for Automotive Transportation – A Feasibility Study," Vol. I, II and III, EPA-460/3-74-012-a, b, c, U. S. Environmental Protection Agency, Ann Arbor, Michigan, July 1974.
5. W. L. Nelson, "Guide to Refinery Operating Costs," The Petroleum Publishing Co., Tulsa, Oklahoma, 1970.
6. W. T. Tierney, E. M. Johnson and N. R. Crawford, "Energy Conservation Optimization of the Vehicle-Fuel Refinery System," SAE Paper No. 750673, Fuels and Lubricants Meeting, Houston, Texas, June 1975.
7. F. H. Kant, A. R. Cunningham and M. H. Farmer, "Effects of Changing the Proportions of Automotive Distillate and Gasoline Produced by Petroleum Refining," EPA-460/3-74-018, U. S. Environmental Protection Agency, Washington, D. C., July 1974.
8. "The Potential for Energy Conservation in Nine Selected Industries – Petroleum Refining," Conservation Paper No. 10, Federal Energy Administration, Washington, D. C., 1975.
9. K. C. Vyas, "Coal and Oil-Shale Conversion Looks Better," Oil and Gas Journal, August 26, 1974.
10. F. L. Hartley, "Oil Shale: Another Source of Oil for the United States," Oil Daily's Third Annual Synthetic Energy Forum, New York, New York, June 10, 1974.
11. A. E. Harak, L. Dockter, A. Long and H. W. Sohn, "Oil Shale Retorting in a 150-Ton Batch-Type Pilot Plant," Report of Investigations 7995, U. S. Bureau of Mines, 1974.
12. D. P. Montgomery, "Refining of Pyrolytic Shale Oil," 155th National Meeting, American Chemical Society, San Francisco, California, April 2-5, 1968.
13. R. L. Wise, R. C. Miller and H. W. Sohns, "Heat Contents of Some Green River Oil Shales," Report of Investigations 7482, U. S. Bureau of Mines, 1971.
14. S. K. Kunchal, "Energy and Dollar Requirements in an Oil Shale Industry," 170th National Meeting, American Chemical Society, Chicago, Illinois, August 1975.
15. T. A. Hendrickson, "Oil Shale Processing Methods," Quarterly of the Colorado School of Mines, Vol. 69, No. 2, April 1974.

16. L. G. Austin, "Note on Rittinger's Law of Grinding," *Transactions of Society of Mining Engineers, Vol. 254, December 1973.*

17. S. R. S. Sastri and K. S. Narasimhan, "Predicting Grindingmill Energy Use," *Chemical Engineering, September 1, 1975.*

18. "U. S. Energy Outlook – Coal Availability," *National Petroleum Council, U. S. Dept. of the Interior, Washington, D. C., 1973.*

19. "Liquefaction and Chemical Refining of Coal," *A Battelle Energy Program Report, Battelle Columbus Laboratories, July 1974.*

20. W. W. Bodle and K. C. Vyas, "Clean Fuels from Coal," *Oil and Gas Journal, August 1974.*

21. S. Akhtar, et al., "Synthoil Process for Converting Coal to Non-Polluting Fuel Oil," *4th Synthetic Fuels from Coal Conference, Oklahoma State University, Stillwater, Oklahoma, May 6-7, 1974.*

22. L. Grainger, "Energy Conversion Technology in Western Europe," *Phil. Trans. R. Soc. Lond., A. 276, 1974, pp. 527-539.*

23. J. H. Gary, "Liquid Fuels and Chemicals from Coal," *Mineral Industries Bulletin No. 5, Colorado School of Mines Research Institute, September 1969.*

24. H. E. Jacobs, J. F. Jones and R. T. Eddinger, "Hydrogenation of COED Process Coal-Derived Oils," *Industrial and Engineering Chemistry, Vol. 10, No. 4, 1971.*

25. H. H. Lowry, "Chemistry of Coal Utilization," *John Wiley and Sons, Inc. New York, New York, 1963.*

26. "Engineering Evaluation of Project Gasoline-CONSOL Synthetic Fuel Process," *R and D Report No. 59, Office of Coal Research, U. S. Dept. of the Interior, Washington, D. C., July 1970.*

27. "Engineering Evaluation and Review of CONSOL Synthetic Fuel Process," *R and D Report No. 70, Office of Coal Research, U. S. Dept. of the Interior, Washington, D. C., February 1972.*

28. A. L. Coun and J. B. Corns, "Evaluation of Project H-Coal," *Contract 14-01-0001-1188, Office of Coal Research, U. S. Dept. of the Interior, Washington, D. C., 1967.*

29. S. Akhtar, J. J. Lacey and M. Weintraub., "The SYNTHOIL Process – Material Balance and Thermal Efficiency," *67th Annual AIChE Meeting, Washington, D. C., December 1974.*

30. H. H. Storch, N. Columbic and R. B. Anderson, "The Fischer-Tropsch and Related Syntheses," *John Wiley and Sons, Inc., New York, New York, 1951.*

31. S. Katell, "10,000 BPD Fischer-Tropsch Synthesis Plant," *Report No. 58-7, U. S. Dept. of the Interior, Bureau of Mines, Washington, D. C., August 1958.*

32. J. C. Hoogendoorn, "Fischer-Tropsch Process," *Clean Fuel from Coal Symposium, Institute of Gas Technology, Chicago, Illinois, September 10-14, 1973.*

33. J. S. S. Brame and J. G. King, "FUEL Solid, Liquid and Gaseous," *St. Martin's Press, New York, New York, 1967.*

34. G. K. Goldman, "Liquid Fuels from Coal," *Noyes Data Corp., 1972.*

35. W. A. Scheller and B. J. Mohr, "Production of Ethanol and Vegetable Protein by Grain Fermentation," *169th National Meeting, American Chemical Society, Philadelphia, Pennsylvania, April 1975.*

36. R. G. Sheehan and R. F. Corlett, "Methanol or Ammonia Production from Solid Wastes by the City of Seattle," *196th National Meeting, American Chemical Society, Philadelphia, Pennsylvania, April 1975.*

37. D. L. Klass and S. Ghosh, "Fuel Gas from Organic Wastes," *Chemtech, November 1973.*

38. A. H. Brown, "Bioconversion of Solar Energy," *Chemtech, July 1975.*

39. C. Marchetti, "Hydrogen and Energy," *Chemical Economy and Engineering Review, Vol. 5, No. 1, January 1973.*

40. R. L. Savage, et al., "A Hydrogen Energy Carrier," *Vol. I, N74-11727, National Technical*

Information Service, Springfield, Virginia, 1973.

41. *M. Steinberg, "A Review of Nuclear Sources of Non-Fossil Chemical Fuels," Energy Sources, Vol. 1, No. 1, 1973.*

42. *J. R. Garvey, "Future Development in Coal-Fired Powerplants," Proceedings of American Power Conference, Vol. XXIX, 1967.*

43. *"Fuels for the Electric Utility Industry 1971-1985," Edison Electric Institute, New York, New York, 1972.*

44. *D. A. Tillman, "Fuels from Recycling Systems," Environmental Science and Technolgoy, Vol. 9, No. 5, May 1975.*

45. *H. W. Schulz, "Cost/Benefits of Solid Waste Reuse," Environmental Science and Technology, Vol. 9, No. 5, May 1975.*

46. *W. Tipler, "Energy Economics of Automotive Power Generation," SAE Paper No. 750761, 1975 SAE Off-Highway Vehicle Meeting, Milwaukee, Wisconsin, September 1975.*

47. *H. B. Jenson, J. R. Morandi and G. L. Cook, "Characterization of the Saturates and Olefins in Shale-Oil Gas Oil," 155th National Meeting, American Chemical Society, San Francisco, California, April 1968.*

48. *H. E. Carver, "Conversion of Oil Shale to Refined Products," Quarterly of the Colorado School of Mines, Vol. 59, No. 3, July 1964, pp. 19-38.*

49. *J. R. Morandi and R. E. Poulson, "Nitrogen Types in Light Distillates from Aboveground and In Situ Combustion Produced Shale Oils," 169th National Meeting, American Chemical Society, Philadelphia, Pennsylvania, April 1975.*

50. *R. E. Poulson, "Nitrogen and Sulfur in Raw and Refined Shale Oils," 169th National Meeting, American Chemical Society, Philadelphia, Pennsylvania, April 1975.*

51. *F. S. Eisen, "Preparation of Gas Turbine Engine Fuel from Synthetic Crude Oil derived from Coal, Phase II, Final Report," AD/A-007 923, National Technical Information Service, Springfield, Virginia, February 1975.*

52. *G. R. Hill and L. B. Lyon, "A New Chemical Structure for Coal," Industrial and Engineering Chemistry, Vol. 54, No. 6, June 1962.*

53. *H. W. Sternberg, R. Raymond and S. Akhtar, "SYNTHOIL Process and Product Analysis," 169th National Meeting, American Chemical Society, Philadelphia, Pennsylvania, April 1975.*

54. *R. H. Wolk, N. C. Stewart and H. F. Silver, "Review of Desulfurization and Denitrogenation in Coal Liquefaction," 169th National Meeting, American Chemical Society, Philadelphia, Pennsylvania, April 1975.*

55. *"1974 General Motors Report on Programs of Public Interest," General Motors Coropration, Detroit, Michigan, April 15, 1975.*

56. *W. J. Schultz, C. E. Miesiak, A. L. Hamilton and D. E. Larkinson, "Credibility of Diesel Over Gasoline Fuel Economy Claims by Association," SAE Paper No. 760047, Automotive Engineering Congress, Detroit, Michigan, February 1976.*

57. *"Motor Trend," January 1976, p. 30.*

DISCUSSION

M. A. Elliott *(Energy Consultants)*

A study like this depends on your original philosophy and I don't mean this in any critical sense, but it seems to me one should look at the recovery ratio of the in situ resource. And if we do this with primary recovery of petroleum, we'd have to multiply these figures by 0.33, or something like that. With some strip mining operations you multiply by 0.8. With room and pillar underground mining, by 0.5. It seems to me that if you want to look at the overall picture, it gets complicated. But you ought to start with the in situ resource and then carry it all the way through to the end use.

Stebar

Recovery ratio is an important consideration. However, in our study, where we were attempting to match potential fuels with engines for optimum energy efficiency, we did not feel that recovery ratios should be considered. If one were to evaluate the overall resource potential, he ought to start with the in situ resource, as you suggest.

E. M. Dickson *(Stanford Research Institute)*

I think this question is directed partly to you and partly to the organizers. The emphasis so far is certainly skewed against the electric vehicle. I wonder if this represents some kind of bias? Our analyses seem to indicate that if you start with coal in the ground, and the goal is miles of transportation delivered, then going through coal, through electricity, and through an electric vehicle to transportation delivered gives an efficiency that would fit into the matrix of your final slide. It would be of the order of 14% or so, and therefore, comparable to that which we are usually experiencing with petroleum. On the other hand, going from coal, through a syncrude, through a distillate process, and into the conventional vehicle gives a number somewhat similar to yours. In other words, it seems as if the electric vehicle is getting a poor deal.

Stebar

We don't want to convey that impression, certainly. We have made similar analyses, as you suggest, but based on our information, as long as fossil fuels are available, energy efficiencies are generally higher with vehicles other than electrics. Now when we get to the point where we're restricted to bio-mass and nuclear sources, the situation changes and it looks to us like you should consider electric vehicles. But we think that's a long way down the road. Of course major advances in battery technology might stimulate interest in electric vehicles sooner.

F. L. Dryer *(Princeton University)*

Can you give us a little more information on the characteristics of your spark ignition and diesel engines? For example, did the diesel engine have an open chamber or a prechamber?

Stebar

The diesel engines were the prechamber type, typical of present passenger car diesels. The gasoline engines were of the conventional variety with oxidizing converters and other familiar exhaust control hardware.

S. S. Penner *(University of California, San Diego)*

I wonder if you would give us your off-the-cuff numbers for the Brayton and Stirling cycles?

Stebar

The combined energy efficiency of the gas turbine cycle engine in its present state of development is considerably lower than for either the spark ignition or diesel engine. We're amused at the very high efficiencies quoted for the gas turbine in some quarters. Remember for our study we've stipulated several constraints; equal performance, equal emission levels, representative driving, etc. We also restricted our evaluations to demonstrated, full-scale vehicles driven the way people typically drive. We did not consider hypothetical gains that might be possible based on theoretical or analytical treatment.

The Stirling is reported to be a very efficient engine cycle, but so far it has not been demonstrated in an actual vehicle driven on the FTP.

W. T. Lyn *(Cummins Engine Company)*

In your conversion conclusion, did you take into account the entire product slate which could have been an energy resource in other applications?

Stebar

Yes, we took into account the quantity and heating value of all useful products.

E. E. Ecklund *(U.S. Energy Research and Development Administration)*

When you play this kind of a game, you've got to be pretty careful to also consider the time frame. In today's environment we know that the internal combustion engine with gasoline is not the best. As a matter of fact, if you play the game along your route you say let's stick with it. We're not going to do anything else. But the game we're in is alternatives. As you consider a longer time frame, then the situation is likely to come out quite differently.

Stebar

That remains to be seen. We are dealing with moving targets in all areas including emission standards and vehicle efficiency. And the efficiency of the spark ignition engine has been improved substantially since 1974 and hopefully can be improved further. Objectively, you don't change to an alternative just for the sake of change. There must be advantages in important areas of concern. You must select the fuel and engine combination that's best for the conditions that will exist.

APPENDIX A

CALCULATION OF AUTOMOBILE FUEL-UTILIZATION EFFICIENCY

For this study, automobile fuel-utilization efficiency was defined as the energy theoretically needed to drive an automobile over a prescribed driving cycle divided by the energy actually consumed. Thus the efficiencies of individual components such as the engine, transmission, etc., were included. The energy theoretically needed was determined by using a vehicle power equation which was integrated over the specified driving cycle. The energy actually consumed was determined from experimental measurements for a vehicle driven on that same cycle. Both the EPA Urban and EPA Highway test cycles were utilized. These cycles were used to determine a composite automobile fuel-utilization efficiency. This efficiency was computed using the following equation:

$$\eta = \frac{0.55\ E_u + 0.45\ E_h}{(\text{fuel heating value})\ (0.55\ FC_u D_u + 0.45\ FC_h D_h)} \qquad (A\text{-}1)$$

where:

η	=	automobile fuel-utilization efficiency
D_u	=	driving distance for the EPA Urban Cycle, 11.99 km
D_h	=	driving distance for the EPA Highway Cycle, 16.48 km
E_u	=	energy theoretically needed for the EPA Urban Cycle, MJ
E_h	=	energy theoretically needed for the EPA Highway Cycle, MJ
FC_u	=	measured fuel consumption for the EPA Urban Cycle, in ℓ/km
FC_h	=	measured fuel consumption for the EPA Highway Cycle, in ℓ/km
fuel heating value	=	32.42 MJ/ℓ gasoline
		36.29 MJ/ℓ diesel 2

The energies theoretically needed for the EPA Urban and Highway driving cycles (E_u and E_h) were obtained from Table A, for the appropriate inertia weight. These energy values were calculated using the following vehicle power equation:

$$P_T = P_{\text{rolling resistance}} + P_{\text{aerodynamic drag}} + P_{\text{inertial}} \qquad (A\text{-}2)$$

where:

P_T	=	total power	in watts (W)
		(if negative, set equal to zero)	
$P_{rolling\ resistance}$	=	$0.04086\ M\ V$	in W
$P_{aerodynamic\ drag}$	=	$0.01284\ C_D A_F V^3$	in W
$P_{inertial}$	=	$0.27755\ M\ A_C\ V$	in W
A_C	=	vehicle acceleration	in m/s^2
$C_D A_F$	=	drag coefficient x frontal area	in m^2
M	=	vehicle mass	in kg
V	=	vehicle velocity	in km/h

Values for V and A_C were dictated by the driving cycle. For $C_D A_F$, it was assumed that the values varied linearly with vehicle weight from $C_D A_F = 0.836$ m^2 for a 907 kg car up to $C_D A_F = 1.115$ m^2 for a 2 268 kg vehicle. These values are typical for current automobiles.

TABLE A

Automobile Energy Requirements for
EPA Urban and Highway Driving Cycles

Vehicle Mass, kg (lbm)	$C_D A_F$, m^2	Energy, MJ							
		Rolling Resistance		Aerodynamic Drag		Inertial		Total	
		Urban	Highway	Urban	Highway	Urban	Highway	Urban	Highway
907 (2000)	0.836	1.087	1.602	0.856	3.087	2.131	2.067	4.074	6.756
1 134 (2500)	0.882	1.358	2.003	0.905	3.259	2.601	2.381	4.864	7.643
1 361 (3000)	0.929	1.629	2.403	0.950	3.431	3.071	2.701	5.650	8.535
1 588 (3500)	0.975	1.901	2.805	0.999	3.603	3.541	3.023	6.441	9.431
1 814 (4000)	1.022	2.174	3.205	1.047	3.774	4.011	3.342	7.232	10.321
2 041 (4500)	1.068	2.445	3.605	1.093	3.946	4.480	3.664	8.018	11.215
2 268 (5000)	1.115	2.717	4.005	1.141	4.118	4.950	3.984	8.808	12.107

SESSION I – SUMMARY

P. S. MYERS

University of Wisconsin

Gentlemen, as one who has helped to run the show, but did not really participate in the preparation, I think that from an unbiased viewpoint I can say that we've had a good session this morning. We've had some very interesting presentations. I think, if it wasn't clear to you before you came, it should be clear to you now that there is a transportation energy problem in the foreseeable future. You heard two different projections this morning on automotive fuel consumption. The first speaker said automotive fuel consumption is going to go up, and the second speaker said it was going to go down, at least if you look at automobiles. One of the speakers indicated that, of the interregionally traded oil, the U. S. would take somewhere between 25 and 30%. I think this raises real questions as to whether we can pay for it. Will world politics permit this? If this is the magnitude of the problem in 1990, what's it going to be like 15 years after that?

I think it is also clear this morning that this intensification of the energy problem is going to bring changes in fuels, and it is probably going to bring changes in engines as well. I think you will hear more about potential changes in fuels as the program develops. It appears that there might be minor or major changes in engines, depending on your viewpoint. I think the message came through loud and clear, however, that fuel conservation is essential. We are going to have to conserve fuel. I think you recognize the meaningful suggestion that perhaps we need a new criterion, miles per unit of original energy, and it was suggested that this ought to include the in situ energy as well as the acquisition energy.

In looking at ways of changing engine and fuel combinations there seemed to be pretty common agreement that you are not going to get the maximum fuel economy with the conventional spark ignition engine and unleaded fuel, and that we are operating, at the moment, with the least optimum combination. It has been suggested that there are significant differences in the economy of engine-fuel combinations. The type of fuel that you make does affect the energy consumption, but probably does not have as much effect as the efficiency of the engine. Only an increase in engine efficiency begins to be of the magnitude required to cover the projected increases of energy consumption. The last paper gave a review of potential fuels and I thought it was interesting to see the range of fuel conversion efficiencies. It ranged from 20 to 95 or 98%. This is quite a range in fuel conversion efficiency. It was also interesting to look at just one of the fuels, coal, to see the range of conversion efficiencies. They ranged from 35% with the Fischer-Tropsch process to as high as 95% for powdered coal. So there's quite a range of efficiencies for a particular fuel. This, then, raises the question of whether we design the engine of the future for the fuel of the future, or do we design the fuel of the future for the engine of the future? There certainly is a question as to which one comes first. I think you will find running through this conference, the interrelationship between the engine of the future and the fuel of the future.

SESSION II

FUTURE AUTOMOTIVE FUEL SOURCES

Session Chairman
M. A. ELLIOT

Energy Consultant
Houston, Texas

COAL AS A SOURCE OF AUTOMOTIVE FUELS

E. L. CLARK

Energy Research and Development Administration (ERDA), Washington, D.C.

ABSTRACT

The conversion of coal to fuels for internal combustion engines was first implemented on a commercial scale nearly fifty years ago. Some thirty years ago, during World War II, over 100,000 barrels per day of high octane fuels were produced from coal in Germany. Cognizant of these historical data, EDRA's Fossil Energy Program includes a major effort to produce transportation fuels from coal.

Coal liquefaction projects currently in the bench-scale and pilot plant stages are aimed at producing primary liquid products at maximum through put and minimum cost. Samples of these products are currently being supplied to bench-scale refining units for determining the effectiveness of state-of-the-art refinery processes to produce synthetic crude oils and specification transportation fuels. Since coal-derived liquids will be small in volume for some years compared to those from petroleum, synthetic crude oils amenable to refinery processing will provide the first introduction to actual use of coal-derived automotive fuels. In addition, as larger quantities of coal-derived liquids become available in 1977 and 1978, refining studies now in progress will provide the techniques for the preparation of sizable samples of transportation fuels for large-scale end-use testing.

Coal gasification projects now in progress are expected to lower the cost of preparing synthesis gas from coal. This will stimulate the production of motor fuels using processes similar to those used now to produce motor gasoline in South Africa. Improved catalysts for such processes are under development.

Many technical and logistic problems are still to be solved before coal-derived automotive fuels become a reality. Large increases in coal production will be necessary and their impact on the economy are under study. The rate of process development and new plant construction and design will depend on Governmental action. There will be differences between petroleum gasolines and those derived from coal. The latter will probably have higher nitrogen content and greater aromaticity with attendant environmental problems. Solution of these will require continuing cooperation between ERDA and the automotive industry.

INTRODUCTION

The history of coal conversion to other fuels includes considerable data on the production of liquids for use in internal combustion engines. Nearly fifty years ago, the first commercial coal hydrogenation plant was built at the Leuna works in central Germany. This plant continued in operation through the second World War, during which time it produced from coal over 15,000 barrels/day of aviation and motor gasoline. Kenneth Gordon, in a paper presented in 1946 before the Institute of Mechanical Engineers, in London, described the process used at Billingham for conversion of Durham coal to gasoline. Gordon also summarized the production of high octane gasoline in Germany during World War II which reached a maximum of 100,000 barrels/day.

The Fossil Energy Program at the Energy Research and Development Administration (ERDA) is cognizant of these historical achievements. It also recognizes the critical importance of transportation fuels which utilize approximately fifty percent of the petroleum barrel. The ERDA Fossil Energy Program structure is influenced by many factors, and a program strategy has evolved to account for these factors. The program seeks to develop a mix of technologies to provide the complex American economy with the enormous amounts of energy in various forms that it requires: natural or synthetic gas for space heating, electric power, and industrial processing; liquid fuels for transportation, electric power, heating and chemical feedstocks; combustible solids for power generation and industrial applications. This paper will limit discussion of the program to those areas most directly involved in potential production of motor fuels.

COAL LIQUEFACTION

The prime objective of coal liquefaction research is to provide technology which is economically feasible and environmentally satisfactory to convert coal to a clean liquid fuel for electric power generation, transportation, and heating for industry and homes. Major ongoing projects in this area are given in Table 1.

Of the five projects listed, three — Solvent Refined Coal (SRC), H-Coal and Synthoil — are most pertinent to the production of liquid fuels. All three have as their objective the production of a "primary" liquid from coal. These so-called primary liquids have economic and process-oriented reasons for their production. In liquefying coal, it is necessary to add hydrogen both to increase the hydrogen/carbon atomic

TABLE 1

Ongoing Coal Liquefaction Projects

Major Projects	Contract Value Millions of $	Contractor	Location	Key Events
• Coal Oil Energy Development (COED)	$21.0	FMC	Princeton, N. J.	Pilot Operations Complete FY 75
• Solvent Refined Coal (SRC)	$41.9	PAMCO	Tacoma, Wash.	Pilot Operations Started Mid FY 75
• H-Coal	$ 8.0	HRI	Trenton, N.J. Cattletsburg, Kentucky	PDU* Runs FY 75 Pilot Plant Decision Mid FY 76
• Clean Coke	$ 6.5	U.S. Steel	Monroeville, PA.	PDU Complete FY 75 Pilot Plant Decision FY 76
• Synthoil	$ 6.9	Foster Wheeler	Bruceton, PA.	RFP for Construction June 75

PDU — Process Development Unit

ratio and to remove undesirable heteroatoms present in the original coal. The cost and thermal efficiency of converting coal to a liquid are dependent largely on the amount of hydrogen used in the conversion. In the three processes mentioned, the hydrogen addition is minimized to improve thermal efficiency and reduce costs. The fuels produced under these conditions are limited to use as boiler fuels for power generation. An examination of the anatomy of these three processes should indicate their potential for producing a variety of liquid fuels.

The SRC process transfers hydrogen to the coal, primarily through a donor solvent which is itself regenerated by molecular hydrogen present in the reactor at pressures of 1,500 to 2,500 psig. Currently, the SRC process is being used to produce boiler fuel, but by changes in operation, the quantity of the hydrogen transferred may be increased to convert coal to distillable oils.

The Synthoil and H-Coal processes are being operated to produce high molecular weight products from coal. Both processes use a solid catalyst within a high pressure reactor. The Synthoil process uses highly turbulent flow of a slurry of coal plus a recycle oil and hydrogen gas over a fixed-bed catalyst. The H-Coal process uses an ebullated bed of catalyst kept in suspension by the flow of a similar hydrogen and coal plus recycle oil slurry. In both cases, reduction of reactor throughput, increased pressure of operation, or maintenance of higher catalyst activity by more frequent renewal could increase hydrogen addition to the coal and produce lower molecular weight liquids more similar to petroleum.

The SRC process is further along in its development than other coal liquefaction processes. The SRC process can use a minimum amount of hydrogen (less than 3,000 SCF/bbl) and is currently being operated under these conditions to produce a material that is a solid at temperatures up to 400°F. This product is essentially ash-free, has a sulfur content of under 0.9 wt.% and a heating value of 16,000 Btu/lb. It is an environmentally acceptable fuel for solid-fuel-fired boilers. Combustion tests are in progress in laboratory burners. A full-scale boiler test is planned for this fuel early in 1976. Preliminary tests are evaluating combustion of this fuel as a liquid-melt for use in boilers or gas turbines. ERDA recently announced the award of a contract to UOP to evaluate use of current refinery techniques for upgrading "primary" coal liquids to specification fuels. SRC product is being evaluated under this contract. Additional tests in other petroleum processing laboratories are planned for products of all these coal liquefaction processes.

The results of this test work will answer an important process engineering problem. To what extent must coal be converted to provide a product which can be handled by conventional refining procedures to provide a complete slate of liquid fuels? The liquefaction of coal as a process is not ideally suited to the use of selective catalysts for producing a specification product. Down-stream processing in which more elegant methods can be used may be able to handle crude, coal-derived oils more efficiently and retain the economic advantages of an efficient primary conversion step. Thus, the feedback from these refining and upgrading tests will be used to adjust coal conversion process conditions to optimize an overall conversion of coal to specification fuels.

We do not expect this effort to establish a single refining scheme or one particular coal liquefaction process as an ideal answer to liquid fuel production. The experience in the petroleum industry has been that the tremendous variations in petroleum feedstocks, production and transportation costs, regional demands for product mix, seasonal cycles, and local prices for competing fuels have resulted in a wide variety of refinery designs. A large number of competing refinery process operations are available to the refinery designer, and even after fifty years of sophisticated technical development, no single ideal process for petroleum refining has emerged.

Coal and other fossil fuel feedstocks are even more variable than petroleum feedstocks, and some of their product markets are extremely susceptible to competition by easily transportable liquid fuels. Thus, the fossil fuel program strategy is to advance a broad spectrum of processes, specifically suited to wide ranges of feedstocks and markets.

The ultimate commercialization of a synthetic fuel process depends very heavily upon process efficiency and cost. Even small increases in process efficiency result in huge savings because of the large throughput of material in these plants. Consequently, it is necessary to keep the program aimed at optimum process configurations. ERDA's Fossil Energy Program conducts continuing evaluations to provide comparisons of the process alternatives being considered. These studies provide program guidance, as well as information for the development of realistic plans for demonstration plant projects.

One such demonstration plant project is in progress. In July 1974, an RFP was issued for a clean boiler fuel demonstration plant. On January 17, 1975, the Office of Coal Research awarded a contract to Coalcon for the phased design, construction and operation of a 2,600 ton/day demonstration plant using a hydrocarbonization process for producing 3,900 barrels/day of liquid product, and 22 million cubic feet daily of pipeline quality gas. This plant will fulfill a near term objective of the coal program to provide a fuel substitute from coal for a portion of higher grade oil and gas used in the generation of electricity and in industrial heating processes.

This plant is the first demonstration plant ERDA has undertaken. It is the last step before commercialization. While the design and engineering is being funded by the Government, the industrial partner will share the construction and operating costs on a 50-50 basis. The plant schedule is being accelerated, and we hope to have the plant begin operations in 1979.

While this demonstration plant will produce liquids for power plant or industrial fuel, these liquids will also be amenable to upgrading procedures currently being tested. We expect to have available, as a result of our test programs, considerable knowledge of the processing steps required to convert coal-derived liquids to automotive fuels. Several months ago, we discussed with representatives of the General Motors Corporation procedures and fuel quantities which would be required for determining the acceptability of transportation fuels prepared from coal. As sufficient quantities of representative primary fuels from coal become available, we plan to use our processing data to prepare adequate quantities of transportation fuels for actual test programs. Fig. 1 indicates the anticipated supply projections for these "primary" fluids from current bench-scale units and future pilot plants. Current supplies are barely enough to provide materials for testing refinery processes. Future supplies should be ample for large-scale testing of both refining processes and actual fuels.

In Fig. 1, we assume that by 1977, a Synthoil unit will be in operation producing some 30 barrels/day of oil. In 1978, provided anticipated progress is made, we expect a large pilot plant using the H-Coal process will start operation and will produce 700 to 1,800 barrels/day of heavy oils. If suitable acceleration can be achieved for our demonstration plant program, it will also contribute to our supply of coal-derived oils. As we are now doing for the SRC plant products, combustion tests will be run and materials will be made available for upgrading to transportation fuels. By that time, we shall have developed suitable refining procedures and facilities for this upgrading. We shall look to the automotive industry for evaluation and testing of these refined, coal-derived fuels.

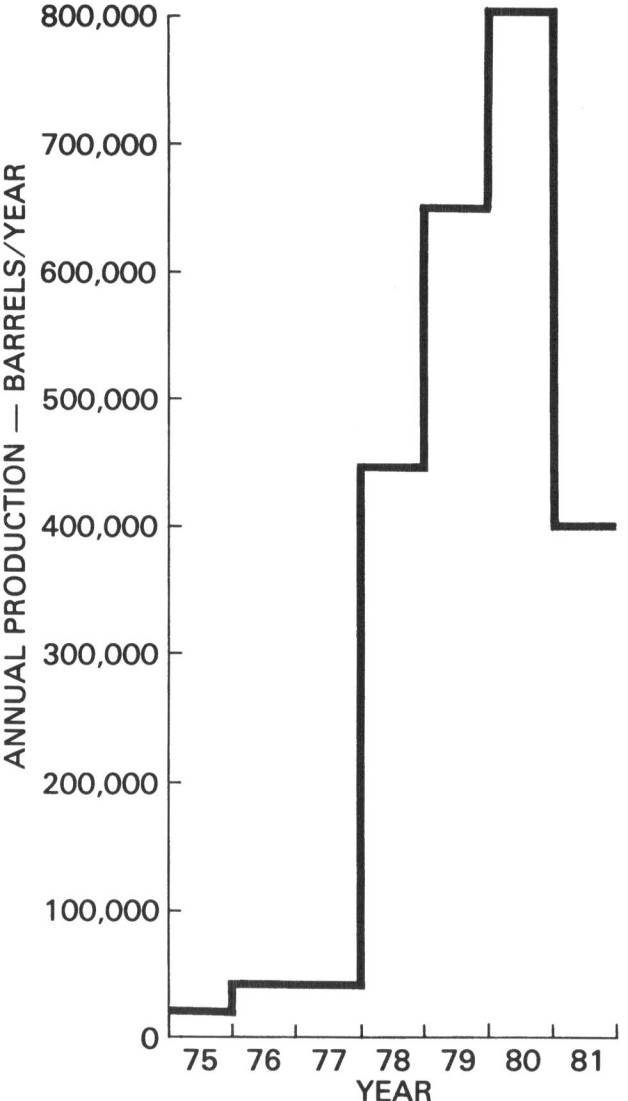

Fig. 1. Projected quantities of primary coal liquefaction products.

HIGH BTU GASIFICATION

The major objective of the high Btu gas program is to provide improved technology for the manufacture of pipeline quality gas from coal; improvements which, firstly conserve coal as they permit greater efficiency, and secondly decrease the cost of manufacture of gas by about 20 percent compared to present technology. The major ongoing projects in this area are given in Table 2.

TABLE 2

Ongoing High Btu Gasification Projects

Major Projects	Contract Value Millions of $	Contractor	Location	Key Events
• CO$_2$ Acceptor Process	2.0	Conoco Coal Dev. Co.	Rapid City, S.D.	Methanation Plant Construction Complete FY 75
• Hygas Process	3.0	Institute of Gas Technology	Chicago, Ill.	Steam Oxygen System Construction Complete FY 75
• Liquid Methanation	4.0	Chemical Systems Inc.	To be Determined	Complete Pilot Plant Construction FY 75
• Ash-Agglomerating Process	1.0	Battelle Columbus	West Jefferson, Ohio	Complete Pilot Plant Construction FY 76
• Steam-Iron Process	6.0	Institute of Gas Technology	Chicago, Ill.	Complete Pilot Plant Construction FY 76
• Bi Gas	13.2	Bituminous Coal Research	Homer City, PA.	Complete Pilot Plant Construction FY 76
• Synthane	9.6	Rust Engineering/ Lumus Corp.	Brucetown, PA.	Complete Construction FY 75

While the main thrust of this portion of our program is to produce a synthetic natural gas, it is of critical importance to producing liquid fuels. For producing pipeline gas, the first step is to react coal with oxygen and steam to produce a mixture of hydrogen and carbon monoxide appropriately called "Synthesis Gas." This gas mixture is a raw material for synthesizing a wide variety of products. It can be reacted with steam over a catalyst to produce pure hydrogen required for coal liquefaction. The hydrogen, thus produced, can also be reacted with nitrogen for production of ammonia. Synthesis gas can be converted to methanol which has possibilities as a component of automotive fuels. Synthesis gas is being directly converted to liquid fuels now by a modification of the Fischer-Tropsch process in a plant being operated by SASOL in South Africa.

Thus, our high Btu gasification program can provide a raw material for direct and indirect production of liquid fuels. We anticipate that by development of economical methods for coal gasification, the coal liquefaction processes previously discussed will become more efficient by access to lower cost hydrogen. We also have the exciting prospect of the synthesis of liquid fuels directly from basic building blocks. Currently, in South Africa, a wide range of petroleum products — from paraffin waxes to motor gasoline — are being produced in this manner. Work is in progress to improve gasoline quality by this process through more selective catalysts. Success will depend on a reduction in the cost of converting coal to a synthesis gas. An important advantage of this synthetic approach is the absence of nitrogen and sulfur compounds in the liquids produced.

Some progress can be reported now on a synthesis gas route to automotive gasoline. As mentioned above, synthesis gas can be converted to methanol. Furthermore, with inexpensive synthesis gas from coal, lower cost methanol should be possible. Under a contract with ERDA, the Mobil Research and Development Corporation is successfully converting methanol in a bench-scale unit to automotive gasoline. The gasoline is highly aromatic and has a Research Octane rating of 90 to 92. Preliminary automotive tests indicate good performance. The process design of a semi-commercial unit is being developed. Economics of commercial production will depend on methanol costs and successful integration of the coal gasification, methanol production, and methanol conversion steps.

To achieve any measure of success in producing significant quantities of automotive fuel from coal, various technical and logistic problems must be solved. The following are included in the solutions to these problems:

1. A suitable capability must be established for increased production of coal.

2. Primary liquefaction and refining processes must be developed to ensure operability at economic levels.

3. Governmental assistance will be required to eliminate some of the economic risks in competing with petroleum.

COAL PRODUCTION

For any appreciable production of liquid fuels from coal, it is obvious that large increases in coal production will be required. These increases will be in addition to increased requirements of coal for power generation and steel production. In planning for the conversion of coal to liquid and gaseous fuels, the problem of coal availability is being considered. Several studies have evaluated constraints on increasing coal production to levels as high as two billion tons/year by 1985. This very high objective would require a threefold expansion from the production of roughly 600 million tons in 1974.

These studies suffer from the deficiencies inherent in all broad-based overviews which evaluate in a "macro" manner and often do not consider "micro" effects which can exert major influences. However, some interesting and meaningful observations can be made. Ample coal reserves exist to support production levels exceeding two billion tons/year. Sufficient labor should be available for strip-mining which is assumed to be utilized for production of 56% of the desired 1985 production. Some labor shortages (as much as 20%) may be anticipated for underground mining required to achieve maximum 1985 production objectives. With adequate development and allocation, sufficient water resources should be available. Transportation, capital and equipment availability were considered and, if reasonable incentives are provided, these necessary ingredients should be available.

However, while the problems of water supply seem non-existent on a regional basis, one finds insufficient supplies when local authorities are contacted. Transportation expansion will require capital assistance to railroads for additional rolling stock. The expansion of capacity to produce mining equipment will require much more detailed study than available in most "broad-brush" evaluations. The latter indicate the need for expansion, but do not supply the details of all the supporting expansions in supplying components such as large bearings and electrical machinery.

It appears that both the production of coal and its conversion will depend on the action or lack of action by governmental agencies. Several bills to provide action are being considered by the Congress. The Administration has revealed several potential programs involving various methods for assistance to industry. One cannot predict at this time which plan might be implemented. From the activity going on, it appears that some plan will be implemented within the coming year. Our program is aimed at providing a variety of routes for accelerating such an implementation.

Our program plan in liquefaction of coal is illustrated in Fig. 2. We show the various types of processes being studied at the left of this schematic. In the center we indicate critical steps common to all coal liquefaction processes. These include separation of heavy oils from solids, conversion of residues to char, and gasification of char and heavy oils to produce the hydrogen required in all liquefaction processes. At

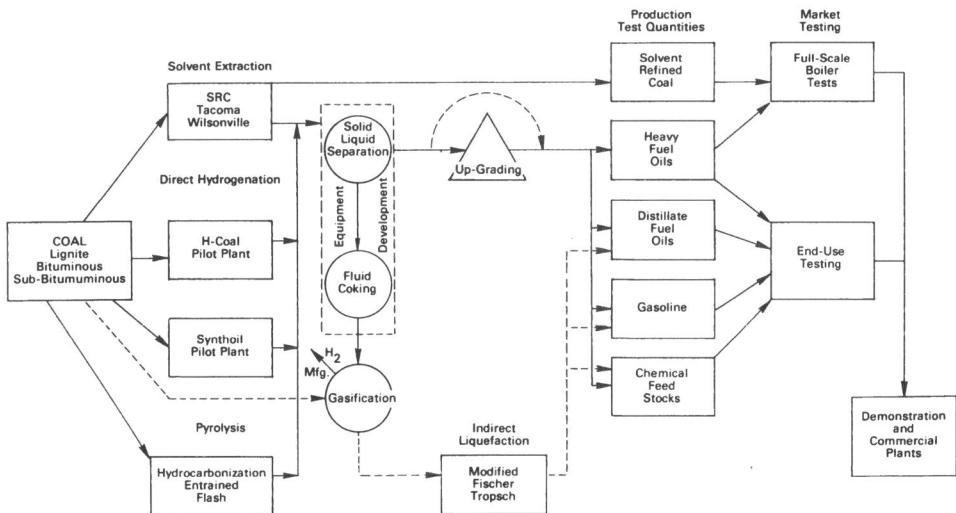

Fig. 2. ERDA coal liquefaction development strategy.

the right, are shown the refining and upgrading studies for conversion of coal liquefaction products to usable, specification fuels. As larger quantities of representative products become available, samples of specification fuels will be prepared for testing. We expect the cooperation of the automotive industry in this future program.

CONCLUSION

Speculation as to the rate of progress of ERDA's coal programs is an uncertain art. However, it appears quite certain that for some time to come, the quantities of coal-derived liquid fuels will be considerably smaller than those produced from petroleum. As a result, the first approaches to upgrading coal-derived liquids would be to make them amenable for acceptance by a conventional refinery together with petroleum crude oils. This approach is being stressed in our refining studies and we are working with petroleum companies and others skilled in evaluating complex refinery feedstocks. Again, these results will be reduced to reality by actual preparation of specification fuels to ensure their acceptability to the products of the automotive industry.

As in the past, we look forward to continual exchanges of information with the engineers who specify engines and their fuels. Through this cooperation we anticipate a logical program to our mutual benefit.

DISCUSSION

H. F. Mason *(Chevron Research Co.)*

What do you mean by "modified" Fischer-Tropsch process?

Clark

Considerable work is going on, most of it proprietary, on the improvement of the Fischer-Tropsch catalyst. There is another advantage now. Traditionally, the philosophy of the Fischer-Tropsch process and the coal-hydrogenation process has been to minimize the production of methane and ethane. This has been partly because, in Europe, there is no market for methane or ethane, except as fuel. In the United States the availability of very inexpensive natural gas has kept demand for methane and ethane down. This situation has changed. We foresee that the new Fischer-Tropsch plant will be an altogether different one. The average molecular weight of the product will be reduced. Before, you could not do that because you would make too much methane. Now, methane becomes an advantage as natural gas prices begin climbing toward $4/million Btu. We now have a chance of modifying the Fischer-Tropsch process to suit the American desire for natural gas. We'll pay a higher price for it, but a hybrid plant which will produce both natural gas and liquid fuels could be extremely advantageous in our coming economy. That's what I mean by "modified."

S. S. Penner *(University of California, San Diego)*

What's your best estimate on dollars per barrel for liquified coal?

Clark

I think it's going to be somewhere between $15 and $25 a barrel, but you can't hold me to the number because any major changes in capital costs could make a drastic change in the cost of the plant and the cost of the fuel. If you look at the economic structure of a synthetic fuel plant, you find that somewhere between 65 and 80% of the product cost is associated with the capital investment in the plant. Coal cost is 15-16%. Labor, catalyst and chemicals together are 5-8%. If we could reduce the capital cost by 20%, we would have a profound effect on the economics of the Fischer-Tropsch process and also the synthetic natural gas process.

J. W. Hodgson   *(University of Tennessee)*

Several people have expressed concern about the possible carcinogenic nature of coal-derived liquids. Would you say something about that?

Clark

There's a tremendous history on this. This history was started by Exxon when they evaluated the carcinogenic nature of cat cracker cycle stock. They did a splendid job. A similar job was done by Union Carbide when they were running their coal hydrogenation facility in South Charleston. I think a reasonable amount of precaution must be taken, and that these precautions are being taken now with heavy petroleum liquids. Fortunately, very few of the volatile materials made in a coal liquefaction plant are carcinogenic. It's only the non-volatile materials. So you have the contact problem and an aerosol problem and there's ample background to deal with these. We are continuing to study it to ensure that in each process we do not make materials that may be unusually carcinogenic. The refining problem, I feel, is under reasonable control. There may be a problem with increased carcinogens in automobile exhaust with a more aromatic fuel, but I'm not familiar with that.

MOTOR FUELS FROM OIL SHALE —
PRODUCTION AND PROPERTIES

J. H. GARY

Colorado School of Mines, Golden, Colorado

ABSTRACT

The oil shales of Colorado, Wyoming and Utah are not shales and do not contain any liquid oil. The base material is a marlstone, very similar to limestone, which is dense and non-porous. The organic material is a solid called kerogen which does not melt and has very limited solubility in such organic solvents as benzene, acetone and ether. It is necessary to heat the kerogen to temperatures near 900°F to cause the kerogen to decompose and give a liquid shale oil, gases and carbon as products.

The organic content of the Western shales is about 14 percent by weight and, upon pyrolysis, yields from 25 to 40 gallons of oil per ton of rock. There are two general methods for converting the solid kerogen to shale oil; one is by mining and retorting the shale above ground and the other is by establishing communications (introducing porosity) in the shale and retorting the shale underground or in situ.

The raw shale oil produced from the retorting is high in unsaturated hydrocarbons, nitrogen and sulfur. In order to produce satisfactory motor fuels more intensive and expensive processing is required than to produce motor fuels from conventional crude oils.

Background will be presented on shale and shale oil properties, retorting and refining methods and properties of finished motor fuels.

CHARACTERISTICS AND LOCATION OF OIL SHALE

The oil shales of Colorado, Utah, and Wyoming are not shales, and do not contain any liquid oil. They consist of a fine-grained non-porous sedimentary rock, similar in many respects to limestone, containing a solid organic material which yields petroleum-type hydrocarbons only after destructive distillation at temperatures of approximately 900°F.

The organic matter in oil shale is only slightly soluble (approximately 2%) in common hydrocarbon solvents, and upon decomposition by heat yields a waxy oil which is a solid at room temperature. This led Professor Crum Brown of Edinburgh University to apply the name "kerogen", from the Greek words meaning producer of wax, to the organic fraction of oil shale.

Significant deposits of oil shale occur in at least eight of the United States (Fig. 1) and 23 foreign countries, but the largest deposits occur in the organic-rich sedimentary rocks of the Green River Formation in Colorado, Utah, and Wyoming. The location of oil shale deposits 10 feet or more thick, and containing 15 or more gallons of shale oil per ton are shown in Fig. 2. These deposits contain an estimated 1.8 trillion barrels of oil and cover approximately 11 million acres. Of these, high grade shales containing 25 gallons or more of oil per ton of rock have a total capacity of approximately 600 billion barrels of shale oil (1). The sections of Colorado, Utah, and Wyoming in which the oil shale deposits occur are areas of low annual rainfall.

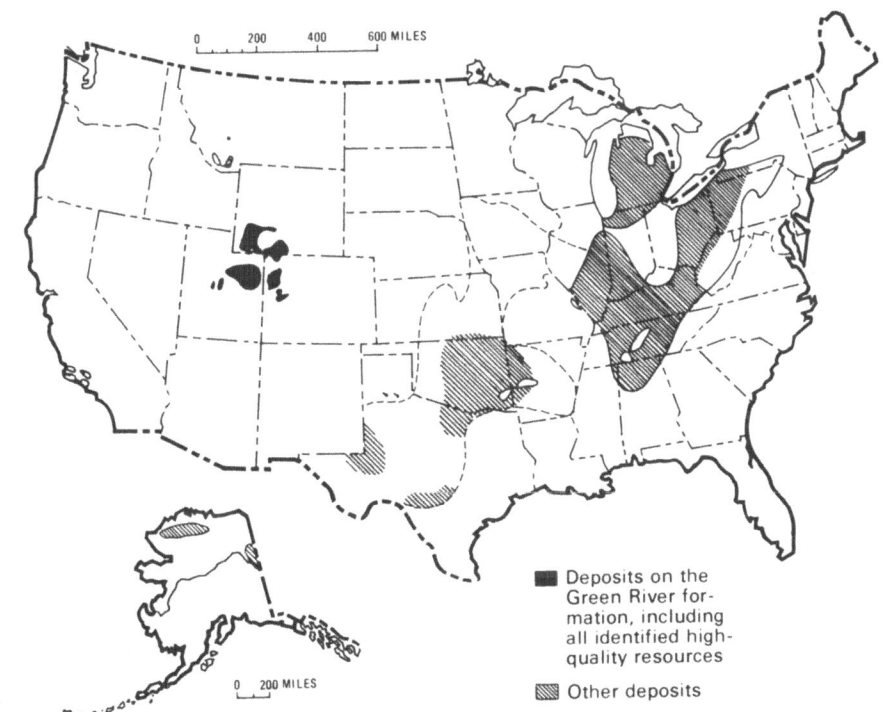

Fig. 1. Principal reported oil shale deposits of the United States.

References pp. 146-147.

The richest of the oil shale zones in the Piceance Creek Basin has long been known as the "Mahogany Ledge" because of its color and the ledging outcrops of the rich oil-shale layer. The Mahogany Zone varies in thickness from less than 75 feet to more than 225 feet and covers an area of more than 770,000 acres. Recent studies have shown that two other rich zones underlie the Mahogany Zone and that near the middle of the Piceance Basin there is a sequence of almost continuous rich oil shale to a depth of more than 1500 feet below the top of the Mahogany Zone (2).

The typical composition of the Colorado and Utah oil shale from the Mahogany Zone containing 25 gallons of oil per ton is given in Table 1 (3, 4).

The organic content of the Western shales averages from 12 to 20 percent by weight and, upon pyrolysis, yields from 15 to 40 gallons of oil per ton of shale.

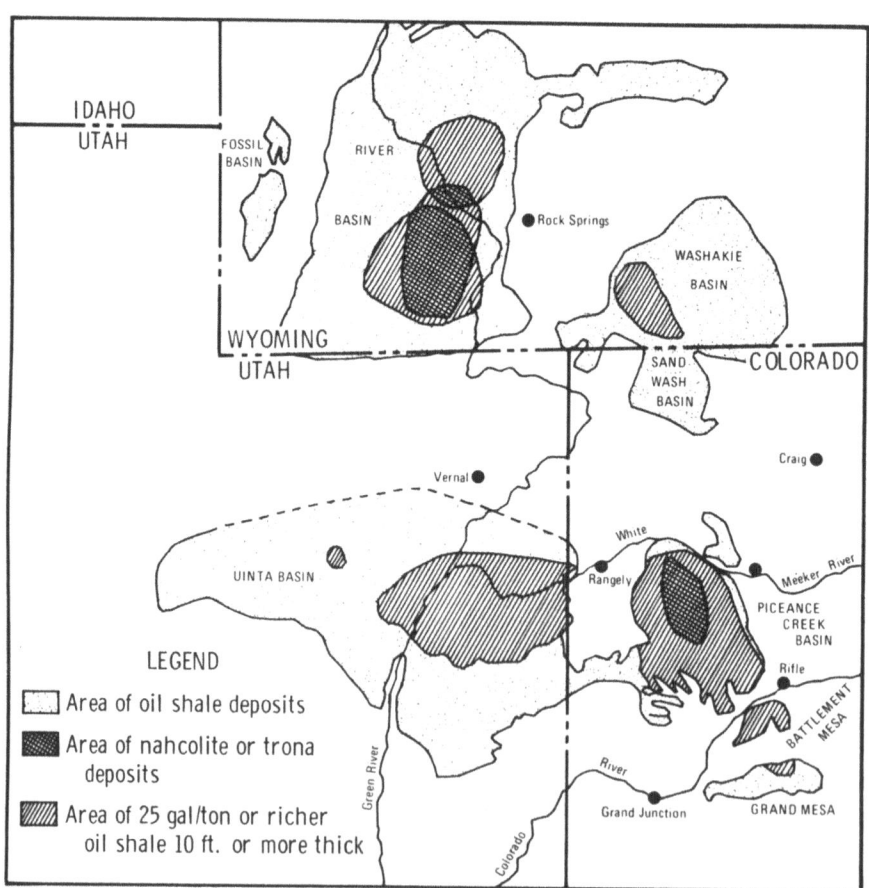

Fig. 2. Colorado, Wyoming and Utah oil shale deposits.

TABLE 1

Typical Oil Shale Composition, Weight Percent

(25 gallons of oil per ton)		
Organic Matter		13.8
Mineral Matter		
Carbonates (dolomite)	41.3	
Feldspars	18.1	
Quartz	11.2	
Clays (illite)	11.2	
Analcite	3.4	
Pyrite	1.0	
		86.2
	TOTAL	100.0

OIL SHALE PROCESSING

The low porosity of the marlstone containing the kerogen and the necessity of heating the shale to about 900°F before liquid or gaseous hydrocarbons are produced require either the mining of the shale and heating in above ground retorts, or the establishing of communication by fracturing, or other means, in the underground deposits so heat can be introduced, and the products formed can be removed. Quite logically, these are classified as *surface* and *in situ* processes.

Surface Retorting – All surface retorting processes require the oil shale to be mined and crushed prior to retorting, and then be conveyed to the retort by belts or vibratory conveyors.

In the retort, the oil shale is heated to temperatures ranging from 900° to 1000°F which thermally decompose the solid kerogen and convert it to gas and oil vapors. The surface retorting processes can be divided into two types, direct and indirect heating. The gas combustion process (including Paraho) is typical of the direct heating processes. The TOSCO process is the only indirect heating process that has been field tested using United States shale in a "semiworks" size plant (1000 tons of shale per day).

The gas combustion retort (Fig. 3) is a vertical, refractory-lined vessel in which the crushed shale, moving continuously downward, is heated to retorting temperature by direct gas-to-solid heat exchange with rising hot gases. Recycle gases entering the bottom of the reactor cool the hot retorted shale and are themselves heated almost to retorting temperature. They mix with injected air at a point approximately one-third of the way up from the bottom. The heat needed for the retorting is produced by burning the recycle gases and some of the carbon residue on the spent shale. These

References pp. 146-147.

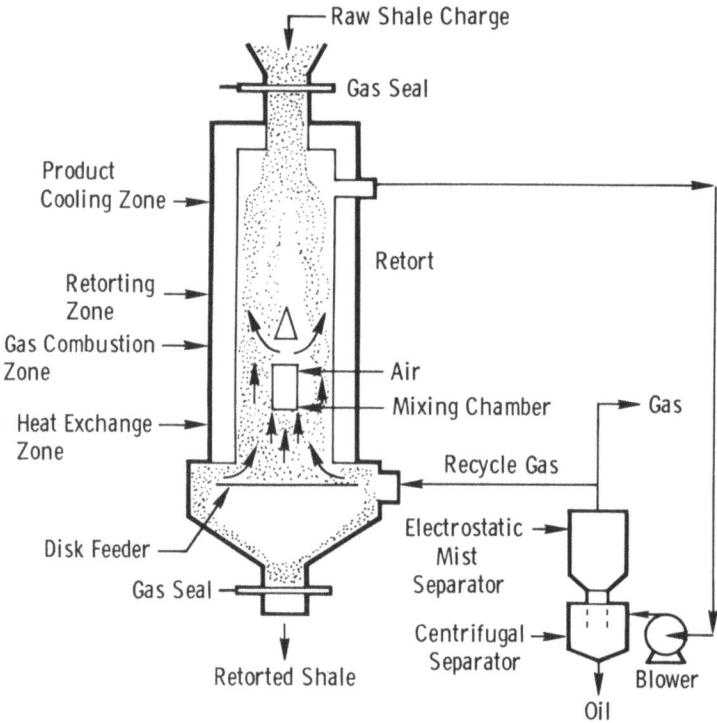

Fig. 3. Recycle gas is mixed with air and burned within the gas-combustion retort. Gases flow upward and shale moves downward.

hot gases heat the shale in the zone immediately above the air injection points to the retorting temperature of 900°F or above, and the kerogen is thermally decomposed to release oil vapor and gas. The hot gases containing the oil vapors are cooled by the entering raw shale as they pass upward and out of the retort to the oil separation system. The oil is separated by passing the cooled gases through mist collectors, and then part of the gas is recycled to the bottom of the retort. The remainder of the gas has a heating value of 80 to 100 Btu per cubic foot, and can be used as a low heating value gas in other areas of the plant.

Water is formed during oil shale retorting at rates from two to five gallons of water per ton of shale processed. This water passes up and out of the reactor with the gas stream, and much of it is condensed with the oil in the mist separation system. The water separated from the oil fraction contains a variety of organic and inorganic compounds which must be removed before the water can be used for subsequent plant or waste disposal (such as wetting spent shale) purposes.

The Paraho kiln is similar to the gas combustion retort, but has a unique raw-feed distributor and a patented discharge grate for the spent shale, and uses a multi-level gas injection system.

The TOSCO process (Fig. 4) uses an indirectly heated rotary-type kiln in which hot ceramic balls transfer heat to the shale. Raw shale is crushed then preheated and conveyed to a kiln feeder by hot flue gases from the ball heater. The flue gas is separated from the shale and the shale fed into the pyrolysis drum with the hot balls. There, it is brought to the retorting temperature of 900°F by conductive and radiant heat exchange with the balls which causes the solid kerogen to thermally decompose and produce a carbonaceous residue, oil vapors and gases. The kiln discharge is passed over a trommel screen to separate the balls from the spent shale. The hot gases are withdrawn to a fractionator where they are partially condensed and separated into water, shale oil, and a fuel gas having a heating value of about 800 Btu per cubic foot.

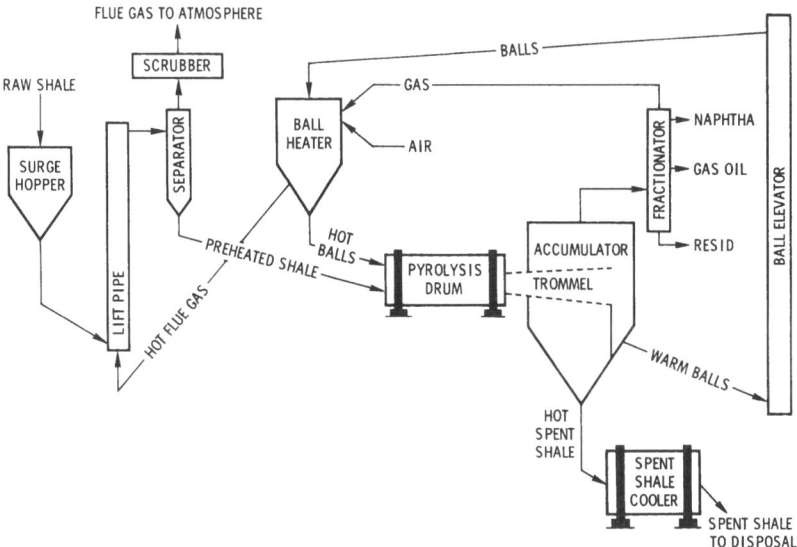

Fig. 4. TOSCO II process.

The separated balls are carried by an elevator to the ball heater in which they are heated to about 1250°F by burning a portion of the fuel gas produced in the process.

The water separated from the oil is similar to the water produced by gas combustion process and the organic and inorganic contaminants must be removed or neutralized before the water can be used elsewhere in the process. Two to five gallons of water are produced per ton of shale processed. A 50,000 barrels per day shale oil plant processing 25 gallon Fisher assay oil shale will produce from 170,000 to 420,000 gallons of process water per day.

In Situ Retorting – A number of petroleum companies and the U. S. Bureau of

Mines have been conducting experiments on the in situ production of shale oil for a number of years. This involves underground heating of the oil shale to retorting temperatures of 850 to 1000°F either by combustion in the formation, or by introduction of hot gases such as natural gas or superheated steam.

The major advantages of in situ processing are that it eliminates the mining, transportation, and crushing of the shale before retorting and the disposal of the spent shale after retorting.

The amount of material that must be handled in surface processing is massive compared to that handled in conventional mining operations. For example, more than 1-1/2 tons of shale must be mined, moved, crushed and retorted to produce one barrel of oil from a 25 gallon per ton analysis shale. For a 100,000 barrel per day surface plant, this means handling 150,000 tons of shale each day.

The largest underground mining operation in the world removes only 60,000 tons of rock per day and the largest open pit operation about 300,000 tons per day.

In addition, in the above-ground shale oil facility, over 100,000 tons of spent shale must be disposed of, as 80% of the mined material is waste. The difficulty of disposal is increased because the volume of the spent shale is about 12% greater than the volume of the shale before it was mined.

To retort oil shale in place, there must be some method of introducing permeability into the shale formation. Also, a method of heat injection must be developed that will cause pyrolysis of the organic matter in oil shale to yield useful products.

Oil shale has a low thermal conductivity (similar to fire brick) and this requires that some method of heat transfer other than conductivity be utilized. Heat flow also governs the ratio of heat effectively used in retorting to the total heat generated, and to fully utilize the fuel capability of the oil shale, efficient methods of heating the shale must be developed.

One method of in situ retorting that has been tested involves drilling wells through the oil shale in a pattern very much like that used in present waterflood or fireflood projects in conventional petroleum reservoirs. These patterns generally consist of a central-injection well surrounded by recovery or producing wells. Horizontal fractures must be created between injection well and recovery wells to provide paths for injecting the combustion air or hot gases and the movement of the shale oil to recovery wells. Spacing of the wells depends upon fracture distance, natural fractures, jointing, bedding planes, and overall economics (4, 5, 6).

Oil shale in place has little or no measurable permeability. Probably more than any other factor, ultimate success of in situ processing is dependent upon developing procedures for inducing satisfactory permeability in the formation. Techniques under consideration include explosive blasting (nuclear or conventional), hydraulic fracturing, electro-fracturing and combinations of fracturing and leaching (6).

Recently, a great deal of publicity has been given to one non-nuclear alternative method that is a combination of underground mining and in situ retorting. This is the Garret Process and requires strata of oil shale with a satisfactory thickness of both in-place shale and overburden. From 10 to 25 percent of the in-place shale is mined by the conventional room-and-pillar method and transported above-ground for surface retorting or disposal. Vertical holes are drilled upward in the ceilings of the underground rooms formed, packed with explosives and detonated. The fragmented rock falls into the mined-out voids. The shale-filled chimneys thus induced are retorted in situ by either combustion or hot gas circulation heating methods. The oil and gases formed are removed for further processing.

SHALE OIL PROCESSING

Crude shale oils are low API gravity, medium-sulfur, high-nitrogen, and highly-unsaturated oils (Table 2). The viscosities and pour points are higher than for many petroleum crude oils of similar gravity. This means that additional processing, such as vis-breaking, catalytic hydrotreating, or hydrocracking, must be performed at or near the retort site in order to reduce the viscosity and pour point to levels permitting the shale oil to be transported by pipeline during periods of cold weather. In general, more intensive and higher cost processing is required to produce satisfactory motor fuels from shale oil than from conventional crude oils.

Vis-breaking is a light-to-moderate thermal cracking operation which breaks the heavy paraffin and olefin molecules into particles of lower molecular weight, thereby lowering the viscosity and pour point of the oil. This serves the dual purpose of permitting the oil to flow satisfactorily during cold weather and reducing pumping

TABLE 2

Properties of TOSCO II Raw Shale Oil and Conventional Crudes *

	Conventional Crudes	TOSCO II Shale Oil
Gravity, °API	15-44	22
Sulfur, Wt %	0.04-4.1	0.8
Nitrogen, Wt %	0.01-0.65	1.9
Ni, ppm	0.03-45	6
V, ppm	0.002-348	3
As, ppm	0-0.030	40
Viscosity, SUS @ 100°F	31-1025	106
Conradson Carbon, Wt %	0.1-11.4	4.6
Bromine Number	negligible	49.5

*From Ref. 7

References pp. 146-147.

cost year around. Vis-breaking at the retort permits the shale oil to be pumped to a refinery for further processing.

Catalytic hydrotreating of the shale oil at the retort location replaces some of the processing done at the refinery if vis-breaking is used at the retort site. It has the disadvantage of requiring a substantial amount of water or portion of the light product to produce the hydrogen needed. However, catalytic hydrotreating not only lowers the viscosity and pour point of the shale oil, but can also reduce the sulfur and nitrogen contents to acceptable levels, and saturate the unsaturated components of the shale oil with hydrogen.

There are a number of processing sequences involving catalytic hydrogenation that can be applied to upgrading shale oil. Both the Final Environmental Statement for the Prototype Oil Shale Leasing Program prepared by the U. S. Department of the Interior (1), and the National Petroleum Council's Task Group on Oil Shale selected a processing scheme involving fractionation, delayed coking, and hydrocracking. This scheme was selected as the example to be discussed.

The shale oil is heated and partially vaporized in a gas- or oil-fired heater and fed to a distillation tower. The distillation tower separates the shale oil into an overhead gas stream, side streams of naphtha and light gas oil, and a heavy oil bottoms (residuum).

The heavy oil bottoms stream is sent to a delayed coking unit where it undergoes a severe thermal cracking process to produce coke and a vapor stream containing gas oil, naphtha and gases. The gas stream is returned to the flash zone of the distillation tower for separation into fractions.

The gas fraction from the distillation tower is sent to the gas processing section where hydrogen sulfide, ammonia, mercaptans, carbon dioxide, and low-boiling sulfides are separated and processed in a sulfur recovery unit.

The naphtha and gas oil fractions from the distillation tower are processed in a catalytic hydrogenation unit which removes the sulfur as hydrogen sulfide, nitrogen as ammonia, and saturates the olefins. The hydrocarbon stream product from the catalytic hydrogenation unit is a low-sulfur, low-nitrogen, storage-stable synthetic crude oil (Table 3). The ammonia and hydrogen sulfide are separated and the hydrogen sulfide is converted to elemental sulfur in a conventional Claus sulfur plant.

AUTOMOTIVE FUELS FROM SHALE OIL

A number of companies have produced finished motor fuels from shale oil, but only a few of the results have been released for publication. The most recent study was one carried out under the direction of the Navy Energy and Natural Resources Research and Development Office. Ten thousand barrels of shale oil produced by a surface process (Paraho) were processed into gasoline, jet fuel, diesel fuel, and heavy fuel oil (8).

TABLE 3

Properties of an Upgraded Shale Oil

Property	Value
Gravity, °API	46.2
Sulfur, wt %	0.005
Nitrogen, wt %	0.035
Pour Point, °F	below 50
Viscosity, SUS @100 °F	40

In addition, the Laramie Energy Research Center, Energy Research and Development Administration, has performed laboratory processing of crude shale oil to produce motor gasoline, diesel, and burner fuels (Table 4). The conclusions summarized here are based on the results determined in both of these studies.

Gasoline can be produced which meets all ASTM D-439 and Military F-46 specifications except that for oxidation stability. The F-46 specification requires a minimum induction period of 480 minutes while induction periods of above 360 minutes were obtained for the shale gasoline. Oxidation stability is a measure of gum formation in storage and usually can be related to the concentrations of unsaturated compounds in the gasoline. Additional hydrogenation under more severe operating conditions should eliminate this problem (3, 9, 10).

In the laboratory, diesel fuels were made which met all ASTM D-975, specifications for No. 1D, 2D, and 4D diesel fuel oils. However, some difficulty was experienced at the Gary Western Refinery in Gilsonite, Colorado due to limitations of the hydrotreating facility. Fuels produced had high wax and high gum contents which prevented the diesel fuels from meeting freeze point and thermal stability specifications. It is believed that with adequate hydrotreating facilities, no problem will be experienced.

Aircraft turbine fuels were produced at the Gary Western Refinery that met all ASTM D-1655 specifications for Jet A and Jet B fuels and military specifications MIL-T-56245 for JP-4 and JP-5 jet fuels except for some freezing points and net heats of combustion. Under more realistic processing conditions, neither of these should be a problem.

SUMMARY

In summary, it is estimated that a shale oil industry can be established which can produce one to one and one-half million barrels per day of shale oil. This is from 5 to 10 percent of the U. S. daily consumption of petroleum.

References pp. 146-147.

TABLE 4

Properties of Synthetic Motor Fuels from Shale Oil

		Diesel Fuel			Jet Fuel	
	Gasoline (8)	1-D (11)	2-D (11)	4-D (11)	JP5/ Jet A (8)	JP4/ Jet B (8)
Distillation, °F						
10%	130	400	442	450	376	232
20%	148	409	454	468	387	250
50%	215	422	484	507	422	292
90%	326	469	537	605	482	368
EP	385	517	554	628	518	405
Dist. Residue, %	1.0	2.0	1.0	2.0	1.0	1.0
RVP, psia (Max)	8.8					2.0
RON	82					
MON	91					
Cu Strip Corrosion	1A				1B	1B
Gum, mg/100ml	2.8					
Sulfur, ppm	30	100	200	200	27	13
Nitrogen, ppm		141	166	202		
Flash point, °F		188	212	265	140	
Pour point, °F		‹ 0	0	20	−26	−91
Gravity, °API		41.9	39.8	39.0	44.0	57.0
Carbon Residue on 10% Resid		0.14	0.16	0.36		
Ash, Wt %		‹ 0.001	‹ 0.001	‹ 0.001		
Vis. @ 100°F, cSt		1.66	2.40	2.96		
Cetane index		48	54	56		
Net heat of combustion, Btu/lb					18,160	18,340
Aromatics, Vol %					21.0	11.7
Olefins, Vol %					4.3	0.4

Technology is available to produce motor fuels meeting existing motor fuel specifications from Western oil shale, but more severe and extensive processing is required than for fuels produced from conventional crude oils. As a result, the synthetic motor fuels will be more costly than today's petroleum based fuels.

REFERENCES

1. U. S. Dept. of Interior, Final Environmental Statement for the Prototype Oil Shale Leasing Program, I2, 1973.
2. Anon., Colorado School of Mines Quarterly, 17 (4), 1922, p. 34.
3. C. M. Frost, H. C. Carpenter, C. B. Hopkins, S. S. Tihen and P. L. Cottingham, U. S. Bur. Mines

Rept. Inv. 5574, 1960, 17 pp.

4. B. F. Grant, *Colorado School of Mines Quarterly, 59 (3), 1964, pp. 39-46.*

5. V. D. Allread, *Colorado School of Mines Quarterly, 59 (3), 1964, pp. 47-73.*

6. J. H. Gary, *World Oil, 161 (2), 1965, pp. 98-101.*

7. E. D. Burger, D. J. Curtin, G. A. Myers and D. K. Wunderlich, *Amer. Chem. Soc., Div. of Petroleum Chem. Preprints, p. 765-775, Chicago Meeting, August 24-29, 1975.*

8. Applied Systems Corp., *Executive Summary Report: "The Production and Refining of 10,000 Barrels of Crude Shale Oil into Military Fuels," Office of Naval Res., Contract N00014-75-0055, June 1975.*

9. P. L. Cottingham, E. R. White and C. M. Frost, *Ind. Eng. Chem., 49 (4), 1957, pp. 679-684.*

10. M. O. Rosenheimer and J. R. Kiovsky, *Preprints, Div. of Petrol. Chem., A.C.S., 12 (4), 1967, pp. B147-B164.*

11. P. L. Cottingham and L. G. Nickerson, *Amer. Chem. Soc., Div. of Petroleum Chem. Preprints, Philadelphia Meeting, April 6-11, 1975, pp. 535-541.*

DISCUSSION

S. L. Meisel *(Mobil R&D)*

You mentioned that for above-ground retorting, considering the water problem, you're probably limited to something like 1,000,000 to 1,500,000 barrels a day of production. Is there a similar limit for in situ production of shale oil?

Gary

The 1,000,000 to 1,500,000 barrels per day limit is not only due to the water. It's also due to some other problems. The primary problem is the large number of people needed to support a mining industry that can mine the required shale. There are not many people out there. If you bring the people in to do the mining, you've got to build towns for them. You've got to build schools. You've got to bring in business, and you end up with a substantial number of people. This means we have to have sanitary facilities, water, and everything else that goes with it. The 1,000,000 to 1,500,000 barrels per day limit is based on these considerations.

Now, if we go to in situ work we're going to be limited by most of the same problems. If we use the Garrett process, the figures vary on how much material has to be mined, but the most optimistic is somewhere between 10 and 15%. That means that's the least we have to mine. If we want a million barrels a day, that means we're still mining 250,000 to 400,000 tons of shale a day, which is a substantial amount. If we could get a true in situ process that involved just drilling wells and then doing something like hydraulic fracturing, presumably we wouldn't necessarily be limited.

J. P. Longwell *(MIT and Exxon)*

If you look at the Canadian tar sands, the amount of mining per barrel of product is comparable to what's being considered here. They're undertaking this big mining operation, and I guess planning to get up to production rates comparable to those you're talking about. As far as very large scale handling of solids, they may be doing some of the pilot work, and they seem to have a faster time schedule.

Gary

They're also doing surface mining so that in some ways it's comparable to shale oil mining. But perhaps they don't have as big a disposal problem as we do with shale. They aren't limited by water. They have quite a bit of water.

THE INFLUENCE OF NUCLEAR ENERGY
ON TRANSPORTATION FUELS

J. L. RUSSELL

General Atomics Company, San Diego, California

ABSTRACT

Nuclear reactions are a potentially unlimited source of energy for the future. However, nuclear energy is not in a form easily used in transportation.

The only significant near-term influence of nuclear energy on transportation fuels is displacement of oil from the generation of electricity so that it becomes available for other uses. Currently in the United States, about nine percent of the electricity generated is from nuclear plants. By 1980, the nuclear portion will be between 15 and 20 percent and rapidly increasing.

In the mid-term (before the year 2000), nuclear energy may be providing process heat to coal gasification and liquefaction plants. Nuclear heat may be used to crack water into hydrogen and oxygen to provide feedstocks for chemical and synthetic fuel industries. Liquid hydrogen may become the fuel for large, long-range air transport. Also, nuclear explosives may be accepted as a means of stimulating production of otherwise uneconomical natural gas fields.

At some time in the future, use of fossil fuels for transportation will become uneconomic because of price pressure from other applications. How soon this will occur depends on many variables. However, nuclearly produced hydrogen, ammonia, and electricity are all potentially viable alternatives to carbon based fuels.

INTRODUCTION

The term "nuclear energy" commonly embraces four technologies – thermal fission reactors, breeder reactors, fusion reactors, and nuclear explosives. These technologies potentially provide essentially unlimited sources of energy for the

References p. 161.

future. With the possible exception of application to nuclear ships, these technologies are not in a form easily used in transportation.

Of these four technologies, only one, thermal fission reactors, is now in the marketplace as a commercial energy source. About 9% of the electricity now being generated in the U. S. comes from these nuclear plants.

The breeder reactor, with hopes of commercialization before the end of the century, is the object of a federally financed technology development program.

Fusion is very promising, but has yet to be demonstrated as a technical reality. Recent developments, however, give hope that scientific feasibility will be demonstrated before the end of this decade. This may make possible a commercial fusion industry early in the next century.

The energy of nuclear explosives has been suggested for many applications including gas and oil stimulation, geothermal stimulation, and shale recovery. Only the gas-stimulation concept has been expressly mentioned in ERDA summary documents (1). The idea of releasing large numbers of nuclear explosives underground has not yet been accepted as an attractive energy alternative.

Fusion and fission technologies are broadly described below. Also, the mildly complex way in which each technology can be made self-sufficient into essentially unlimited fuel reserves will be described. Finally, some of the technologies being considered for applying nuclear energy to areas other than electricity production are discussed.

THERMAL FISSION

The only significant near-term influence of nuclear energy on transportation fuels is to generate more electricity with nuclear power, thereby releasing oil for transportation. According to recent ERDA forecasts for the U. S. (2), the current 9% of the electricity generated from nuclear plants will grow by 1985 to around 28%. This will amount to about 10% of the total primary energy consumed in the U. S. and will be an oil equivalent of around 4.7 million barrels per day that would not otherwise be available for transportation. These same forecasts suggest that the total U. S. energy consumption during that time will increase by around 35%. Clearly, during this period nuclear energy will relieve but cannot completely solve the anticipated energy deficit. However, with favorable resolution of the current regulatory, financial, and public acceptance problems, the nuclear industry could supply most of the U. S. electrical energy requirements by the end of the century. In round numbers, this industry has produced some 60 electric power stations and has orders for around 150 more to be delivered over the coming 10 to 12 years. About one-third of these reactors are boiling water reactors (BWR), the other two-thirds being pressurized water reactors (PWR).

There is one high-temperature gas-cooled reactor (HTGR) in the U. S. The HTGR, because of its higher temperature capability, may provide the key to expanding the use of nuclear energy into fields other than electricity production.

BREEDER REACTORS

Thermal reactors consume principally the uranium isotope U-235, which is present in normal uranium only at about 0.71% by weight. The other isotope, U-238, is only minimally utilized. On the other hand, breeder reactors, by converting the U-238 to plutonium, use all the uranium as fuel. They can also convert thorium to U-233, which is also fissionable. The various kinds of reactors then could feed fuel to each other in complex ways, with all the uranium and thorium eventually being used, rather than the fraction of a percent used by present thermal reactors. Thus it might be feasible to mine granite, or perhaps even the oceans, for the trace concentrations of uranium. In this way, the fission energy supply could be measured in millions of years.

Two kinds of breeder reactor are receiving significant research attention: the liquid metal fast breeder reactor (LMFBR); and the gas cooled fast reactor (GCFR). The former is the object of the primary effort of the U. S. government to develop a commercial breeder reactor before the end of the century. The GCFR is at present an advanced technology back-up.

The role of the GCFR in the nuclear energy economy can be explained by considering the three phases from 1970 to 1990, from 1990 to 2020, and after 2020. The first enrichment-fueled thermal converter phase is shown schematically in Fig. 1. The large numbers of LWRs (Light Water Reactors) will result in a growing plutonium stockpile. The thorium-fueled HTGR will grow in importance.

Introduction of the plutonium-fueled fast breeders will lead to a transition phase from an enrichment-fueled toward a breeder-fueled economy as shown in Fig. 2. The plutonium supply available from LWRs will allow rapid introduction of fast breeders. As the proportion of breeders increases and the energy growth rate slows, the excess fuel production capability of fast breeders will provide U-233 fuel for advanced thermal converters, HTGRs, and possibly LWBRs (Light Water Breeder Reactors). The requirement for uranium enrichment will level off and phase out as the associated LWRs are retired from service.

As shown in Fig. 3, after about 2020 a stable breeder-fueled advanced thermal converter phase can be reached in which high gain, low doubling time breeders provide both plutonium fuel for themselves and for energy growth, and U-233 fuel for advanced thermal converters. Dependence on uranium enrichment will be eliminated and the full energy potential of uranium and thorium ores will be available. Gas-cooled reactors, GCFR and HTGR, have superior characteristics to fill the roles of high gain breeder and advanced thermal converter.

References p. 161.

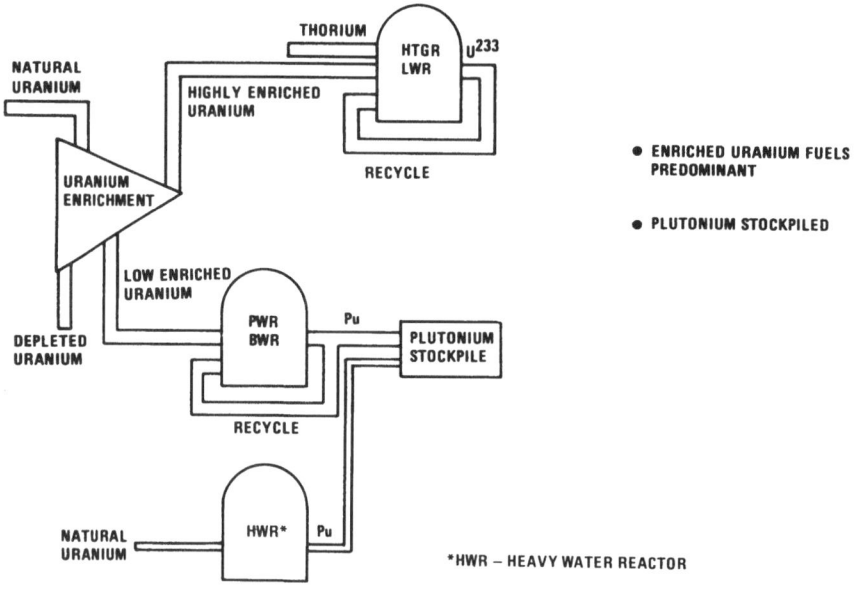

Fig. 1. Nuclear economy 1970-1990.

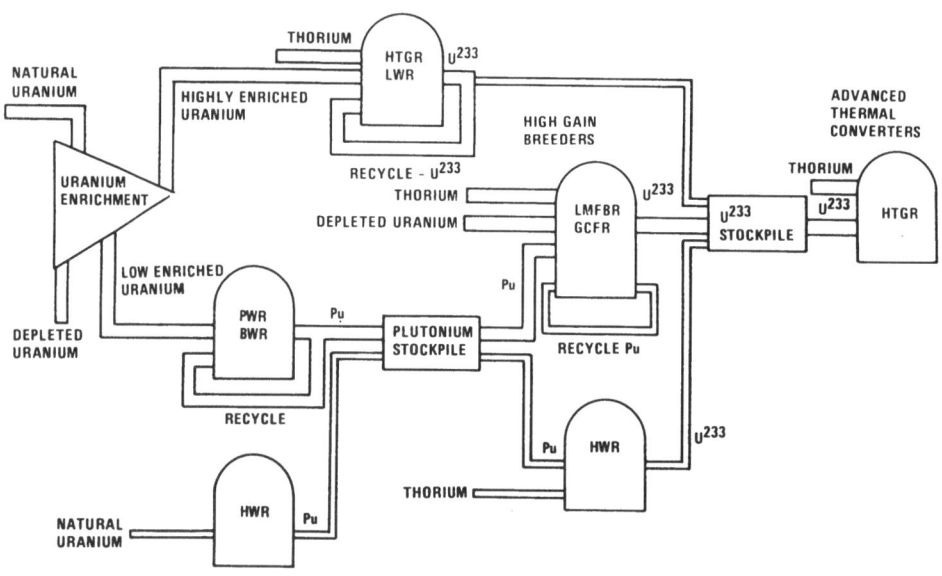

Fig. 2. Nuclear economy 1990-2020. Transition phase from enrichment-fueled to breeder-fueled.

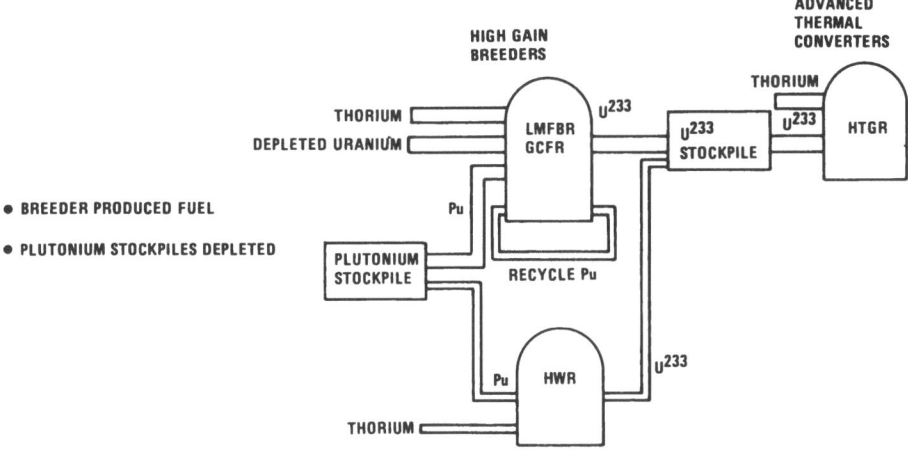

Fig. 3. Nuclear economy after 2020.

The number of HTGR plants that can be fed by one FBR depends on the HTGR conversion ratio and the FBR breeding ratio. The high breeding ratio of the GCFR is advantageous in such a combination. In a symbiotic breeder/converter economy, Fig. 4 shows that combining a GCFR with a breeding ratio of 1.47 and an HTGR with a conversion ratio of 0.87, three HTGRs can be fed by one similar-sized FBR.

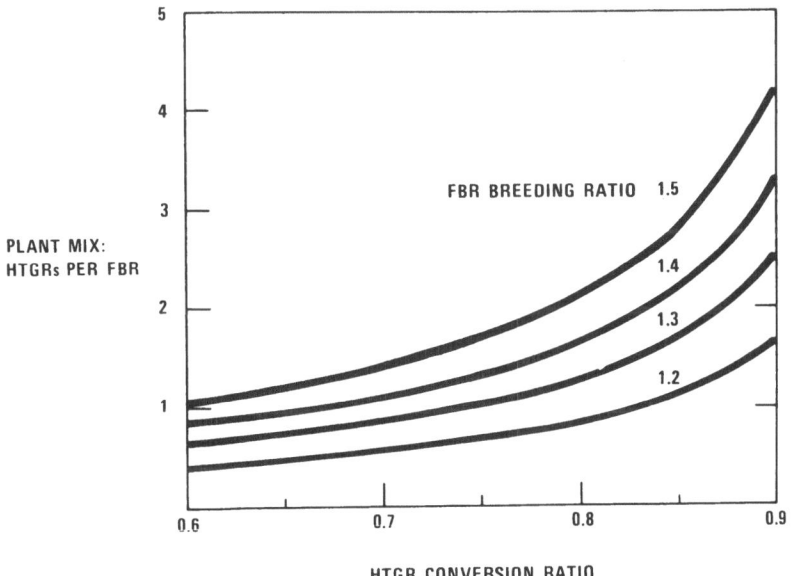

Fig. 4. HTGR and FBR combination in a self-sustaining system at equilibrium.

References p. 161.

Self-sustaining breeder/converter complexes can be operated such that some of the fissile material generated by the breeder is reserved for inventory of new electric generating capacity rather than consuming all of it as make-up for existing plants. System growth is thus achieved by reducing the number of converters supported by one breeder.

It can be seen in Fig. 5 that while low-gain breeders may yield a maximum self-sustained generating capacity growth rate of about 3.5%, a high-gain breeder may yield a maximum growth rate as high as 9% per year.

This figure demonstrates that the early rapid growth of a breeder industry cannot fuel itself. The plutonium must come from the LWR inductor. For example a typical LWR would produce some 0.06 kg/year of Pu, per MW(t*). This could provide for a 10% annual growth rate of breeders totaling about half the power of the LWRs assigned to this duty, assuming an out-of-pile inventory fraction of 1/3 and a breeder rating of 1 MW/kg fissile (in core). The total number of LWRs in fact coexistent with this breeder-HTGR association thus could far exceed the latter's demands for the most rapid expansion that practical building would allow.

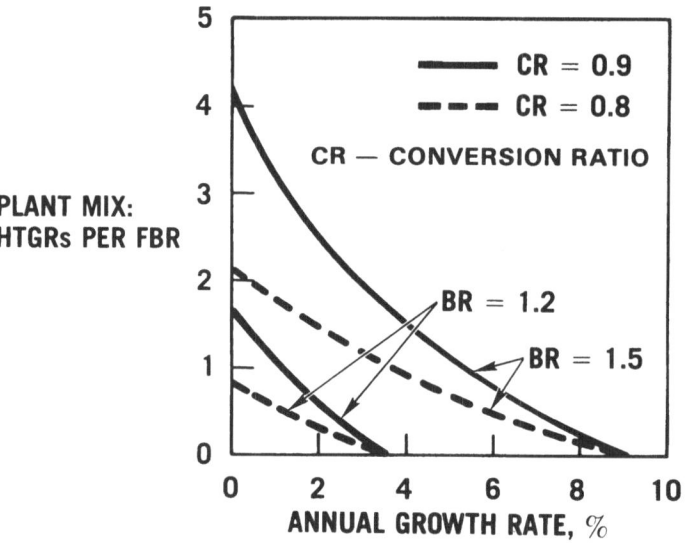

Fig. 5. Growth potential of symbiotic systems.

Thermal

The nuclear efficiency of the thermal reactors also influences the number that can be supported by a breeder. Thermal reactors with high conversion ratios (such as HTGR) can supply a significant portion of their own fuel requirements. Low-gain breeders will not achieve self-sufficiency at high growth rates of nuclear electrical generating capacity. Significant growth rates in this instance will require continued use of expensive U-235 enrichment processes to provide fuel for thermal reactors. At lower growth rates the industry can achieve self-sufficiency once the appropriate overall plant mix is reached. The appropriate plant mix is defined such that the surplus fissile material produced by the breeders is used as make-up fuel for thermal reactors and as inventory for new thermal and fast reactors according to the prevailing growth rate. With increasing nuclear performance of the reactors and with varying annual growth rates of the nuclear industry, the overall fissile balance can be maintained by proper adjustment of the plant mix. The ranges of attainable growth rate and plant mix are rather limited if the breeding ratio of the breeder is small. High breeding ratios, however, result in larger ranges within which self-sufficiency is achievable.

FUSION

Fusion derives its energy from the combining of light elements to form heavier and more stable ones. It is generally conceded that the only such reaction that presently conceivable machines are likely to be able to handle is the D-T reaction. Deuterium (D) is a heavy isotope of hydrogen, H-2, while tritium (T) is an even heavier isotope of hydrogen, H-3. These combine in a fusion process to form helium and to release a neutron and a great deal of energy. Deuterium is present in ordinary hydrogen (in water for example) at 0.015% by weight. It is relatively simple to remove from water, making its supply almost unlimited. Tritium on the other hand, is unstable and has a 10.8-year half-life. It is not found in nature, but it can be produced by neutron irradiation of another abundant element, lithium. Therefore, by properly arranging lithium around the fusion reactor to absorb the fusion-produced neutron, it can breed enough tritium to feed itself. This procedure leads, again, to a fuel with a supply measured in millions of years.

Over a dozen government-sponsored and private laboratories across the U. S. are seeking to verify the scientific feasibility of controlled thermo-nuclear fusion. All are or have been experimenting with a least one of the four following confinement systems:

- Tokamak-like toroidal
- Magnetic mirror, open
- Theta pinch, pulsed
- Laser fusion

Although all these systems are being applied to the familiar problems of plasma

References p. 161.

heating and confinement, their use is also yielding valuable information on fusion engineering and technology. In 1952, Project Sherwood, the nation's first effort to control for peaceful purposes a thermonuclear reaction, was created. Project Sherwood produced the whimsical Perhapsatron, the Columbus, the Linear Dynamic Pinch, the Astron, and other plasma-study devices. Its major programs were:

- The Pinch Program
- The Stellarator Program
- The Magnetic Mirror Program

Project Sherwood also produced the Scylla theta pinch device, in which plasma temperatures above the ignition level were truly reached for the first time. And as ironic as it may seem, perhaps the most important of Project Sherwood's discoveries is that no obstacle to prevent our someday attaining the final goal of controlled thermonuclear fusion has ever been uncovered.

Also, much valuable work has and is being accomplished at the ERDA-supported Princeton Plasma Physics Laboratory and the Lawrence Livermore Laboratory. Within the past two years, Princeton's Adiabatic Toroidal Compressor Tokamak has proved for the first time that plasma temperature could be increased with neutral beams and also with compression. The world's largest mirror machine is the 2X II device at Lawrence.

The Oak Ridge National Laboratory also started plasma research for fusion applications in 1952, but it was not until 1955 that extensive study began there. Today, the ORNL ORMAK device, which has demonstrated size scaling for Tokamaks, is testing high-power neutral-beam heating.

There is a high probability that the program at General Atomics, begun in 1957, will be the one to first demonstrate the plasma conditions of temperature, density, and confinement time required to "break even." This is scheduled to occur in 1978. Fig. 6 shows a conceptual drawing of the Doublet III which is now being developed for this important scientific demonstration.

Table 1 shows the presently anticipated milestones for the U. S. fusion program as presented by ERDA (1).

Once fusion becomes a practical, commercial heat source, sometime in the next century, it can be used in the same way as fission reactors for generation of electricity and for process heat. There probably will be some special applications of fusion not presently foreseen that use some of the special features of the fusion process.

FUELS

Most primary energy is not used for making electricity. Some 70% is used directly for transportation fuels, industrial process heat, and domestic applications. Near the

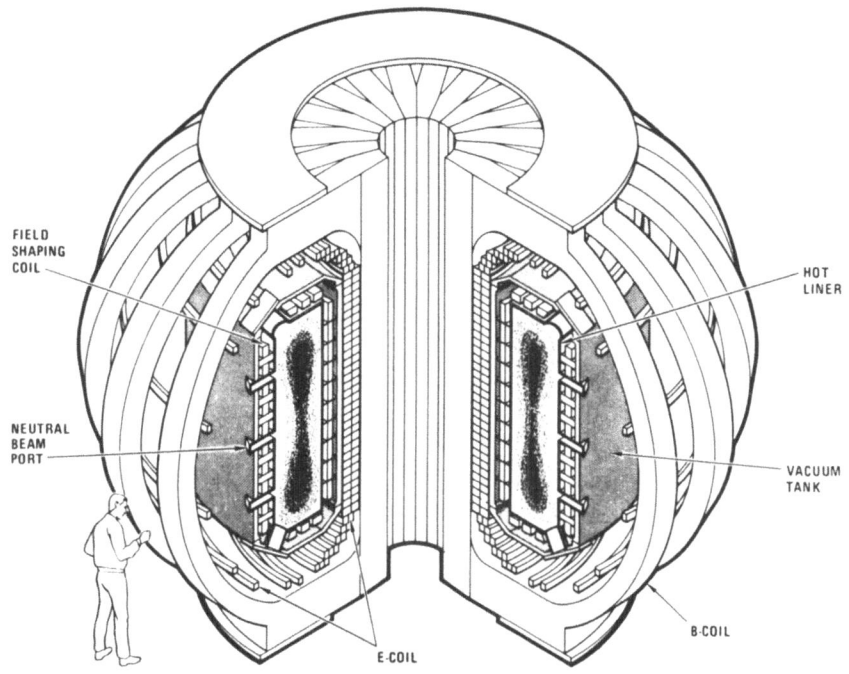

Fig. 6. Doublet III — artist's conception.

TABLE 1

Fusion Power Major Milestones

Milestone	Expected Date
Scientific Feasibility (Doublet III)	1978 – 1980
D-T Burning Experiments (TFTR-TCT)	1980 – 1982
Experimental Power Reactor I	$\sim$ 1985
Experimental Power Reactor II	$\sim$ 1990
Demonstration Power Plant	$\sim$ 1995
Commercial Power Plants	> 2000

end of the century, nuclear energy might begin to play a significant role in transportation fuels in that it could be used to make synthetic fuels. The processes by *References p. 161.*

which these synthetic fuels are made are very energy intensive. Nuclear energy can supply some, or for one process all, of this energy.

Coal gasification, a steam-plus-carbon reaction, is now done by heat supplied from the burning of fossil fuels.

The hydrogen-plus-carbon method of coal gasification (3) begins with steam methane reforming at 1200° to 1600°F, a heat-range well suited to today's HTGR technology. The second step of this method is the reaction between the hydrogen and the carbon to form methane. The sum of the reactions in this process is equal to the reaction shown below for steam plus carbon.

$$2C + 2H_2O + HEAT \rightarrow CH_4 + CO_2$$

The hydrogen-plus-carbon process may be traced through the HTGR core and reactor vessel (Fig. 7). Helium flows downward through the core and is heated to 1620°F. It passes through radial ducts to reformers and upward among the reformer tubes where heat is transferred to the steam-methane mixture. The helium then flows through the adjacent steam generators where it gives up additional heat. Then it passes to the helium circulator where it is compressed and discharged through the upper horizontal ducts back into the core.

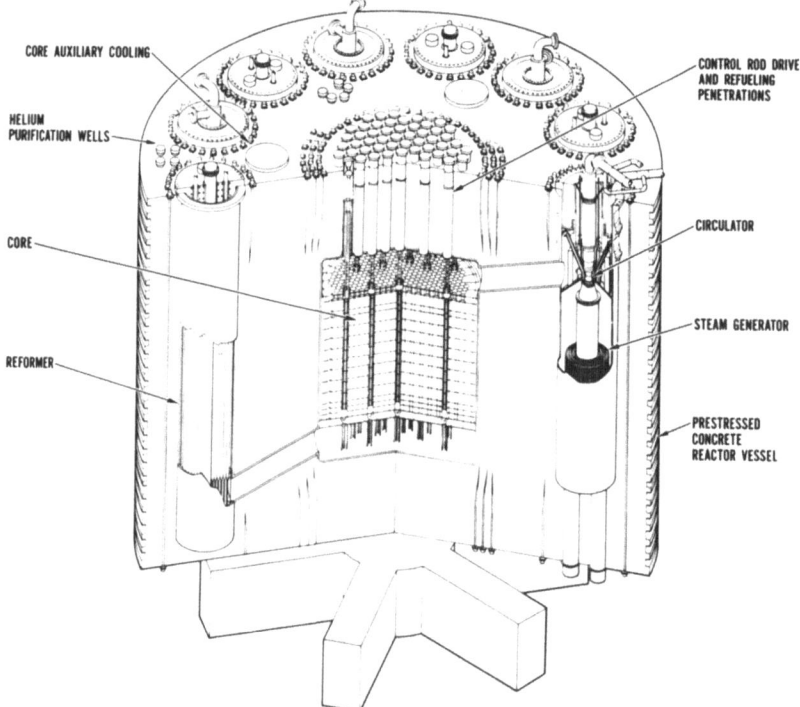

Fig. 7. HTGR process heat source.

Most features of the process-heat HTGR are identical to those of the HTGR used for electric power generation.

The reformer is the one new piece of equipment in this HTGR configuration. Its design probably will differ significantly from that of the steam generator, but experience and technical information gained in the latter's development will be valuable to the reformer design.

The safety implications of coupling a nuclear with a chemical processing plant will require careful technical analysis. Operational requirements and limitations of the two plants must be identified, and a suitable control and safety system must be developed.

The large commercial HTGR presently offered to utilities has a heat rating of 3000 MW(t). Table 2 shows typical values for a coal-gasification plant using this size HTGR core.

TABLE 2

HTGR Coal Gasification Plant
Typical Values Anticipated

Reactor Heat	3000 MW(t)
Coal Feed	12×10^6 tons/yr
Gas Produced	630×10^6 scf/day
Water Consumed	576×10^3 gallons/day

The process flow diagram, Fig. 8, showing the helium heat transfer loop in the coal gasification plant, is applicable to an HTGR producing hydrogen as an end product. For applications requiring high-purity hydrogen, reformer operating conditions can be altered to decrease the methane fraction in the final product. High-purity hydrogen is not always needed, however, in such processes as coal gasification or in the manufacture of ammonia.

The feedstock from which the hydrogen is produced could be natural gas, petroleum, or coal.

The HTGR steam generators produce the steam consumed in the reaction and in driving the steam turbines. Even though the plant output may be 100% hydrogen, i.e., no net electric power produced, substantial power is required for hydrogen compressors and other in-plant uses.

As carbon reserves diminish, or perhaps as economics and/or ecological considerations dictate, hydrogen may become an important transportation fuel. The development of an efficient and economical process for making hydrogen from water

References p. 161.

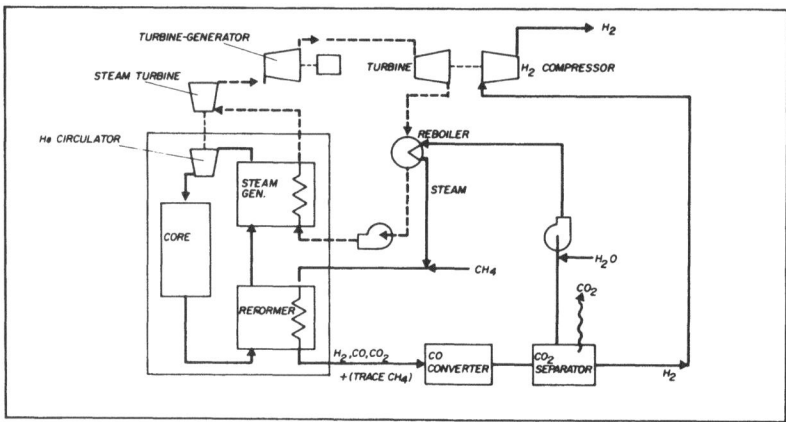

Fig. 8. Process flow for HTGR coal gasification plant.

and nuclear heat (water splitting) lies before us (4). The degree of success in developing it will determine the extent to which hydrogen and hydrogen compounds as intermediate energy forms will be used.

In the early 1960s, the U. S. Army sponsored research on the use of energy from a nuclear reactor to decompose water into hydrogen and oxygen. The hydrogen could be combined with nitrogen from the air to produce ammonia, which could be used as a gasoline substitute. Water was to be decomposed by either electrolysis or a closed thermochemical cycle (water splitting). The latter would avoid the high capital costs and low overall efficiencies associated with electrolysis. An adequately efficient thermochemical cycle was not developed, however, before the work was terminated in 1964.

Direct thermal decomposition of water by heat occurs only at such high temperature that it is impractical to consider this as an engineering process. But, if the water is reacted with certain chemical compounds in a specified sequence, the decomposition of water into hydrogen and oxygen occurs at much lower temperatures. Heat is added at some of the various steps and the chemicals are recycled. The significant point is that the temperatures required in the various steps are much lower than the impractically high temperature required for the direct thermal decomposition. In fact, temperatures for some recently proposed cycles may be as low as 1000 to 1200°F.

Promising results from the water-splitting program of the Euratom Laboratory at Ispra, Italy, have caused renewed interest in the thermochemical cycles. Dozens of different processes of this kind have been published recently. An example is the one proposed at Ispra, shown in Table 3, which illustrates how the chemicals are recycled.

Computer studies indicate the existence of several hundred similar thermochemical

TABLE 3

Ispra Mark I Cycle

$CaBr_2 + 2H_2O \rightarrow Ca(OH)_2 + 2HBr$	1350°F	Endothermic
$2HBr + Hg \rightarrow HgBr_2 + H_2$	480°F	Exothermic
$HgBr_2 + Ca(OH)_2 \rightarrow CaBr_2 + H_2O + HgO$	212°F	Exothermic
$HgO \rightarrow Hg + 1/2 \ O_2$	1100°F	Exothermic

cycles. Research at several laboratories around the world is now directed toward the problem of selecting efficient cycles and developing them to commercial applicability.

Heat to drive the reactions can come from a nuclear reactor. It would appear that in the relatively near future, an essentially unlimited supply of hydrogen may be technologically and economically possible to produce for use as a chemical fuel. The most significant limitation on its availability, assuming the economic barriers are overcome, will be the time and effort required to build a nuclear-chemical industry.

REFERENCES

1. "A National Plan for Energy Research, Development and Demonstration," ERDA-48, Vol. 2 of 2.
2. "Nuclear Power Growth, 1974-2000," WASH-1139 (74), February 1974.
3. R. N. Quade, "High-Temperature Gas-Cooled Reactors for Process Heat," Power Engineering, April 1973.
4. J. L. Russell, Jr., "Nuclear Water Splitting and the Hydrogen Economy," Power Engineering, April 1974.

DISCUSSION

S. L. Meisel (Mobil Research and Development Corp.)

We agree, of course, that hydrogen is the perfect fuel. All it does is produce water. Someplace I read that you actually do get a small amount of byproduct hydrogen peroxide and I wonder if you would like to comment on this?

Russell

That's not my field. I try to make the hydrogen, but there are several people here who are experts in that field. My feelings are that you have to have very good instruments to measure it. Would anyone like to volunteer?

P. C. T. deBoer *(Cornell University)*

We have not been able to find any evidence that this is really a problem.

S. Gratch *(Ford Motor Company)*

How much energy would be required to build those 1,350 nuclear reactors?

Russell

About 1/10 what they put out over their lifetime.

SESSION II — SUMMARY

M. A. ELLIOTT

Energy Consultant, Houston, Texas

I'm not going to attempt to summarize the entire discussion. What I am going to do is get a little bit of my own brand of truth in here.

The one thing that I think came through in all of the presentations was that there is a real logistics problem involved in the establishment of a synthetic fuels industry. Not only are nuclear reactors suffering delays, but liquefaction plants are in competition with gasification plants, and there isn't enough pressure vessel construction to suit everyone's needs. We don't need to belabor this point. But this is something that is very important.

In addition, there are terrific institutional constraints. Just an example of what these are costing you. In the case of existing coal gasification technology, the cost of gas in two years has gone up $1 per million BTU's. This is at the Texas Eastern Four-Corners plant. That plant's cost is now $850,000,000. By the time it's built it's going to a billion. You will never get one of these plants built when there's that much capital involved until: (a) the regulatory agencies take a reasonable attitude, and (b) the government guarantees the loan during construction. These are some facts of life that we all ought to recognize when we look at the real world of establishing a synthetic fuels industry in this country.

The other point that I think came out loud and clear was the question of handling a lot of material. This is something that people who have never been in the coal business or in the mining business really don't understand. It's the reason, for example, that you can't save much money on a coal gasification plant by new technology because the coal gasifier is only 15-20% of the capital investment. You've got coal handling, coal preparation, gas clean-up, and all the down-stream facilities that are common to everything. So here again is a problem that arises simply because you're handling a lot of material.

Now, the final comment on my brand of truth, which, as you see was culled from the papers today, is that we better get on with building some kind of a plant, because we need a yard stick. What are you going to compare new technology with if you've never built a Lurgi gasification plant in the real world? In the real world that plant went up from $350,000,000 to $850,000,000 in two years. Costs are changing day-to-day. Let's get on with it and build one.

SESSION III
AUTOMOTIVE UTILIZATION
OF INTERMEDIATE—TERM FUTURE FUELS

Session Chairman
J. B. HEYWOOD

Massachusetts Institute of Technology
Cambridge, Massachusetts

CHARACTERISTICS OF CONVENTIONAL FUELS FROM NON-PETROLEUM SOURCES — AN EXPERIMENTAL STUDY

R. W. HURN

Energy Research & Development Administration (ERDA), Bartlesville, Oklahoma

ABSTRACT

Experimental data concerning physical properties and engine performance characteristics are reported for a gasoline derived from oil shale and for a gasoline containing a coal-derived component. The paper includes brief descriptions of the materials from which the fuels were derived.

In general, the use of gasolines from shale or coal did not result in unusual short-term effects on engine emissions, engine behavior, or fuel economy.

A more definitive assessment of performance and engine effects in long-term use will require much more extensive testing. This in turn requires increased availability of test fuels from a synthetic fuels development program.

INTRODUCTION

Non-petroleum resources currently are forecast to provide steadily increased amounts of transportation fuels beginning in the middle 1980's. While the conversion technology remains to be fully developed, it is probable that liquid fuels from supplemental sources will be refined from crude materials that will be produced either from retorting oil shale or from partial hydrogenation of coal. To produce refined liquids, these supplemental "crudes" may either be used separately as refinery feedstock, or fed to the refinery as a mixed charge of natural and synthetic crudes.

Products from natural and from synthetic crudes may not be readily distinguishable. However, without question, refinery *process* requirements and the applicable technology will be affected by changes in the composition of the refinery charge caused by a switch to, or inclusion of synthetic crudes. In addition to having an impact on the applicable process technology, the synthetic material — shale or coal-derived — also may impart unique characteristics to the refined *product*.

In a technological sense, of the two effects — that pertaining to process requirements and that pertaining to product characteristics — the former probably far outweighs the latter. Even so, effects related to product characteristics could be highly significant in properly accommodating engines and related accessory systems to the fuels that will be available. It is this later area that this discussion will be directed.

As a first approach to an experimental study of conventional fuels derived from non-petroleum sources, work has been done with a gasoline produced from shale oil and with one produced from a coal-derived synthetic crude (syncrude). Both gasolines were produced using technologies and process parameters believed appropriate for synthetic fuels processing. However, because validation experience and test data were lacking, the choice of process steps was not necessarily optimum. In fact, certain of the resultant fuel properties clearly were unacceptable under current fuel quality criteria; therefore, no significance should be placed upon the absolute values of the fuel properties. Instead, these exploratory findings are expected only to outline the quality of fuel differences that may result from use of a non-petroleum resource, and to provide a basis for estimating the severity of effects that may be encountered.

PREPARATION OF FUELS

The crude shale oil from which the shale gasoline was refined was produced from shale mined from the Naval oil shale reserve, Anvil Points, Colorado. Approximately 15,800 tons of shale rock were retorted in a 56-day continuous run to yield just under 10,000 bbl of crude shale oil. This crude compares with natural domestic crude as follows:

	Shale crude	Natural crude
Gravity, °API	15.1	15-44
Sulfur, wt. percent	0.065	0.04-4
Nitrogen, wt. percent	1.89	0.01-0.7

In the subsequent refining operation, the crude shale oil was fed to a coker/fractionator to yield naphtha, light gas-oil, heavy gas-oil, heavy fuel-oil, coke, and gas. Of these products only the naphtha was used for gasoline production via hydrotreating and reforming. Total refinery input of 9,956 bbl of crude shale yielded 5,761 bbl of product as follows:

> 725 bbl of gasoline
> 454 bbl of JP-4 turbine fuel
> 650 bbl of JP-5/Jet A turbine fuel
> 1,167 bbl of No. 2 diesel fuel

2,765 bbl of heavy fuel-oil

Operations and results in the production of the shale-derived fuels are described in detail in a report to the primary sponsor of the project, the Navy Energy and Natural Resources R&D Office.*

The coal-derived gasoline was refined from a syncrude produced by the COED** process using a Utah coal feedstock. The hydrogenated syncrude was a relatively low-sulfur material of 19.4 API gravity. Subsequently, the syncrude was fractionated to yield test fuels: (1) a straight-run naphtha of 400° F end point (11.8 vol. percent of charge), (2) a 400° to 600° F straight-run distillate (37.3 vol. percent of charge), and (3) 600+ °F bottoms (50.9 vol. percent of charge). Some properties of the fractions are given in Table 1.

TABLE 1

Composition and Properties of Utah Syncrude Fractions

Boiling range	Syncrude as received	IBP-400°F	400-600°F	>600°F +
Yield, vol. percent of crude	100.0	11.8	37.3	50.9
RVP at 100° F, psia	—	1.0	—	—
API Gravity, 60° F	19.4	37.2	22.4	14.2
ASTM Distillation, °F:	D-86	D-86	D-86	D-1160
IBP	195	178	410	644
10 percent	396	257	460	691
50 percent	635*	324	505	801
90 percent	—	379	568	956
End point	—	410	606	—
Sulfur, wt. percent	0.07	0.12	0.03	0.02
Nitrogen, wt. percent ...	0.25	0.25	0.39	0.37

*Cracking observed above 700° F, distillation discontinued.

The heavy fraction (>600°F) was further processed in a fluid catalytic cracker to yield: (1) a 400°F end-point catalytic naphtha, (2) a 400° to 600°F light catalytic gas-oil. The catalytic naphtha then was blended with other refinery stocks to produce a full-boiling range gasoline with a nominal Reid vapor pressure (RVP) of 9 psia and a nominal clear research octane number (RON) of 96.

* The Production and Refining of 10,000 bbl of Crude Shale Oil into Military Fuels. Applied Systems Corporation, Vienna, VA 22180, June 1975.

**Char Oil Energy Development. The FMC Corporation, Princeton, NJ.

SHALE OIL — DERIVED GASOLINE

Physical and Chemical Properties — In brief, the distillation and hydrocarbon type characteristics of the shale oil-derived gasoline (Table 2) are not unlike those of many commercial gasolines. However,

TABLE 2

Properties of Shale Oil - Derived Gasoline

API Gravity, 60° F	87
ASTM Distillation, °F:	
IBP	114
10 percent	125
30 percent	158
50 percent	194
70 percent	261
90 percent	328
End point	387
Reid Vapor Pressure, psia	10.5
Sulfur, wt. percent	0.024
Nitrogen, wt. percent	0.03
Oxygen, wt. percent	0.52
Hydrocarbon types (FIA), vol. percent:	
Aromatics	28.0
Olefins	14.0
Saturates	58.0
Motor Octane No. (MON)	76.0
Research Octane No. (RON)	81.1
Stability data:	
ASTM induction period, min	1440+*
ASTM existent gum, mg/100 ml	25
Accelerated gum (4 hr), mg/100 ml:	
2.0 lbs DMD per 1,000 bbl	32

* 40 lbs of pressure was lost during the entire 24-hr period; however, at no time did it lose 2 lbs in a 15-minute period.

anti-knock quality and fuel stability of this particular gasoline are poor and totally preclude its use as a commercial motor fuel in the form produced. Additional or alternative refinery processing would be required to yield a product acceptable for use in automobiles of current design. While assessment of the applicable or appropriate refinery technology is beyond the scope of this discussion, it appears that available technologies should be adequate to correct the product deficiencies.

With regard to the above, the Navy's contractor observed "these fuels tended to exhibit storage and thermal instabilities. In addition, the fuels contained a high wax content, high particulate matter, and high gum content. It is believed that a higher pressure in the hydrogenation stage (about 2,000-3,000 psi), along with clay treatment of the final products, would reduce or eliminate some or most of these problem areas"(1).

In connection with engine tests of the shale-derived gasoline, its poor octane quality (76 MON, 81 RON) combined with the gum and stability problems make it doubtful that extended tests would provide useful information. With respect to costs involved in any process adjustments to upgrade the product, one can only speculate that a cost penalty is involved. However, even a gross cost estimate is premature without further definition of the process requirements.

Engine Test Results — Although the shale gasoline may not be suitable for extended engine work, data were taken to provide information on emissions and fuel economy. For this work, non-catalytic converter equipped cars were selected in order to observe engine-out emissions. Results are summarized in Table 3.

TABLE 3

Emissions and Fuel Economy Results With Shale — Derived Gasoline

| Fuel | Vehicle | Emissions, g/mile | | | | Fuel Economy, mpg | |
		CO	HC	NO$_x$	Aldehydes	Urban Cycle	Highway
Shale gasoline	1974 Model,	25.8	2.2	2.6	0.14	11.2	17.4
Indolene	351-CID	25.9	2.3	2.7	0.15	10.9	17.7
Shale gasoline	1975 Model,	7.4	1.3	2.7	0.18	12.3	21.1
Indolene*	318-CID	8.7	1.2	2.5	0.17	12.2	20.6

Represents one analysis, all other data represent the average of 3 replicates.

None of the observed differences are consequential. Engine knock, associated with poor octane quality, was the only problem with the shale-derived gasoline.

GASOLINE DERIVED FROM COAL

Physical and Chemical Properties — The light straight run naphtha fraction from distillation of the Utah syncrude was found to be an exceptionally complex material of approximately 55 percent paraffins, 14.5 percent olefins, and 30.5 percent aromatics. Iso- and cycloparaffins accounted for about 85 percent of the paraffins and for the extraordinary complexity of the fuel. The material had reasonably good octane quality and tetraethyl lead (TEL) response:

	Clear	+0.5 ml TEL	+3.0 ml TEL
RON	84.5	86.9	92.4
MON	78.1	79.8	83.6

Both sulfur (0.12 wt. percent) and nitrogen (0.25 wt. percent) were very high and the need to lower these levels would be given appropriate consideration in any plan to utilize the straight run naphtha in an integrated process scheme. Less than two gallons of the naphtha were obtained; it was not used for further process or blending.

Of the syncrude derivatives the preferred material for use in gasoline was the light catalytic naphtha having the properties shown in Table 4.

TABLE 4

Properties of the Light Catalytic Naphtha

Reid Vapor Pressure, psia 4.6	Research Octane No. (RON)
API Gravity, 60°F 48.4	Clear 94.8
ASTM Distillation, °F:	+0.5 ml TEL 97.7
IBP 113	+3.0 ml TEL 100.4
10 percent 162	Motor Octane No. (MON)
30 percent 207	Clear 84.0
50 percent 253	+0.5 ml TEL 86.1
70 percent 305	+3.0 ml TEL 89.5
90 percent 360	Yield:
End point 402	As volume percent of Fluid
Sulfur, wt. percent 0.01	Cat. Cracker Charge 38.1
Nitrogen, wt. percent 0.057	As vol. percent of
	Syncrude 19.4

To produce a full-boiling range gasoline as might be done in commercial practice, the light cat. naphtha was blended with refinery stocks as shown in Table 5. This produced a gasoline of well-balanced volatility and favorable anti-knock quality. The very small amount of fuel that was available was expended in engine tests to obtain fuel economy and performance data that could be compared with a reference fuel (Indolene).

Results of these tests showed no significant differences between the emissions and fuel economy characteristics of reference and test fuels. The data show somewhat elevated CO and HC emissions for the fuel containing coal-derived components (Table 6). These elevated emissions levels are believed to reflect real differences between the synthetic and reference (natural) fuels as evidenced both by continuous and by bag

TABLE 5

Properties of Cat-Cracked Fuel Blend

	Gasoline blend	Components			
		Butanes	Reformate	Alkylate	Cat. Cracked Naphtha from Pilot Plant
Vol. percent of blend :	100	6	30	20	44
API Gravity, 60°F	52.0	–	40.3	71.1	48.4
Reid Vapor Pressure, psia	9.6	76	6.2	7.3	4.6
ASTM gum, g/ml	–	–	1	2	4
Bromine No.	14	–	1.2	1.2	30
ASTM Distillation, °F:					
IBP	90	–	114	96	113
10 percent	150	–	240	152	162
50 percent	262	–	305	215	253
90 percent	363	–	374	288	360
End point	419	–	438	390	402
Research Octane No. (RON)					
Clear	95.6	–	96.5	93.5	94.8
0.5 cc TEL.	98.5	–	98.7	98.5	97.7
3.0 cc TEL	102.6	–	101.3	103.6	100.4
Motor Octane No. (MON)					
Clear	86.2	–	86.9	91.1	84.0
0.5 cc TEL	89.0	–	88.8	97.1	86.1
3.0 cc TEL	93.2	–	92.0	101.4	89.5

TABLE 6

Emissions and Fuel Economy

— Gasoline Containing Coal-Derivative Components —

(1975 Model Vehicle — 318-CID Engine)

Fuel	Emissions, g/mile				Fuel Economy, mpg	
	CO	HC	NO$_x$	Aldehydes	Urban Cycle	Highway
Coal Derivative Gasoline	11.7	1.6	2.7	0.19	12.7	21.0
Indolene	8.7	1.2	2.5	0.17	12.4	20.6

emissions measurements. However, the differences may be attributable to slight difference in the stoichiometry of the fuel mixtures; therefore they may or may not reflect an effect of combustion characteristics, per se. In any event, the differences are so small that they cannot be absolutely established by the few tests that were possible with the limited amount of fuel available.

CONCLUSIONS

The study was severely limited by the small amount of coal-derived material that was available, and by the unsuitability of the shale-derived material for extended engine testing. Nonetheless, some useful observations were made:

- Both shale gasoline and a gasoline containing 44 percent of a coal-derived material were used satisfactorily in short-term engine tests. As compared with results observed using a typical Mid-continent petroleum-derived gasoline, the change to synthetic fuel did not significantly affect emissions, fuel economy, or engine performance.

- The stability of the shale gasoline was inadequate for extended engine tests, but the deficiency is judged correctable by appropriate choice of available refining technology and/or process adjustment.

- The naphtha from catalytic cracking (51 percent yield) of a coal syncrude was found to be an excellent gasoline blending stock. The naphtha was usable as a 44 vol. percent component of a finished, 96 RON unleaded gasoline (remainder: 6 percent butanes, 30 percent reformate, and 20 percent alkylate).

- Engine testing of significant scale and duration will be required to determine and describe the requirements for mutual adaptation of new automotive engines and synthetic fuels. Multiple-barrel lots of test fuels will be needed even for the exploratory phases of such testing. Provision for production of such test fuels should be a prime element in all phases of a synthetic fuels development program.

DISCUSSION

F. B. Parks *(General Motors Research Laboratories)*

I noticed you measured aldehydes in the exhaust from both synthetic fuels. Did you expect higher aldehydes with the exhaust from synthetic products?

Hurn

No. We measure aldehyde emissions with all of our test fuels. We not only measure aldehydes but we measure PNA in the fuel and we are attempting to get some estimate of PNA in the exhaust. We measure aldehydes routinely, except it's not all that routine. We also look at some of the higher oxygenates for most of the work we do.

S. S. Penner *(University of California, San Diego)*

In the shale products, you indicated initially very high sulfur and nitrogen content. Could you give us some idea of the cost of reducing their content?

Hurn

No. I'm not even sure that those within my organization, or even in any other organization whose interest lies in that area, would be willing to give you cost information. I think that the process development is at such an early stage that these cost values probably would be premature.

S. L. Meisel *(Mobil R & D)*

I thought the sulfur distribution in your product from coal was rather interesting. The sulfur was high in the light ends but pretty low in the heavy ends. Is that typical? Do you have an explanation?

Hurn

I have no explanation. I am as interested as you. I have no idea why they showed up in the light product, I'm pursuing that. I don't know whether it's a fluke, or whether it's to be expected.

W. J. McLean *(Cornell University)*

I noticed along the same lines that quite a bit more of the nitrogen comes through in a coal-derived fuel than in a shale-derived fuel. Do you see that in the exhaust products also?

Hurn

We have attempted to develop a relationship between oxides of nitrogen in the exhaust and the nitrogen in material of origin. We've been unsuccessful.

J. B. Heywood *(Massachussetts Institute of Technology)*

Let me add to that. I did some rough calculations at breakfast. The amount of nitrogen in these gasolines is sufficiently small so that you wouldn't detect that effect on the nitric-oxide emissions.

B. W. Joseph *(General Motors Research Laboratories)*

I'd like to know in what form the nitrogen is in the shale?

Hurn

I'm unable to answer that. Can anyone help?

H. C. Huffman *(Union Oil Company of California)*

The nitrogen compounds are largely pyridines and quinolines.

J. P. Longwell *(MIT and Exxon)*

I agree that if you take a gasoline that matches current specifications you'll probably have difficulty distinguishing the performance from petroleum-derived gasolines. It might be interesting, for example, in these coal-derived liquids to pay particular attention to the high-boiling materials because some of the aromatics might be different from the normal aromatics you get with gasoline. You might have more multi-ring aromatics which will correlate with problems such as deposits and the increase in octane number requirement with time and also the carcinogenicity of exhaust products.

Hurn

We are looking at these factors and we are working very closely with Alex Mills and the people in ERDA's coal research area. I'd like to make another point. Certainly we are in no way attempting to make gasolines that just mirror the compositional characteristics of conventional gasoline. We're very conscious of the need to produce the fuel as it would be produced most economically from the synthetic product, economically taking into account the end use. This is the combined efficiency on the processing and in the vehicle.

APPLICATION OF A NEW COMBUSTION ANALYSIS METHOD IN THE STUDY OF ALTERNATE FUEL COMBUSTION AND EMISSION CHARACTERISTICS

J. A. HARRINGTON

Ford Motor Company, Dearborn, Michigan

ABSTRACT

A new combustion analysis method is described, and results obtained in an application are presented. The application selected is a study of the effects of EGR (Exhaust Gas Recirculation) on the combustion of methanol in a spark-ignited engine. The combustion analysis method is based upon a measure of the variation of the logarithm of the product (pressure) x (volume)$^{\overline{\gamma}}$ with time. The quantity $\overline{\gamma}$ (an effective specific heat ratio for gases in the engine combustion chamber) is determined readily from the slope of log P versus log V curves. Measurements of log (PV$^{\overline{\gamma}}$) provide for a simplified determination of combustion parameters such as ignition delay interval, burn rate and burn duration.

The log PV$^{\overline{\gamma}}$ measurement method is applied in a single cylinder engine study of the effects of EGR flow rates on the combustion-emission characteristics of methanol fuel. The results obtained for methanol are compared to those for a gasoline at corresponding engine operating conditions and EGR levels. Methanol exhibits shorter ignition delay intervals and burn durations than gasoline. Ignition delays and burn durations for both fuels are extended as engine load is decreased, as the fuel-air mixture is made lean and as the EGR level is increased. With increasing spark advance, ignition delay periods for both fuels are extended while burn durations are shortened. At the same equivalence ratio, air mass flow rate, speed, spark timing, and EGR rate, engine power output with methanol is about 10 percent higher than with gasoline. Under similar conditions but with constant power output, volume based fuel consumption with methanol is about twice that with gasoline, and energy utilization is more efficient. Methanol exhibits lower indicated specific CO emissions at Φ_{FA} values greater than ~ 1.05, and higher CO emissions at leaner mixture ratios. NO

References pp. 208-210.

emissions with methanol are lower than those with gasoline at common operating conditions, mixture ratios and EGR levels. HC emissions with methanol are generally higher than those with gasoline at corresponding non-misfiring mixture ratios and EGR levels. Lean misfire limits with methanol are extended by about 0.10 equivalence ratios relative to those of gasoline at all EGR levels investigated. Methanol has the greater tolerance to EGR.

INTRODUCTION

The energy shortage has stimulated research related to the production and use of fuels derivable from non-petroleum resources such as coal or shale. Coal, being the most abundant resource, could supply fuel needs for the long term. A variety of synthetic fuels ranging from alcohols like methanol to fuels much like todays gasolines might be produced (1). At present however, we are not in a position to identify the optimum synthetic fuel. Even so, because of the long lead times required to produce alternative fuels and/or the engines to use them, the advantages and disadvantages of the potential alternative fuels need to be identified now. From a user's standpoint, it is important that fuel evaluations include measurements of fuel economy, performance, emission characteristics, and combustion characteristics. Such information is essential for the optimal design of future combustion systems or engines which will be expected to operate with the alternative fuels.

Several studies on the use of alternative fuels in internal combustion engines have been carried out. These studies, for the most part, have centered around the use of alcohols and alcohol gasoline blends (2-6) because the alcohols are known to be derivable as synthetic fuels and sufficient quantities are available for test purposes. Other fuels studied include ammonia (7), and blends of ethers, water or hydrogen with gasoline or alcohols (8-11). Measurements of fuel economy, performance and emissions are usually reported in such studies. However a more complete fuel evaluation could be made if measurements of the fundamental combustion characteristics of the fuels are included. With information on combustion characteristics such as ignition delay periods (flame kernel development times) and combustion periods in addition to fuel economy, performance and emission data, the inter-relationships of such quantities could be quantified.

While measurements on fundamental combustion parameters are important to fuel and/or engine evaluations, they are seldom made because the known analysis methods are difficult to implement, and their use requires extensive calculations. The methods most frequently used for combustion analysis are based on measurement of the variation of cylinder pressure with time during combustion or of flame position versus time (12-18). Pressure versus time information is obtained using standard transducers. Flame position versus time measurements are somewhat more difficult to obtain as they require combustion chamber modifications to accommodate ion gap probes or windows through which flame photographs can be taken. These measurements can yield quantitative results, but data interpretation is not straightforward, and extensive calculations normally are required to determine burning gas properties, ignition delay periods, burn rates and combustion intervals.

In this paper, a new combustion analysis method is described, and results obtained in its application to study exhaust gas recirculation (EGR) effects on the combustion of methanol in a single cylinder engine are presented. Part I of the paper deals with a description of the combustion analysis method. The method is based on a measure of the logarithm of the product (pressure) x (volume)$^{\bar{\gamma}}$ where $\bar{\gamma}$ represents an effective specific heat ratio for gases in the combustion chamber. Rationale for use of such measurements is established by reference to conventional knowledge of engine processes and to thermodynamics. The instrumentation required for measurement is described, and instrument performance characteristics are illustrated. Part II of the paper presents experimental results obtained in a single cylinder engine study of the effects of EGR flow rate on the combustion-emission performance characteristics of methanol. Results obtained with methanol at various speeds, loads, spark timings, and EGR rates are compared to those with gasoline at corresponding operating conditions.

COMBUSTION ANALYSIS METHOD

In the ideal Otto cycle (19,20), the compression and expansion strokes of an engine are represented as isentropic processes. Combustion and heat rejection are taken as constant volume processes occurring at top and bottom center piston positions respectively. Real engines only approximate this cycle; however, noncombustion portions of the compression and expansion strokes are nearly isentropic and can be represented mathematically by the equation:

$$PV^{\bar{\gamma}} = C \qquad (1)$$

P = absolute pressure,

V = cylinder volume,

$\bar{\gamma}$ = the effective specific heat ratio of gases in the combustion chamber and

C = a constant for each process.

In a motored engine, the value of C or $PV^{\bar{\gamma}}$ is about the same for both the compression and expansion processes. In a firing engine however, these two processes are separated by a highly nonisentropic combustion process which causes the value of $PV^{\bar{\gamma}}$ to change. Because of the pressure rise due to combustion, the value of $PV^{\bar{\gamma}}$ at post combustion conditions is higher than the precombustion value. Changes in the

References pp. 208-210.

value of $PV^{\overline{\gamma}}$ that take place during combustion are of particular interest for combustion analysis because they reflect the details of the process.

THEORY

Equations relating $PV^{\overline{\gamma}}$ changes to combustion characteristics are summarized in this section. From a first law analysis of the combustion process in engines, incremental changes in the mass of fuel-air mixture burned at any time can be written as:

$$dM_b = \frac{1}{E_c} \left| \frac{d(PV^{\gamma}) + PV^{\gamma}d(PV^{\gamma}d(\ell n(\overline{MW})}{(\gamma\text{-}1)\ (V^{\gamma\text{-}1})} + dQ_{H.L.} \right| \qquad (2)$$

Here E_c represents the heat of combustion per unit mass of fuel, γ represents the specific heat ratio $(C_p(X,T)/C_v(X,T))$, $\overline{MW}$ is the molecular weight of gases in the combustion chamber and $dQ_{H.L.}$ is the heat transfer loss. The relationship for the mass fraction of mixture burned at any time t, is:

$$\frac{M_b}{M_T}\Bigg]_t = \frac{\displaystyle\int_{initial}^{t} \left| \frac{d(PV^{\gamma}) + PV^{\gamma}d(\ell n(\overline{MW}))}{(\gamma\text{-}1)\ (V^{\gamma\text{-}1})} + dQ_{H.L.} \right|}{\displaystyle\int_{initial}^{final} \left| \frac{d(PV^{\gamma}) + PV^{\gamma}d(\ell n(\overline{MW}))}{(\gamma\text{-}1)\ (V^{\gamma\text{-}1})} + dQ_{H.L.} \right|} \qquad (3)$$

If the symbol γ is replaced by $\overline{\gamma}$ to indicate that heat losses will be incorporated into this parameter and if the quantities $V^{\overline{\gamma}\text{-}1}$, γ and $\ell n\ (\overline{MW})$ are taken to be constants during combustion, the equation can be integrated with the following result:

$$\frac{M_b}{M_T}\Bigg]_t \approx \frac{(PV^{\overline{\gamma}})_t - (PV^{\overline{\gamma}})_{initial}}{(PV^{\overline{\gamma}})_{final} - (PV^{\overline{\gamma}})_{initial}} \qquad (4)$$

This approximate relationship for mass fraction burned is quite similar to one developed by Blizzard and Keck (21) and is about equivalent to that used by Rassweiller and Withrow (12). Even though the accuracy of equation 4 is limited by the assumptions made, its use in relating pressure time records to engine combustion characteristics provides valuable information on the details of combustion.

An alternative relationship for mass fraction burned is obtained if the heat release due to combustion is expressed in terms of entropy (i.e. $dS_c = Q_c/T$). In this case, the incremental change in mass burned at any time is written:

$$d(M_b) = \frac{\overline{C}_v(X,T)\, x(M_T)}{(dS_c/dM_b)} \left| d(\ell n PV^\gamma) + d(\ell n \overline{MW}) \right| \qquad (5)$$

where $C_v(X,T)$ is the constant volume specific heat of gases, M_T is the total mass of burnable mixture, and dS_c/dM_b represents the incremental change in entropy per unit mass of mixture burned. The mass fraction of mixture burned at any time t can be written:

$$\left[\frac{M_b}{M_T} \right]_t = \frac{\displaystyle\int_{initial}^{t} \frac{\overline{C}_v(X,T)}{dS_c/dM_b} \left| d(\ell n PV^\gamma) + d(\ell n \overline{MW}) \right|}{\displaystyle\int_{initial}^{final} \frac{\overline{C}_v(X,T)}{dS_c/dM_b} \left| d(\ell n PV^\gamma) + d(\ell n \overline{MW}) \right|} \qquad (6)$$

If the symbol γ is again replaced by $\overline{\gamma}$ (as in equation 4) and if the quantities $\overline{C}_v(X,T)/(dS_c/dM_b)$ and $\ell n\,(\overline{MW})$ are assumed to be constants, the integrated form of equation 6 becomes:

$$\frac{M_b}{M_T} \simeq \frac{\ell n(PV^{\overline{\gamma}})_t - \ell n(PV^{\overline{\gamma}})_{initial}}{\ell n(PV^{\gamma})_{final} - \ell n(PV^{\overline{\gamma}})_{initial}} \qquad (7)$$

Both the above equation and equation 4 are approximate relationships describing mass fraction burned in terms of the pressure and volume. Limitations common to the two equations are that heat transfer effects must be compensated for by an adjustment to the value of γ (as $\overline{\gamma}$) and that $\overline{\gamma}$ values are treated as constants. A more critical assumption common to these equations is that combustion is treated as a simple heat addition process with no change in composition and with no real gas effects. In equation 4, the added assumption that $V^{\overline{\gamma}-1}$ is a constant is valid only if combustion takes place very near to the top dead center (TDC) piston position. In equation 7, the assumption that $C_v(X,T)/(dS_c/dM_b)$ is a constant is likely to be valid only in special cases.

It is not certain which of the approximations is more valid, but a fully detailed assessment of such relationships is beyond the scope of the present report. The relationships were given to show that changes in the value of $PV^{\overline{\gamma}}$ or of $\ell n\, PV^{\overline{\gamma}}$ during

References pp. 208-210.

combustion do reflect upon important features of the process. From the standpoint of instrumentation it is irrelevant whether the basic measurement takes the form of $PV^{\bar{\gamma}}$ or $\ell n\, PV^{\bar{\gamma}}$ since the resulting data could be interpreted according to either of the equations, or according to more exact relationships. Measurements on the variation of either quantity with time (or crank angle) during engine operation therefore could be used to estimate burning rates or to define ignition delays and burn durations.

Ignition delays can be taken as the interval between spark firing and a point where $\ell n\, PV^{\bar{\gamma}}$ (or $PV^{\bar{\gamma}}$) increases to some chosen fraction of the difference between final and initial values. In the present case, ignition delays are taken as the interval between spark firing and the point where $\ell n\, PV^{\bar{\gamma}}$ reaches 5 percent of the difference between final and initial values. Combustion intervals then begin at the end of the delay period and are terminated at a point where $\ell n\, PV^{\bar{\gamma}}$ approaches its final value. Since this final value will be approached asymptotically, an "effective" combustion interval is defined as the interval during which the value of $\ell n\, PV^{\bar{\gamma}}$ increases from 5 to 95 percent of the difference between its final and initial values.

In applying the above concepts to combustion analysis in engines, measurements of the value of $\bar{\gamma}$ and of the variation of $\ell n\, PV^{\bar{\gamma}}$ with time are to be made. Since precombustion portions of the compression stroke and post combustion portions of the expansion stroke are "approximately isentropic," $\bar{\gamma}$ values for chosen portions of the engine cycle can be determined from the slopes of ℓnP versus ℓnV curves. Measurements on ℓnP and ℓnV are also required to obtain $\ell n(PV^{\bar{\gamma}})$ since this quantity can be expressed as

$$\ell n(PV^{\bar{\gamma}}) = \ell nP + \bar{\gamma}\ell nV \tag{8}$$

Thus, the only measurements needed to determine $\bar{\gamma}$ values, ignition delays, combustion periods, and burn rates are pressure and volume. Logarithms of pressure and volume signals are readily obtained with analog electronic circuits and the $\bar{\gamma}$ value can be determined as mentioned above. With $\bar{\gamma}$ values known, the sum indicated by equation 8 can be made electronically to obtain a $\ell n(PV^{\bar{\gamma}})$ signal. Where a signal corresponding to $PV^{\bar{\gamma}}$ is desired, the antilog of $\ell n(PV^{\bar{\gamma}})$ can be generated electronically.

INSTRUMENTATION FOR MEASUREMENT OF COMBUSTION CHARACTERISTICS

Pressure — Pressure measurements are straightforward and can be obtained using standard experimental methods with calibrated transducers(14). In the present application, a water-cooled model 601A Kistler pressure transducer together with a Kistler model 504 charge amplifier were used. Because signal drift problems are sometimes encountered with long term use of piezoelectric transducers (continuous over 4 to 8 hours), a mechanism for automatically resetting the transducer signal to a known level for each engine cycle has been developed. A gate signal generated at the end of the engine exhaust stroke is used to close a fast electronic switch which then causes a resetting of the transducer signal to the known level. Resetting is accomplished with the electronic circuit shown in Fig. 1.

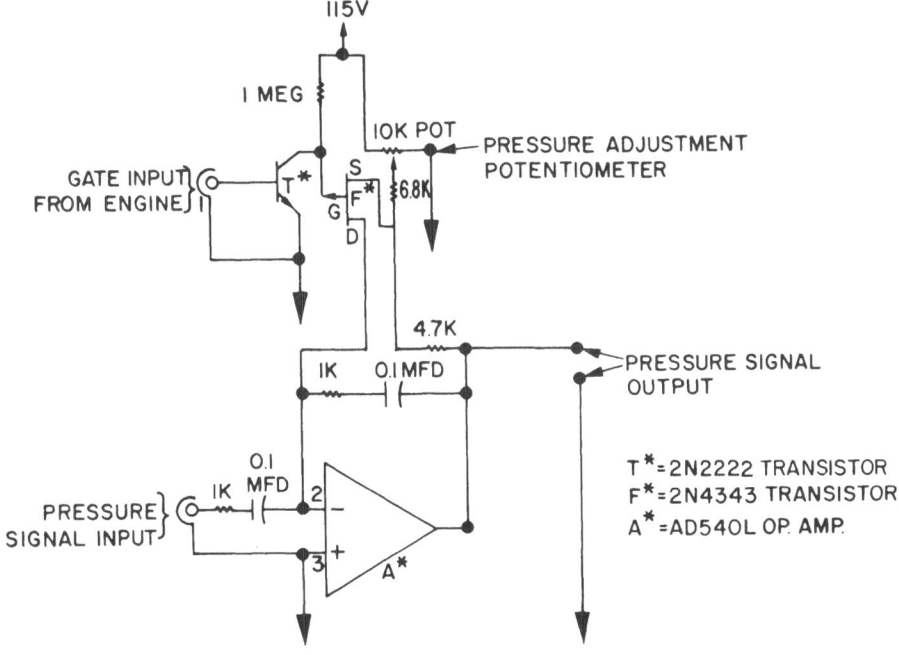

Fig. 1. Pressure reset circuit.

The logarithm of the pressure signal is obtained via a standard electronic log circuit.

Volume — Measurement of cylinder volume as a function of time or of crank angle during engine operation is based on the use of a newly developed device. One device for volume determination is known to be commercially available (Tektronix), but its design is specific to an engine with a piston connecting rod length to crankshaft turn ratio (L/R) of 4:1. Because use of the commercially available device at L/R ratios other than 4:1 gives rise to errors in volume, most volume determinations are based on calculations rather than experimental measurements. In the present work, accurate analog volume signals are essential and as a result a new optical-mechanical device for volume determinations has been developed. The device consists of a crankshaft mounted cam, a constant brightness light source, and an optical detector as illustrated in Fig. 2. The cam is designed to simulate the piston position variation with crank angle or time as the crankshaft rotates. The light source and detector are mounted rigidly to the engine at opposite sides of the cam. As the cam rotates about its axis (crankshaft axis), the amount of light reaching the detector varies directly with the normal distance between the piston surface at any crank angle and the TDC position.

References pp. 208-210.

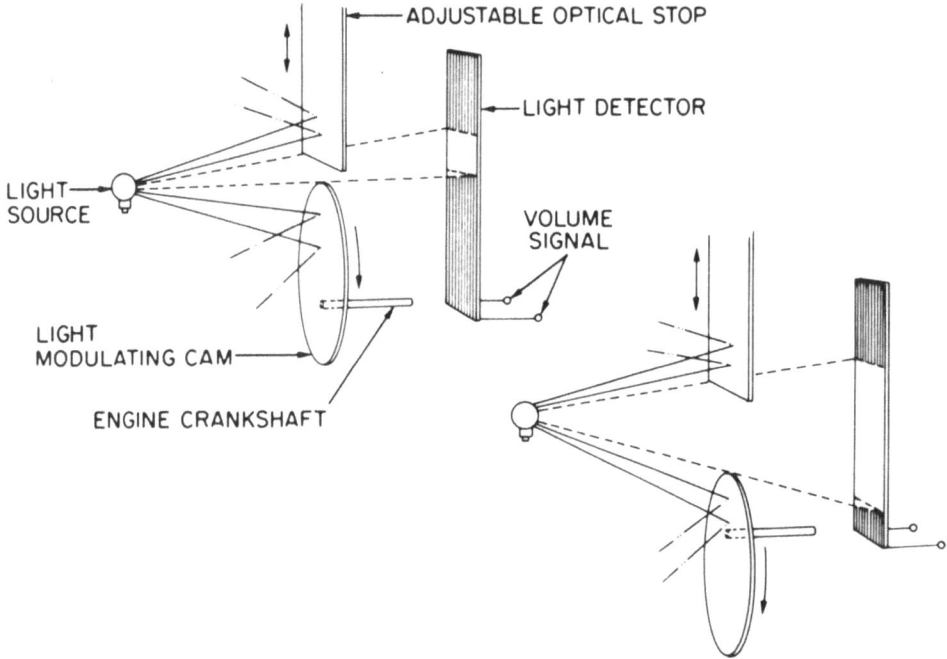

Fig. 2. Schematic of engine combustion chamber volume indicator.

This provides a measure of displacement volume as a function of time or of crank angle. To account for the clearance volume (piston at TDC), the appropriate constant amount of light over and above the light level mentioned above is allowed to reach the detector. The amount of "added light" is adjustable in accordance with compression ratio. The total signal generated by the light detector will therefore be proportional to combustion chamber volume at any crankshaft position and compression ratio.

In making an analysis to determine the "optical" cam profile, it was found that a circular disc spun off axis would be appropriate. With the disc spun off center by a distance $\triangle R$ the motion of the disc edge exactly duplicates the piston motion in an engine (i.e. the piston motion as caused by a slider crank mechanism). The distance $\triangle R$ is defined as

$$\triangle R = \frac{R}{L} \times r_d \tag{9}$$

where r_d = disc radius and R/L = ratio of engine crankshaft turn radius to the piston connecting rod length.

A signal corresponding to the logarithm of volume is obtained via a standard electronic log circuit. A signal for the product of $\overline{\gamma}$ and the logarithm of volume is then obtained by suitable amplification of the logarithm of volume signal.

Logarithm of PV$^{\overline{\gamma}}$ – The logarithm of PV$^{\overline{\gamma}}$ is obtained quite simply by adding the logP and $\overline{\gamma}$ logV signals.

Instrument Performance — Development of the new method for volume determinations with an analog signal output makes it possible to generate accurate curves of P versus V, logP versus logV and $\log PV^{\overline{\gamma}}$ (or $PV^{\overline{\gamma}}$) versus time or crank angle on line as the engine operates. The curves can be displayed on an oscilloscope and recorded on film or recorded on magnetic tape for subsequent computer processing. A block diagram showing one instrumentation arrangement for an oscilloscope display of data is given in Fig. 3. The response characteristics of the device and representative experimental results are given below.

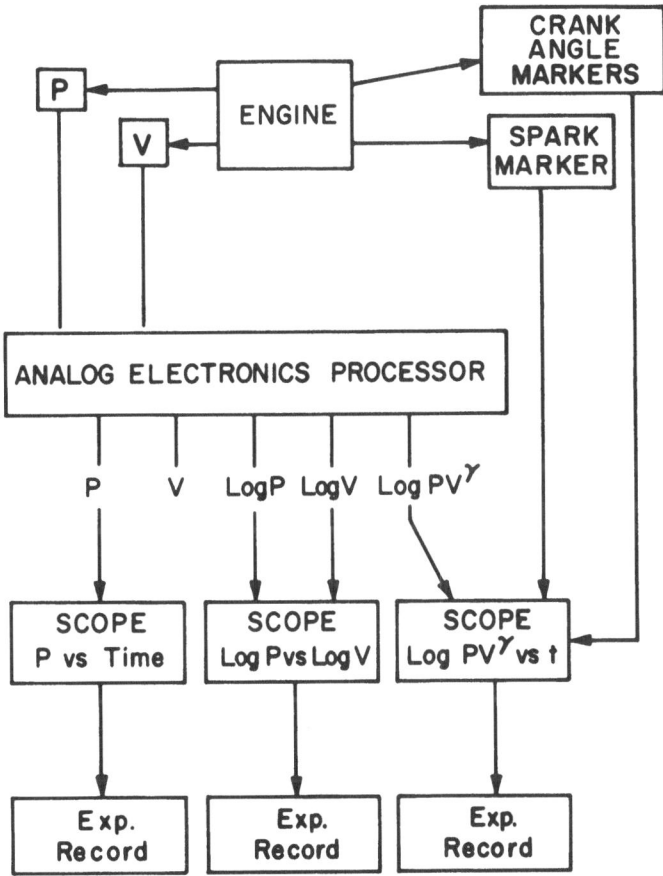

Fig. 3. Block diagram of combustion analysis system.

The initial check out of the instrument consisted of a calibration of the pressure transducer, the volume measuring device and the associated circuitry. In all cases the signal was linearly related to the quantity measured.

Experimental pressure versus volume and log pressure versus log volume diagrams are obtained when the appropriate instrument signals are used to drive the vertical

References pp. 208-210.

and horizontal sweeps of an oscilloscope as the engine operates. Records of oscilloscope traces are shown in Figs. 4 and 5. Fig. 4 shows P vs V (indicator) diagrams obtained for a Waukesha variable compression ratio engine operating at 1000 RPM with a manifold pressure of 750 mm Hg. The upper P vs V curve was obtained under motoring conditions while the lower curve was obtained with the engine firing (Indolene clear fuel, 15°BTDC spark timing). Curves such as these are useful in engine cycle analysis to account for the work associated with pumping, the combustion process, etc. Fig. 5 shows logarithmic P vs V diagrams obtained at the same engine operating conditions. The upper and lower records again correspond to motoring and firing conditions. The square wave pulse at the end of the exhaust stroke results from the action of the pressure reset circuit. The $\bar{\gamma}$ values for the compression and expansion strokes are determinable from the slopes of the curves. For the example shown, $\bar{\gamma}$ at the end of compression was about 1.29 while that after combustion at the beginning of expansion was about 1.24.

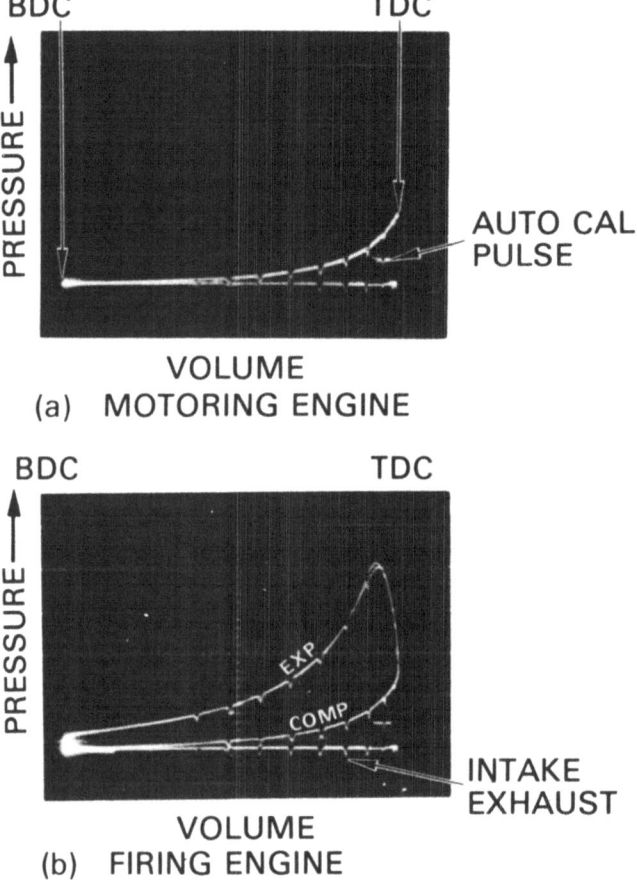

Fig. 4. Pressure versus volume record.

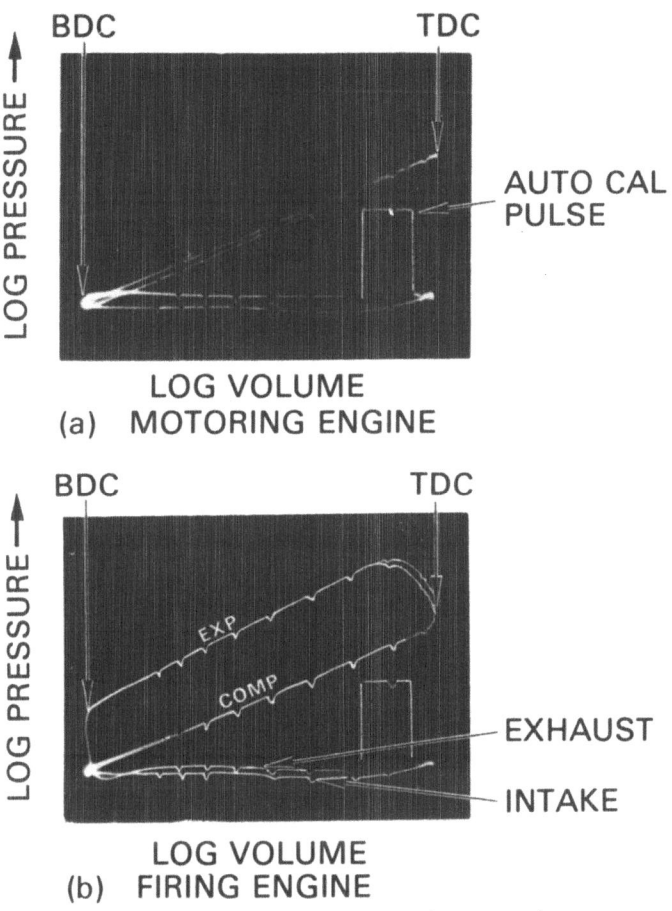

Fig. 5 Log pressure versus log volume record.

Experimental log$PV^{\overline{\gamma}}$ versus time records ($\overline{\gamma}$ set at 1.265) for motoring and firing conditions are shown in Fig. 6. Operating conditions were the same as those given above. Crank angle markers at ten degree intervals and a spark firing marker are superimposed on each curve. Spark firing is indicated by a "step" in the oscilloscope trace. The upper curve demonstrates the "nearly isentropic" nature of the compression and expansion processes under motoring conditions. The lower curve shows how combustion influences the value of log $PV^{\overline{\gamma}}$. At the moment of spark firing, the log $PV^{\overline{\gamma}}$ value corresponds to the motored engine value. After a short delay period during which the flame kernel develops (the ignition delay period, usually of the order of 2 milliseconds), the value of log $PV^{\overline{\gamma}}$ increases and approaches a new higher level constant value at which time combustion is completed. The instantaneous slope of a log$PV^{\overline{\gamma}}$ curve reflects the burning rate. Initially the burn rate is low; it increases with time, and passes through a maximum and then it again decreases toward zero as combustion is completed. This kind of measurement provides data in a

References pp. 208-210.

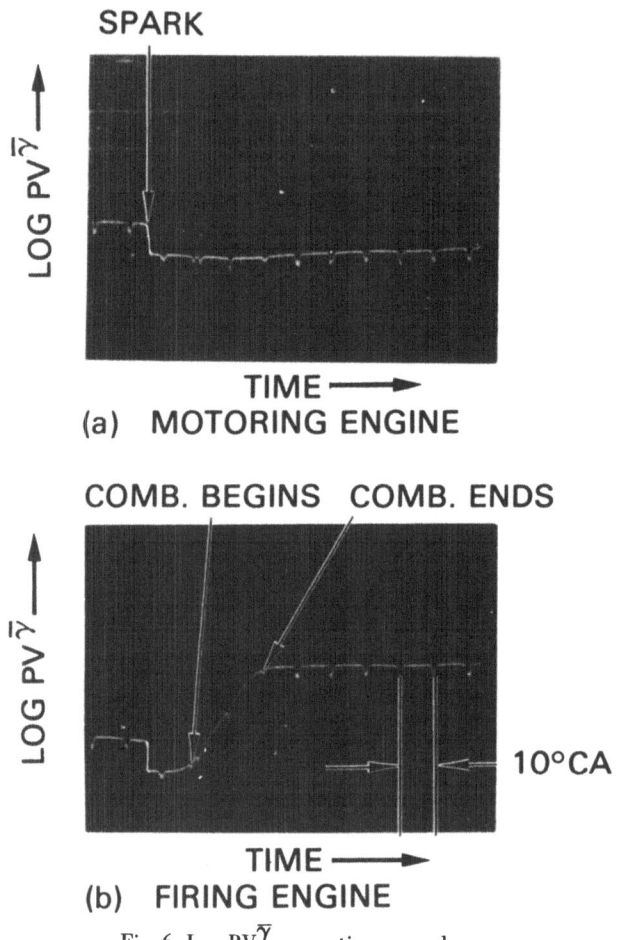

Fig. 6. Log $PV^{\overline{\gamma}}$ versus time record.

form conducive to a greatly simplified determination of ignition delays, burn rates and combustion intervals. Such measurements should be of value in experimental fuel evaluations and/or the testing of engine systems.

The advantages of a $\log PV^{\overline{\gamma}}$ measurement over conventional pressure measurements become apparent when one compares the respective experimental data curves obtained, and considers the effort required to determine combustion parameters. Fig. 7, for example, shows a pressure versus time curve and a $\log (PV^{\overline{\gamma}})$ versus time curve. Both contain features that reflect the details of combustion. With the former, however, it is not apparent when combustion starts, how fast it progresses or when it ends. To get this information is rather tedious as it is necessary to accurately determine the variation of chamber volume with time, to synchronize this with the pressure-time data, to determine the value of $\overline{\gamma}$ and then to evaluate $PV^{\overline{\gamma}}$ or $\log PV^{\overline{\gamma}}$

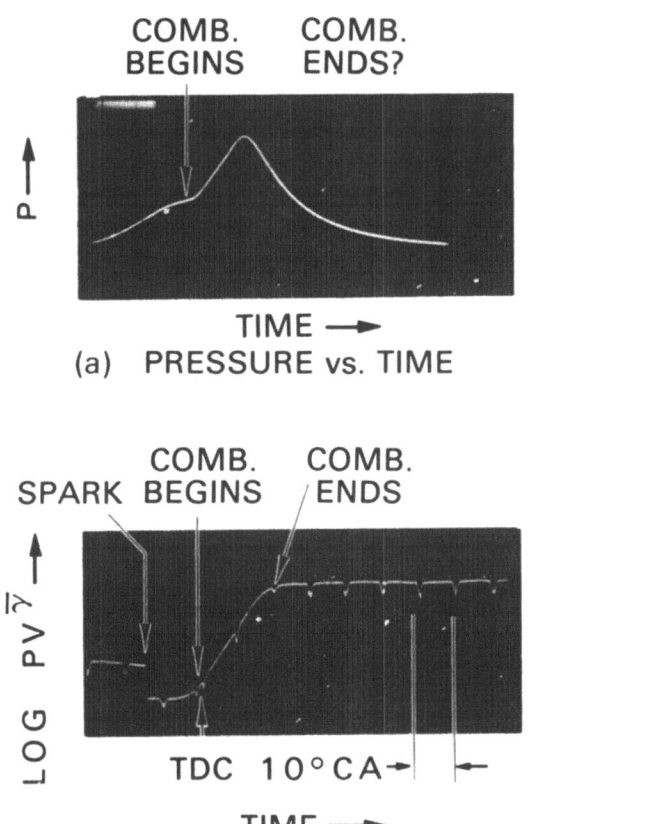

Fig. 7. Pressure versus time and log [(pressure) x (volume)] versus time records of combustion.

over the full time interval of combustion. In contrast to this, the logPV$^{\overline{\gamma}}$ versus time curve provides data in a form conducive to a greatly simplied determination of combustion parameters. Instantaneous burn rate characteristics can be estimated by a simple measurement of curve slope and the points at which combustion begins and ends are readily identified. In the remainder of this paper, the logPV$^{\overline{\gamma}}$ measurement method is applied in a single cylinder engine study of EGR flow rate effects on the combustion-emission characteristics of methanol.

COMBUSTION AND EMISSION CHARACTERISTICS OF METHANOL AS INFLUENCED BY EGR

Several studies on the use of alcohols such as methanol and ethanol as motor fuels have been carried out. Early work (2,3) shows that alcohols can be used as motor fuels and that this use results in a power increase and a fuel economy penalty. Later

References pp. 208-210.

studies (4,5), which also include emissions measurements, confirm the early observations and indicate that alcohol use leads to reduced emissions. Recent studies(6, 22) show agreement with the previously reported performance measurements, but only a partial agreement with respect to emissions. A recent methanol study by the author(6) shows that NO emissions are reduced while HC and CO emissions with methanol are comparable to or higher than those with gasoline at corresponding engine operating conditions. The same study, which included measurements on combustion characteristics, also showed that methanol burns faster than gasoline. Of the known studies with methanol, none has included measurements of EGR flow rate effects.

In the present work, the combustion, emissions and performance characteristics of methanol and Indolene clear gasoline, as influenced by EGR, are measured using a single cylinder engine. Results obtained with the two fuels are compared at common values of engine speed, mixture equivalence ratio, spark timing, EGR, and air flow to establish relative characteristics.

Experimental Equipment — A single cylinder Waukesha variable compression ratio engine couples to a 50 HP dynamometer was utilized. The cylindrical flat head combustion chamber was fitted with transducers and with quartz windows for optical observations of the combustion process. A conventional ignition system was used. Standard instrumentation was used to measure engine speed and spark timing as well as coolant, lubricant, and exhaust gas temperatures. Engine torque was measured via a load cell which was connected at the dynamometer torque arm. Intake air for the engine was provided by a calibrated-regulated choked flow orifice system which was built for use with the engine. The temperature of air introduced to the engine was controlled by use of electric heaters. The fuel system was designed for dual fuel operation, and consisted of high pressure stainless steel fuel supply tanks, metering valves, calibrated rotameters, and a modified carburetor. Fuel was introduced at the carburetor venturi via either of two fine bore (0.02 in. dia.) capillary tubes which were used in place of the normal "main jet" system.* Control over fuel flow to the engine was obtained by varying fuel system pressure and by needle valve adjustments.

Exhaust gas emissions were sampled by a 25 inch long stainless steel probe with sample inlet holes distributed over its length. The probe was positioned in the exhaust pipe with the upstream probe tip located two feet from the exhaust port. The remainder of the sampling system was as illustrated in Fig. 8. The sample line leading to the hydrocarbon (HC) analyzer was heated and maintained at about 190°F to avoid water condensation, possible unburned methanol condensation, and general hydrocarbon hangup problems. At a point near the HC analyzer, the sample line was split into a second stream to provide sample gas for the CO, CO_2, NO and O_2 analyzers. This second sample line contained a two-stage cold trap, particulate filters and

---
* *Previous work showed this type of carburetor modification to reduce significantly fuel flow oscillations normally observed with carburetor use on single cylinder engines.*

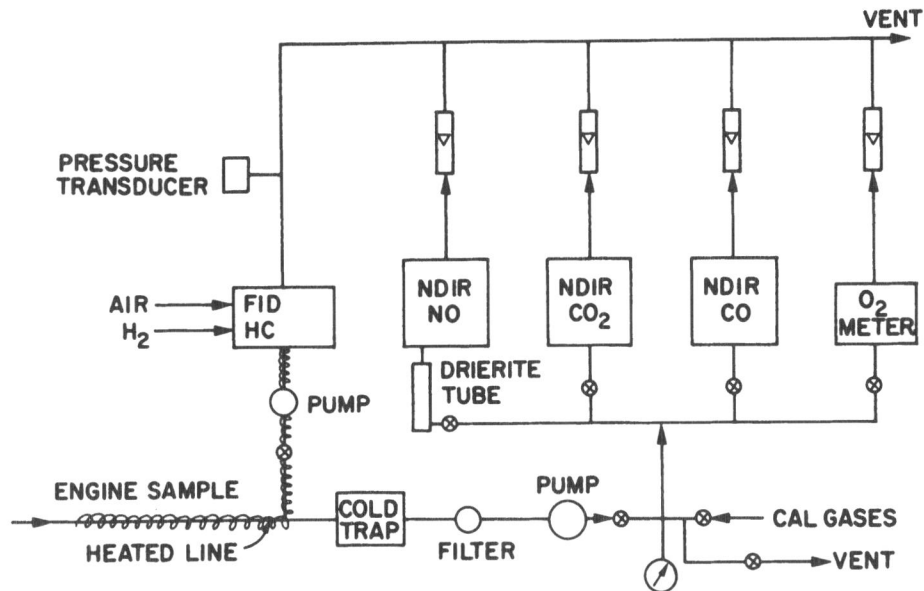

Fig. 8. Schematic of analytical instrument arrangement.

desiccants to assure complete drying of the gases prior to introduction into the analyzers. Dry basis measurements on NO, CO and CO_2 were made using non-dispersive infrared analyzers. A polarographic analyzer was used for O_2. A wet basis hydrocarbon analysis, subsequently converted to a dry basis, was made using hot portions of the gas sample in conjunction with a heated flame ionization detector (FID). FID response to exhaust HC for both fuels was assumed to be the same*, and as a result the reported HC concentrations for methanol exhaust represent lower limit values.

EGR was introduced just below the carburetor throttle plate with rates being determined by a comparison of CO_2 levels in the engine intake system to those in the exhaust. The sampling and analysis system used for intake CO_2 measurements was similar to that used for dry basis measurements on exhaust gases. A low range (0 to 5%) CO_2 analyzer was used. EGR rates are expressed in terms of an EGR index which is defined as:

$$(EGR)_I = \frac{(CO_2)_{Intake}}{(CO_2)_{Exhaust} - (CO_2)_{Intake}} \times 100 \qquad (10)$$

where CO_2 concentrations are dry basis values.

_____

*Manufacturers literature(23) states that FID response to methanol is about 50% of that for paraffinic hydrocarbons. Instruments to determine relative amounts of unburned methanol and "normal" hydrocarbons in exhaust gases were not available at the time these experiments were carried out.

References pp. 208-210.

Combustion analysis was achieved using instrumentation described previously in this report. Analyzer signals were displayed on oscilloscopes and recorded on Polaroid film for subsequent analysis. At each run condition, log (PV^γ) data for 10 consecutive engine cycles were collected, analyzed and then averaged to account, at least in part, for cyclic variations in the combustion process.

Experimental Procedure — Evaluation of combustion, emissions and performance characteristics was accomplished by operating the engine over a range of conditions. Either fuel flow or spark timing was varied while all other parameters were held constant. Operating conditions along with engine specifications are listed in Table 1. Indolene clear gasoline, the current Federal Test Procedure certification fuel, was used to establish a baseline for comparisons. Fuel properties are summarized in Table 2. For a given test, the engine was motored at the chosen speed, and air flow to the engine was set. The ignition system was turned on, and spark timing was adjusted to a chosen value. Fuel flow was then initiated and adjusted to give the chosen equivalence ratio. After steady state engine operation was achieved, the values of fuel and air flow, torque, emissions and combustion characteristics were recorded. Back-to-back runs at corresponding operating conditions were made with each fuel. Selected runs were periodically repeated to assure baseline repeatability.

TABLE 1

Engine Parameters and Operating Conditions

Engine displacement, in^3	37.33
Compression ratio	8.0
Oil temperature, °F	120 ± 5
Coolant temperature, °F	150 ± 10
Coolant mass flow, gals/min	0.4
Inlet air temperature, °F	100 ± 3
Speed, rpm	600, 1000, 1500
Spark timing, °BTDC	0, 15, 30, 45
Mass air flow range, lbs/hr	9 to 35
Manifold pressure, in Hg	19, 24, 28
Equivalence ratio range	1.2 to ∿ 0.6
Exhaust Gas Recirculation (Index)	0 to ∿ 25

Equivalence ratio determinations were based on the known stoichiometric air to fuel ratios of fuels and the measured air and fuel flows, as well as on a computer analysis of exhaust gas emission data. Φ_{FA} is defined as the ratio of actual fuel-air ratio to the stoichiometric fuel-air ratio. To obtain Φ_{FA} values based on exhaust analyses for methanol, an existing computer program for hydrocarbon fuels was modified for use with an arbitrary C-H-O fuel. Input data for the program are the measured CO, CO_2, O_2, HC and NO concentrations in the exhaust and the atomic H/C and O/C ratios of the fuel.

TABLE 2

Properties of Indolene Clear and Methanol

Property	Indolene Clear	Methanol
H/C, O/C	1.87, 0	4.0, 1.0
Specific Gravity, 60°F	0.743	0.796
Stoichiometric A/F	14.6	6.46
Research Octane	96.8	106
Motor Octane	88.2	92.0
Lead Content, g/gal	0.017	0
Boiling point, °F at 1 atm	FBR*	148
Heat of combustion, Btu/lb @ 77°F	∼19,000	859
Latent heat of vaporization, Btu/lb @ 77°F	FBR*	473
FIA		
Vol, % aromatics	32.8	−
Vol, % olefins	2.7	−
Vol, % parafins	64.5	−
Distillation, °F		
IBP	88	−
10%	119	−
30%	171	−
50%	213	−
70%	239	−
90%	299	−
EP	387	−

Full Boiling Range Fuel

Combustion and Power Characteristics — Combustion and power output characteristics obtained for engine operation with Idolene clear gasoline and with methanol are *References pp. 208-210.*

illustrated in Figs. 9 through 16. In each of the three part figures, ignition delays (0 to 5 percent burn), combustion intervals (5 to 95 percent burn) and engine torque values, respectively, are given as a function of mixture equivalence ratio.

Mixture ratio and EGR flow rate effects on these characteristics are shown in Figs. 9 and 10. Ignition delays for both fuels increase with increasing mixture leanness and with increasing EGR. Methanol ignition delays at common values of Φ_{FA} and EGR_I are shorter than those of Indolene clear fuel by 10 to 25 percent. The largest difference in ignition delays tends to occur at the lower EGR levels and at lean mixture conditions. Combustion intervals for the two fuels also increase with leaner mixtures and increasing EGR. Combustion intervals with methanol are shorter than those with Indolene clear at all common EGR flow rates and Φ_{FA} values. With no EGR, methanol combustion intervals are from 8 to 20 percent shorter and with an EGR index of about 18 they are nearly 50 percent shorter. Engine power output for both fuels also decreases with increasing leanness and EGR flow. Without EGR, methanol use results in about 10 percent more power than Indolene clear; at an EGR index of 18, power with methanol is about 25 percent greater.

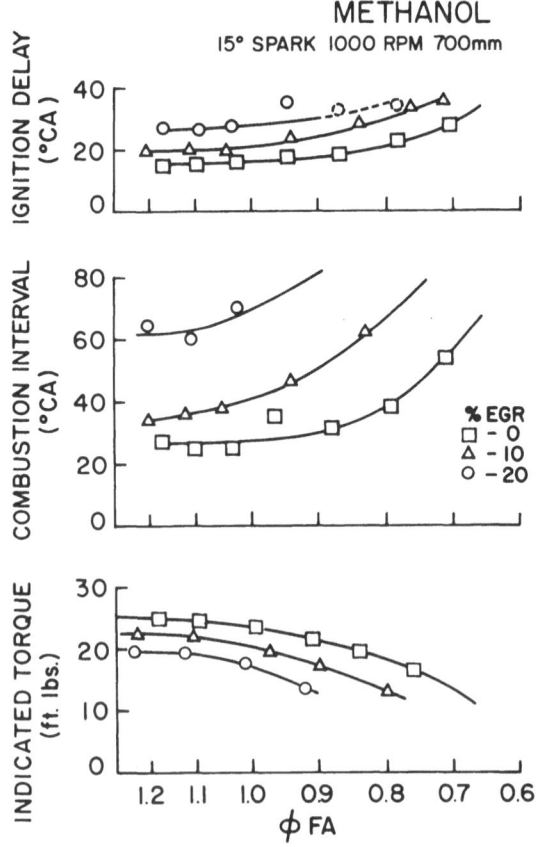

Fig. 9. Influence of Φ_{FA} and EGR on combustion and power.

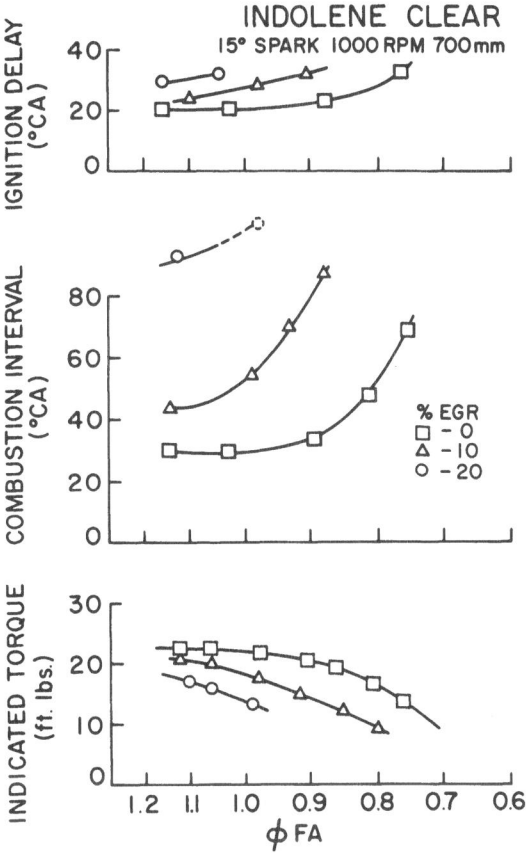

Fig. 10. Influence of Φ_{FA} and EGR on combustion and power.

Spark timing and mixture ratio effects are shown in Figs. 11 and 12. As spark timing is advanced relative to TDC, ignition delay periods for both fuels are extended while combustion intervals are shortened. This same trend in the variation of these combustion parameters with spark timing persists at all EGR levels investigated. A comparison of results for the two fuels at an EGR index of 10 shows methanol ignition delays and combustion intervals to be 19 to 26 percent shorter than those of Indolene clear. At other EGR levels, methanol ignition delays and combustion intervals are from 10 to about 50 percent shorter than those of Indolene clear. Largest differences occur at the higher EGR levels. The influence of spark timing changes on engine power at various Φ_{FA} and EGR values is similar to that observed for gasoline. Methanol use results in a higher power output than Indolene clear at each of the spark timings used. At a spark advance of 45° BTDC (EGR$_I$ $\simeq$ 10), methanol produces about 5 percent more power, and at a timing of 0° BTDC it produces nearly 20 percent more power. The fact that power output with both fuels

References pp. 208-210.

increases continuously as spark timing is advanced from 0° to 45° BTDC indicates MBT (minimum advance for best torque) timings at this EGR level to be greater than 45° BTDC. At the zero EGR level, MBT spark timings for methanol are retarded relative to those for Indolene clear by 2 to 5 degrees in crank angle.

Mixture ratio and throttle setting (mass air flow) effects on the combustion characteristics are shown in Figs. 13 and 14. Ignition delays for both fuels decrease with increasing load and with decreasing mixture leanness. Methanol ignition delays are 15 to 30 percent shorter than those of Indolene clear fuel. Combustion intervals for the two fuels also tend to decrease as air flow is increased. A comparison of results shows combustion intervals for methanol to be from 25 to about 50 percent shorter than those for Indolene clear. The largest difference occurs at the lightest loads and leanest mixture ratios investigated. Similar trends in data are observed at other EGR levels. At the zero EGR level, for example, methanol combustion intervals are 8 to 30 percent shorter than those with Indolene clear. Power output for the fuels increases with increasing airflow as expected. At an EGR_I of about 10 and an airflow of 23.2 lb/hr, power with methanol is 10 to 30 percent higher than with Indolene clear; at an airflow of 16 lb/hr the power with methanol ranges from 5 to 200 percent higher.

Fig. 11. Influence of Φ_{FA} and spark timing on combustion and power.

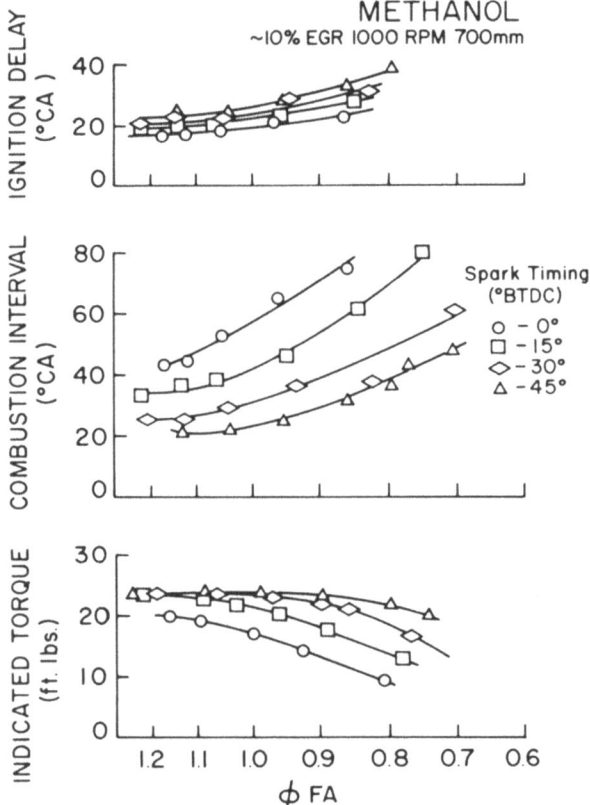

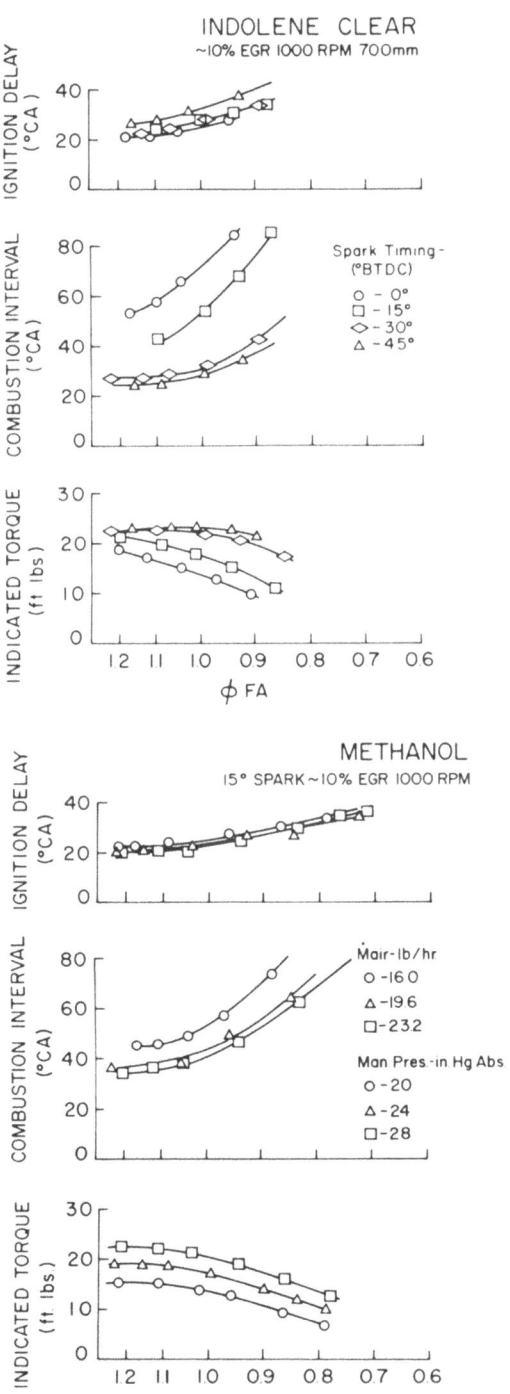

Fig. 12. Influence of Φ_{FA} and spark timing on combustion and power.

Fig. 13. Influence of Φ_{FA} and air flow on combustion and power.

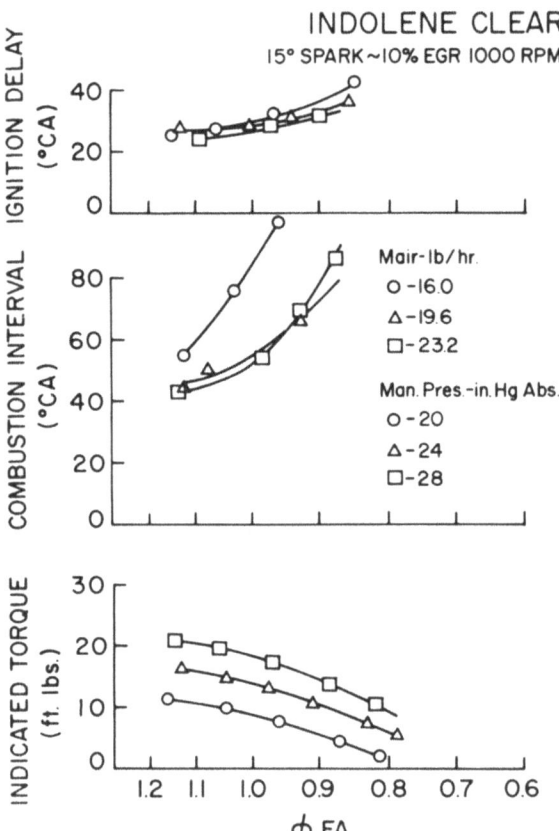

Fig. 14. Influence of Φ_{FA} and air flow on combustion and power.

Engine speed effects on combustion characteristics and power are illustrated in Figs. 15 and 16. For comparison purposes, results obtained at a spark timing of 15° BTDC, an EGR_I of about 10, and a constant mass air flow per cycle ($\sim 0.77 \times 10^{-3}$ lb/cycle) were used. As speed is increased, ignition delay intervals for both tend to increase. A fairly large increase in ignition delay accompanies the speed change from 600 to 1000 rpm while ignition delays at 1000 and 1500 rpm are about the same. Ignition delays with methanol are shorter than those with Indolene clear at each speed. At 600 rpm, methanol ignition delays are about 30 percent shorter and at 1000 and 1500 rpm they are about 12 percent shorter. The effect of speed on combustion intervals differs somewhat for the two fuels. As speed is increased from 600 to 1000 rpm, combustion intervals for both fuels increase, with methanol having the shorter intervals by about 28 percent. However, as speed is increased from 600 to 1000 rpm, combustion intervals for both fuels increase, with methanol having the the shorter intervals by about 28 percent. However, as speed is increased from 600 to 1000 rpm combustion intervals for both fuels increase, with methanol having the shorter intervals by about 28 percent. However, as speed is increased from 1000 to

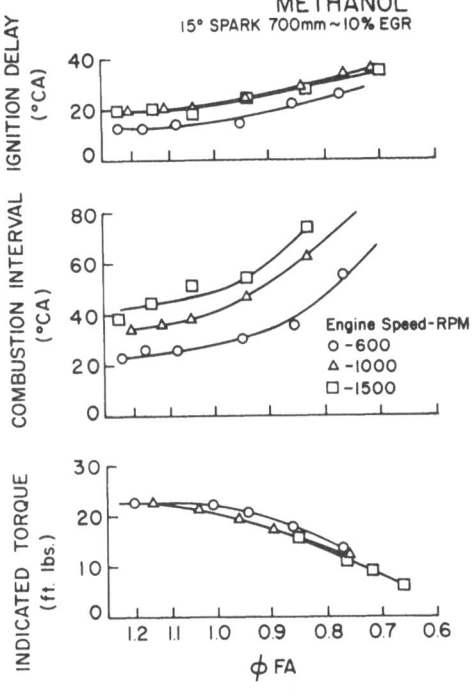

Fig. 15. Influence of Φ_{FA} and engine speed on combustion and power.

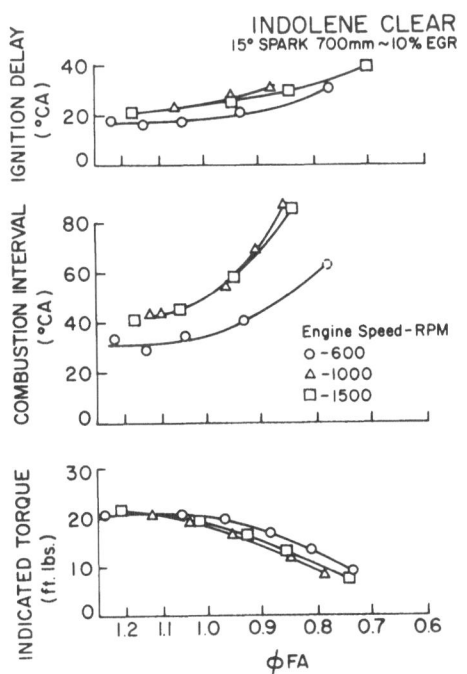

Fig. 16. Influence of Φ_{FA} and engine speed on combustion and power.

1500 rpm, Indolene clear combustion intervals do not change while those of methanol continue to increase in value. As a result, at 1500 rpm, combustion intervals for methanol are equal to those of Indolene clear at fuel rich conditions and are less than 16 percent shorter at the lean mixture ratios. Even though relative combustion interval values change, methanol continues to produce more power than Indolene clear.

Fuel Consumption – Fuel consumption characteristics for the two fuels, illustrated in Figs. 17 and 18, show indicated specific fuel consumption (lb/ihp.hr) and indicated specific energy consumption (BTU/ihp.hr) as a function of Φ_{FA}. Fuel consumption for methanol is seen to be about a factor of two higher than that for Indolene clear at each common value of Φ_{FA}, spark timing and EGR_I (Fig. 17). The corresponding indicated specific energy consumption with methanol, however, is equal to or lower than that with Indolene clear (Fig. 18). At a spark timing of 15° BTDC and an EGR_I

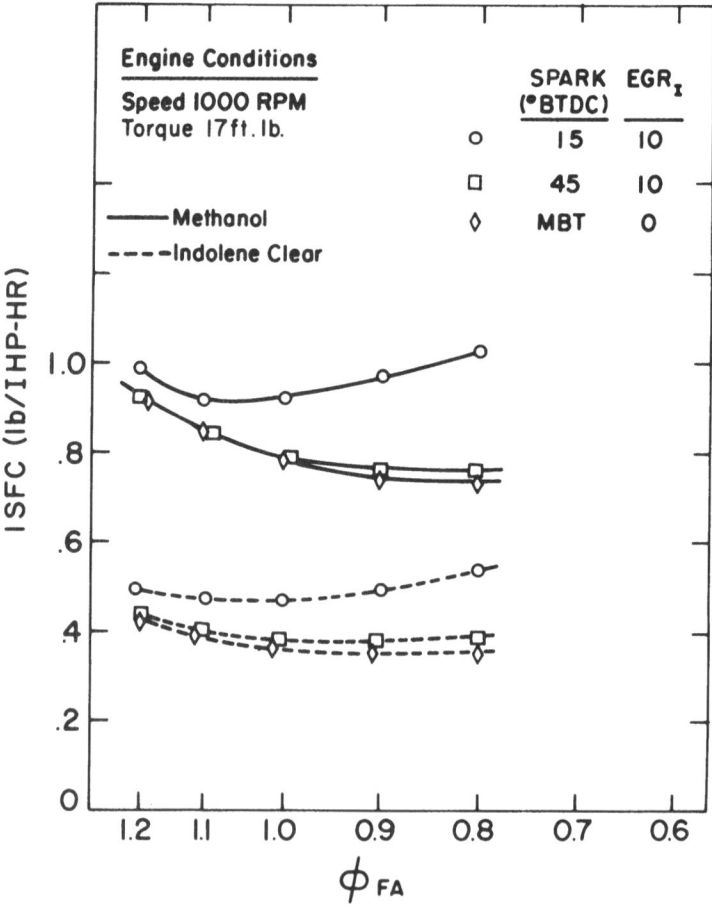

Fig. 17. Influence of Φ_{FA} and spark timing on fuel consumption.

of 10, energy consumption with methanol is 12 to 15 percent lower. At a spark timing of 45° BTDC with an EGR_I of 10, energy consumption with methanol ranges from 3 percent lower at a Φ_{FA} of 1.2 to 10 percent lower at a Φ_{FA} of 0.8.

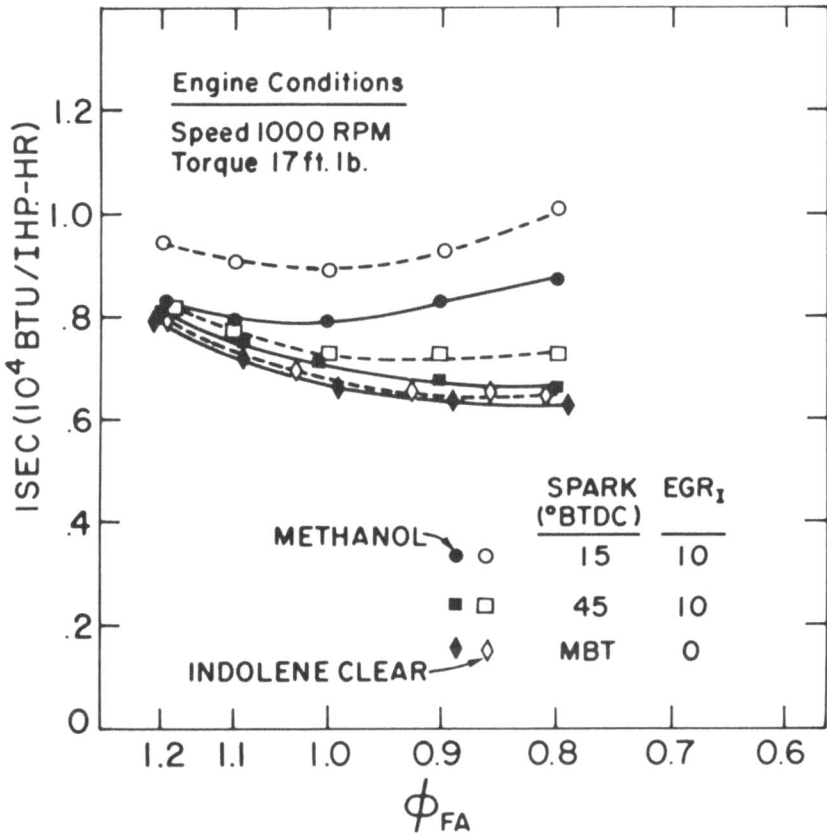

Fig. 18. Influence of Φ_{FA} and spark timing on indicated specific energy consumption.

Emission Characteristics — Emission characteristics for engine operation with methanol and Indolene clear are shown in Figs. 19 through 26.

CO emissions (concentration basis) are shown in Fig. 19. As in the case of gasoline, CO concentration with methanol is a function of Φ only and is not affected by changes in speed, air flow, EGR or spark timing. At mixture ratios richer than $\Phi = 1.05$, CO concentrations with methanol are lower than those of Indolene clear; however, for leaner air-fuel mixtures they are higher by almost a factor of two. In Fig. 20, indicated specific CO emissions (ISCO) are shown. Relative ISCO values for the two fuels at a common EGR_I and Φ_{FA} are about the same as found on a concentration basis.

References pp. 208-210.

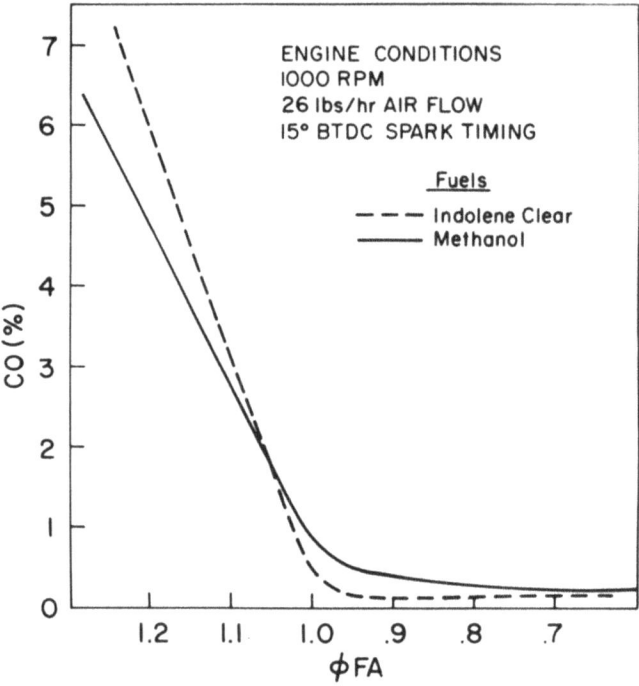

Fig. 19. Influence of Φ_{FA} on CO emissions.

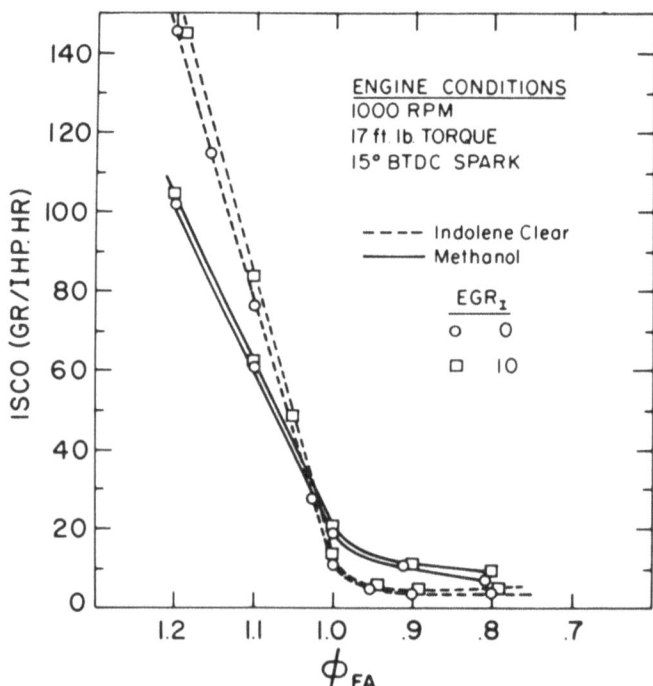

Fig. 20. Influence of Φ_{FA} and EGR on indicated specific CO emissions.

Emissions of NO for both fuels are influenced by Φ_{FA}, air flow (throttle setting), EGR, speed, and spark timing in accordance with patterns known to exist for gasoline. Figs. 21 and 22 show the influence of Φ_{FA} and EGR on NO concentration. NO levels with both fuels are similarly affected by changes in EGR_I and Φ_{FA}. The concentrations of NO with methanol are usually lower than those with Indolene clear at any common value of Φ_{FA}, air flow, EGR_I, speed and spark timing. The largest difference in NO concentrations for the fuels occurs at lean mixture ratios with no EGR. Fig. 23 shows the variation of indicated specific NO emissions (ISNO) with

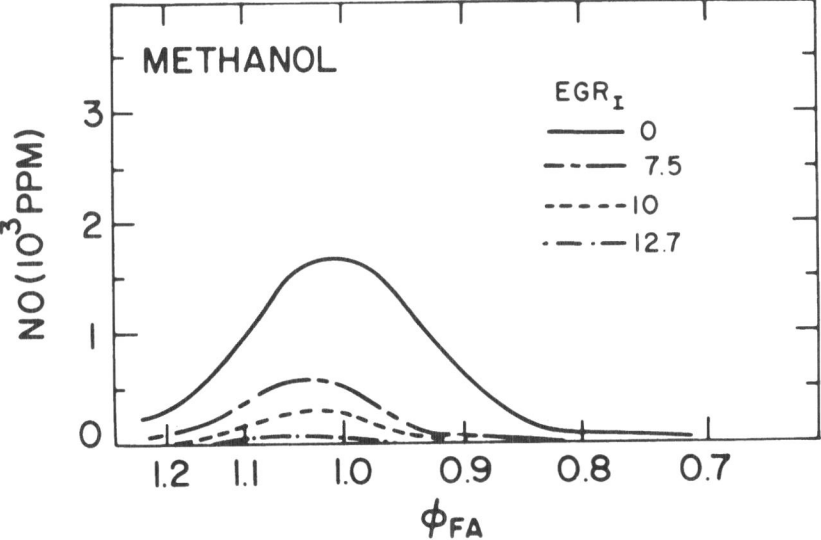

Fig. 21. Influence of Φ_{FA} and air flow on NO emissions.

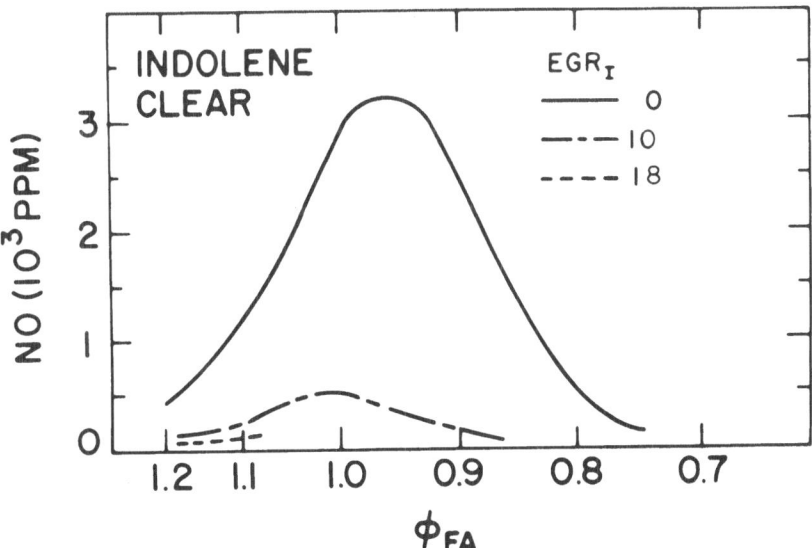

Fig. 22. Influence of Φ_{FA} and air flow on NO emissions.

References pp. 208-210.

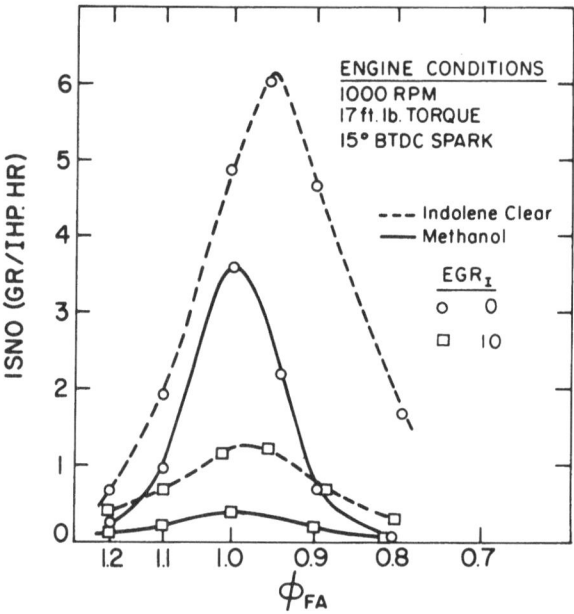

Fig. 23. Influence of Φ_{FA} and EGR on indicated specific NO emissions.

Φ_{FA}. Methanol use results in the lower ISNO values at common Φ_{FA} and EGR_I levels. At fuel rich conditions ($\Phi_{FA} > 1.05$) with zero EGR, ISNO values with methanol are within a factor of 2 of the corresponding Indolene clear values. At fuel lean conditions with no EGR, the ISNO values with methanol are about a factor of 10 lower. With EGR ($EGR_I = 10$ for both fuels), indicated specific NO emissions with methanol are a factor of 3 to 4 lower than with Indolene clear.

HC emissions (concentration basis) for the two fuels also vary with speed, air flow, Φ_{FA} and spark timing in a manner similar to that for gasoline.* With EGR addition, HC levels for Indolene clear vary about as expected; however, those for methanol deviate from the "normal gasoline" behavior. Figs. 24 and 25 show the results. HC concentrations with methanol at fuel rich conditions decrease with increasing EGR while those with Indolene clear do not. Fig. 26 shows the variation of indicated specific HC emissions (ISHC) with Φ_{FA} for operating conditions identical to those of Fig. 23. The ISHC values for the two fuels at the zero EGR level are seen to differ by less than 15 percent at any common Φ_{FA} value. At an EGR_I of 10 and at fuel rich conditions ($\Phi_{FA} > 1.0$) methanol use results in ISHC levels that are lower than with Indolene clear, at Φ_{FA} values between 1.0 and 0.85 they are higher, and at still leaner mixtures they become lower again due to the onset of misfiring with Indolene clear.

---

* *Recall that HC levels for methanol represent lower limit values.*

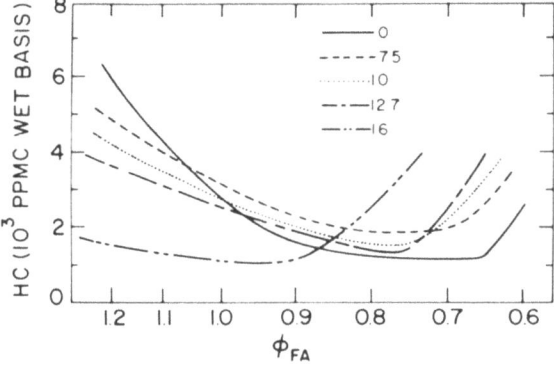

Fig. 24. Influence of Φ_{FA} and EGR on HC emissions.

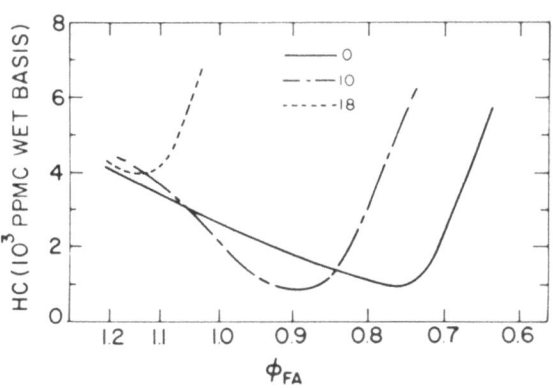

Fig. 25. Influence of Φ_{FA} and EGR on HC emissions

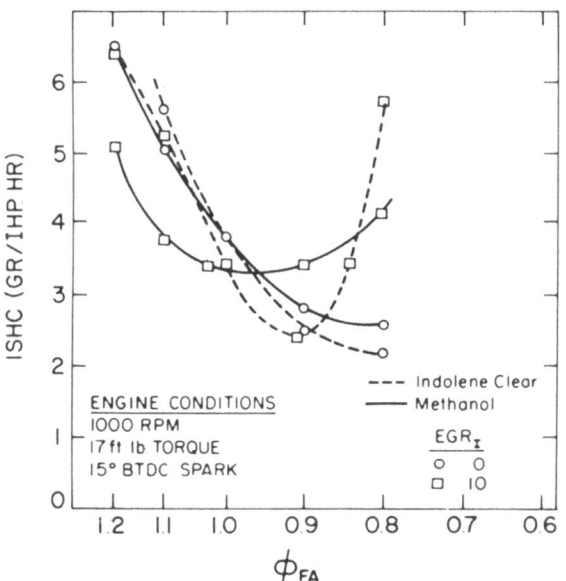

Fig. 26. Influence of Φ_{FA} and EGR on indicated specific HC emissions.

Lean limits, as indicated by a sharp rise in HC emissions with decreasing Φ_{FA} values, vary as expected with increasing EGR. The amount of "lean limit enrichment" per unit increase in EGR index is nearly identical for the two fuels although zero EGR lean limits differ (Fig. 27). Methanol lean limits are extended by about 0.1 units in Φ_{FA} relative to those for Indolene clear at the same EGR levels.

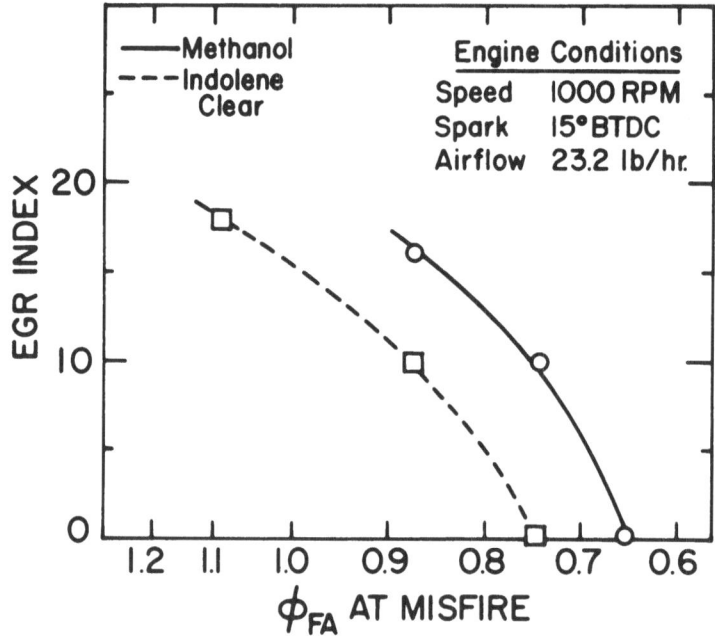

Fig. 27. Variation of Φ_{FA} at misfire with EGR index.

Other Observations — During these experiments, measurements on intake manifold temperatures and exhaust gas temperatures (at port) were also made. When operating with methanol, intake manifold temperatures were depressed by as much as 70°F relative to the 100°F carburetor inlet temperature. With Indolene clear fuel, intake manifold temperature depression was on the order of 15°F.

Exhaust gas temperatures with methanol were 20°F to 70°F lower than with Indolene clear fuel. Generally, exhaust gas temperatures for both fuels increased with speed, spark retard, and mass air flow at any given value of Φ_{FA} and EGR. Exhaust gas temperatures decreased with increasing EGR.

Discussion of Results — Results obtained in applying the $logPV^{\overline{\gamma}}$ method in a study of the combustion-emission characteristics of methanol show that methanol usually burns faster than Indolene clear (combined effects of ignition delay and combustion intervals) at common values of engine speed, Φ_{FA}, spark timing, EGR, and air flow. This fast burn characteristic of methanol is expected to be reflected in relative MBT spark timing settings for the fuels. At EGR_I values of 10 or more, MBT spark timings were not extablished however, because torque maxima were not reached at our most advanced spark timing setting. At the zero EGR level, MBT spark timings for methanol were found to be retarded relative to those for Indolene clear. This zero EGR result is consistent with published MBT spark timings(5,6). Previous work on methanol does not include the use of EGR, and with the exception of a study by the author(6), does not include measurements on burn characteristics.

Power output and fuel consumption data obtained in the present study are consistent with those of earlier (zero EGR) studies(3,5,6). At the zero EGR level and at a constant air flow to the engine, methanol use results in a 10 percent power increase relative to Indolene clear (at identical Φ_{FA}, spark timing, etc.). As EGR is added, the relative power levels produced by the fuels vary somewhat, but methanol continues to produce the higher power. Fuel consumption (SFC) with methanol is about a factor of two higher than with Indolene clear. The difference in fuel consumption rates for methanol and gasoline is large but not unexpected because the heating value of methanol is less than half of that of Indolene clear (Table 2). Where fuel usage is expressed in terms of energy rather than volume, methanol fares much better. Energy consumption with methanol is generally 5 to 10 percent better than with Indolene clear. The actual magnitude of the difference in energy consumed depends on engine operating conditions and EGR levels.

The concentration of CO emissions with methanol and with Indolene clear was observed to be a function of Φ_{FA} only. With methanol, indicated specific CO emissions at fuel rich conditions are lower than those with Indolene clear, and at fuel lean conditions are as much as 50 percent higher. NO emissions for both fuels vary with operating parameters in accordance with patterns known to exist for gasoline. Indicated specific NO emissions with methanol are lower than those with Indolene clear at any common operating point. For operation at fuel rich conditions with zero EGR, indicated specific NO emissions with methanol are within a factor of two of those with Indolene clear; at fuel lean conditions, particularly those approaching misfire limits, NO emissions with methanol are as much as a factor of 10 lower. At an EGR_I of 10, NO emissions with methanol are about a factor of four lower than with Indolene clear at each Φ_{FA} value. The CO and NO results at zero EGR are consistent with those of the earlier studies made with EGR(5,6).

HC emissions with both fuels also vary about as expected with operating parameters, mixture ratio and EGR. An exception to the normal HC variation with increasing EGR appeared with methanol at fuel rich mixture ratios where an increase

References pp. 208-210.

in EGR resulted in decreased HC emissions. At the zero EGR level, indicated specific HC emissions with methanol (as measured) are about the same as with Indolene clear. At an EGR_I of 10, methanol use at fuel rich mixture ratios results in HC emissions that are as much as 25 percent lower, and at lean mixture ratios, HC emissions that are up to 35 percent higher. Considering the fact that the present HC measurements for methanol represent lower limit values, it would appear that methanol use is of no advantage with respect to HC emissions. EGR addition results in an expected lean limit enrichment with both fuels. The amount of enrichment per unit increase in EGR is about the same with both fuels. Methanol lean limits are extended by about 0.1 units in Φ_{FA} relative to those with Indolene clear at each of the common EGR levels used.

CONCLUSIONS

The instrumentation developed to provide a direct measure of $\log(PV^{\overline{\gamma}})$ versus time or of $PV^{\overline{\gamma}}$ versus time is a valuable new tool for the analysis of combustion in engines. Its application simplifies the determination of combustion parameters such as ignition delay interval, burn duration and burn rate.

Overall results of this study indicate that methanol use in internal combustion engines is of no significant advantage with respect to emissions. Much lower NO emissions can be achieved with methanol, but HC and CO emissions tend to be equal to or higher than those with Indolene clear. Methanol use results in a slight power increase but at a significant fuel economy penalty. From the standpoint of energy utilization within the engine, methanol is somewhat better than Indolene clear.

ACKNOWLEDGMENT

The author would like to thank Messrs. E. H. Schanerberger and A. D. Colvin of Engineering and Research Staff, Ford Motor Company for assistance in the collection and analyses of experimental data and for contribution to the design and fabrication of combustion analysis instrumentation.

REFERENCES

1. R. T. Johnson, "Energy and Synthetic Fuels for Transportation," SAE Paper No. 740599, presented at SAE West Coast Meeting, Anaheim, California, August 1974.
2. J. A. Bolt, "A Survey of Alcohol as a Motor Fuel," SAE Special Publication 254, New York, 1964.
3. E. S. Starkman, J. H. Newhall and R. D. Sutton, "Comparative Performance of Alcohol and Hydrocarbon Fuels," SAE Special Publication 254, New York, 1964.

4. R. K. Pefley, M. A. Saad, M. A. Sweeney and J. D. Kilgroe, "Performance and Emission Characteristics Using Blends of Methanol and Dissociated Methanol as an Automotive Fuel," Intersociety Energy Conversion Engineering Conference, Paper No. 719008, SAE New York, 1971, p. 38.

5. G. D. Ebersole and F. S. Manning, "Engine Performance and Exhaust Emissions: Methanol versus Isooctane," SAE Paper No. 720692, presented at National West Coast Meeting, San Francisco, August 1972.

6. J. A. Harrington and R. M. Pilot, "Combustion and Emission Characteristics of Methanol," SAE Paper No. 750420, presented at SAE Congress, Detroit, Michigan, February 1975.

7. R. F. Sawyer, E. S. Starkman, L. Muzio and W. L. Schmidt, "Oxides of Nitrogen in the Combustion Products of an Ammonia Fueled Reciprocating Engine," SAE Paper No. 680401, presented at SAE Mid Year Meeting, Detroit, Michigan, May 1968.

8. R. W. Reynolds, J. S. Smith and I. Steinmetz, "Methyl Ethers as Motor Fuel Components," presented at 168th National Meeting of ACS Division of Petroleum Chemistry, Atlantic City, New Jersey, September 1974.

9. W. J. Most and J. P. Longwell, "Single Cylinder Engine Evaluation of Methanol-Improved Energy Economy and Reduced NO_x," SAE Paper No. 750119, presented at SAE Congress, Detroit, Michigan, February 1975.

10. M. W. Dowdy, J. F. Stocky and T. G. Vanderburg, "An Examination of the Performance of Spark Ignition Engines Using Hydrogen Supplemented Fuels," SAE Paper No. 750027, presented at SAE Congress, Detroit, Michigan, February 1975.

11. R. F. Stebar and F. B. Parks, "Emission Control with Lean Operation Using Hydrogen-Supplemented Fuels," SAE Paper No. 740187, presented at SAE Congress, Detroit, Michigan, February 1974.

12. G. W. Rassweiler and L. L. Withrow, "Motion Pictures of Engine Flames Correlated with Pressure Cards," SAE Transactions, Vol. 38, May 1938, pp. 185-204.

13. B. D. Peters and G. L. Borman, "Cyclic Variations and Average Burning Rates in an S. I. Engine," SAE Paper No. 700064, presented at Automotive Engineering Congress, Detroit, Michigan, January 1970.

14. D. R. Lancaster, R. B. Krieger and J. H. Lienesch, "Measurement and Analysis of Engine Pressure Data," SAE Paper No. 7500026, presented at Automotive Engineering Congress, Detroit Michigan, February 1975.

15. S. Currey, "A Three Dimensional Study of Flame Propagation in a Spark Ignition Engine," SAE Paper No. 62-452B, Presented at Automotive Engineering Conference, Detroit, Michigan, January 1962.

16. G. A. Harrow, P. L. Orman and G. B. Toft, "The Effects of Engine Operating Variables on the Time of Flame Propagation in a Spark Ignition Engine," J. Inst. Petroleum, Vol. 4, No. 475, pp. 204-214.

17. E. S. Starkman, F. M. Strange and T. J. Dahm, "Flame Speeds and Pressure Rise Rates in Spark Ignition Engines," SAE Paper No. 59-83V, Presented at SAE International West Coast Meeting, Vancouver, B.C., August 1959.

18. E. S. Starkman and G. S. Samuelson, "Flame Propagation Rates in Ammonia-Air Combustion at High Pressure," 11th Symposium on Combustion, The Combustion Institute, Pittsburgh, Pennsylvania, 1967, pp. 1037-1045.

19. L. C. Lichty, "Combustion Engines Processes," McGraw-Hill, New York, 1967.

20. E. F. Obert, "Internal Combustion Engines and Air Pollution," Intext Educational Publishers, New York, 1973.

21. N. C. Blizzard and J. C. Keck, "Experimental and Theoretical Investigation of Turbulent Burning Model for Internal Combustion Engines," SAE Paper No. 740191, Presented at Automotive Engineering Congress, Detroit, Michigan, February 1974.

22. J. M. *Colucci, presentation at the NSF-EPA Sponsored Symposium-Workshop on Alternative Fuels, Ann Arbor, Michigan, October 1974.*
23. *Beckman Instruction No. 1306 A, "108A and 109A Hydrocarbon Analyzers," Beckman Instruments, Inc., Fullerton, California, 1966.*

DISCUSSION

F. L. Dryer *(Princeton University)*

There's an interesting alternative to EGR in the case of methanol, and that's the addition of water in solution with methanol. Have you considered that? In that case, the volumetric efficiency of the engine should go up rather than down or stay nearly the same.

Harrington

Yes. We have considered that, but we haven't done any work on it yet. We are aware of the work that has been done by Exxon.

W. E. Bernhardt *(Volkswagenwerk AG)*

You indicated that HC emission is greater with methanol. How did you measure HC?

Harrington

The hydrocarbons were measured using a heated sample system with a heated flame ionization detector. We lumped hydrocarbons into essentially one group. We are aware that the exhaust of a methanol-fueled engine contains a fair amount of unburned methanol. We treated the unburned methanol as being a standard hydrocarbon and also assumed that the flame ionization detector response to what we call hydrocarbons in the case of methanol, and in the case of Indolene clear, were the same. As a result, the figures that I showed represent lower limit values for methanol.

F. L. Dryer

There are ways of experimentally correcting for the flame ionization response, which happens to be lower in the case of methanol, by using a catalytic flame ionization system. I'm wondering whether you should include unburned methanol emissions as a hydrocarbon emission. It turns out that methanol is not very photochemically reactive.

R. W. Hurn *(U.S. Energy Research and Development Agency)*

I would like to respond to Dryer's comment on not counting methanol as HC emission. Fred suggested that we shouldn't count it because it is photochemically unreactive. In this context of interest in methanol, that's true. But the question also arises, should the unburned methanol be counted for its toxic properties. We should also look at the other products. Most of us here know that aldehydes are of concern. I just wanted to be sure that interest in all methanol *combustion* products was not overlooked.

N. A. Henein *(Wayne State University)*

Have you observed any difference in cycle-to-cycle variation with methanol as compared to Indolene?

Harrington

Yes, as a matter of fact, when we operate with methanol, we observe a slightly reduced cyclic variation as compared to Indolene. I don't know the reason. It may be just the fact that we have a considerably greater volume of fuel in the chamber to achieve the same conditions.

R. C. Schwing *(General Motors Research Laboratories)*

Have you considered the engine efficiency and power gain that can be achieved because of methanol's properties?

Harrington

I think you're referring to the volumetric efficiency of the engine. In our case, the air flow rate to the engine was regulated so that even though the temperature of the charge was decreased and one *could* have got more charge in, we regulated this and maintained exactly the same air flows. Methanol would have benefits at wide-open throttle conditions.

J. P. Longwell *(MIT and Exxon)*

Did you see any evidence of preignition reactions?

Harrington

Yes. Whenever we operated at a mode that would give rise to knock with Indolene clear, we generally saw preignition with methanol. Another point is that we have not yet observed any knock with methanol.

J. B. Heywood *(Massachusetts Institute of Technology)*

One of the trends which interested me was as you go to higher speed, ignition delay and combustion duration with methanol became longer than that for the Indolene. Do you have any qualitative explanations for that trend?

Harrington

Purely speculative. It would indicate that possibly the influence of turbulence on combustion reactions differs for these fuels.

J. P. Longwell

I want to continue on the line of John Heywood's question. One piece of speculation might be that, as you increase rpm, the time for preignition reactions is reduced. Therefore, you might have less preignition reactions at the higher rpm. This might account for the fact that Indolene and methanol tend to come together because preignition decomposition of methanol would probably produce mostly CO.

Harrington

In some of our experiments, as we approached misfire conditions, we noticed a rise in CO emissions. This may support that sort of idea.

W. M. Scott *(Ricardo Consulting Engineers)*

I'd like to ask you a question on technique. You mentioned your method of determining the volume of the cylinder for your modeling purposes. I believe a photomechanical device plotted it. Would you like to comment first on how important the accurate measurement of volume is? Next describe the mechanism you used and then tell us how accurately you think it does describe the distances.

Harrington

I think that the device we used exactly duplicates the motion of a piston in an engine. It turns out that you can generate a signal corresponding to the volume by spinning a circular disk off-axis.

Scott

What do you do about the effect of bearing clearance that's affecting the position of the piston relative to the crank?

Harrington

We don't account for that. I don't think you could account for that if you used digital methods for generating volume signal either. It's one of the uncertainties of a measurement or of a calculation.

Heywood

The GM Research people, of course, have done a fair amount of work on evaluating accuracy required in volume measurements, particularly relating volume and crank to pressure traces. Their conclusion was you need a fairly high accuracy, on the order of 1% or so, to get a good match on calculated indicated mean effective pressure to what you determine experimentally.

F. L. Dryer

I would like to add a further comment about toxicity vs. photochemical smog generation with methanol. The question of toxicity of methanol still, I believe, is quite unresolved. It turns out that the toxicity levels that are presently being quoted are quite a bit higher than those that are of critical concern where one would be typically concerned with photochemical smog generation. The medical evidence that has been used to establish those toxicity levels is still very unclear. In fact, there is quite a bit of indication that methanol in inhalation is not absorbed.

ENGINE PERFORMANCE AND EXHAUST EMISSION CHARACTERISTICS OF A METHANOL-FUELED AUTOMOBILE

W. E. BERNHARDT and W. LEE

Volkswagenwerk AG, Wolfsburg, Germany

ABSTRACT

Laboratory and road tests showed methanol to be a very attractive, clean-burning alternative fuel for automobiles with relatively minor problems which can be overcome. A number of VW production vehicles have been converted to methanol operation through the use of an exhaust-heated intake manifold combined with a heating feature using engine coolant and, of course, a modified carburetor. Tests indicated that more power is obtained with methanol because its higher heat of vaporization cools the mixture entering the engine much more than gasoline. This increases the air-fuel mixture density and the mass flow. The gain in power output with pure methanol is about 10%.

When the vehicle is operated on pure methanol, it needs some form of cold starting aid for ambient temperatures below 8°C. There are several possibilities for improving cold starting and warm-up, such as adding volatile starting additives to methanol, using special "cold start" substances (e.g. butane, methyl ether, gasoline) which are sprayed into the intake air during starting, or employing a small flame preheater in the intake manifold.

Vapor lock is not a problem when pure methanol is used. Furthermore, tests with cars modified to run on methanol indicated acceptable to good cold-start driveability.

Fuel economy was measured during exhaust emission tests, driveability tests, and specific fuel economy tests. Because of methanol's lower energy content, mass specific fuel consumption is noticeably greater than that with gasoline. However, fuel consumption related to consumed energy is considerably lower than that with gasoline. This means that methanol burned more efficiently than gasoline. At 2,000 rpm and wide-open-throttle, a 17% increase in brake efficiency has been observed.

Automobile exhaust emissions and air pollution can be reduced by use of methanol fueled engines. Carbon monoxide (CO) emissions from the methanol fueled engine correspond approximately to those from the gasoline engine. However, tests on a VW PASSAT 4-cylinder engine at WOT and various engine speeds showed that it is possible to reduce CO emissions from the methanol fueled engine especially at low engine speeds as compared to gasoline.

When methanol is used as engine fuel, a significant reduction in nitrogen oxides (NO_x) emissions is possible. Furthermore, very low levels of hydrocarbon (HC) emissions were observed for methanol. Only about 10% of the organic emissions measured with the FID are hydrocarbons, as demonstrated by gas chromatographic techniques. Thus, methanol fueled automobiles are environmentally sound with regard to hydrocarbon emissions.

At the same compression ratio, aldehyde emissions from a methanol fueled engine are noticeably higher than from a gasoline fueled engine. However, aldehyde emission can be reduced by increasing the compression ratio, controlling the combustion process and by adding up to 10% water to methanol.

Polynuclear aromatic hydrocarbon emissions, some of which are regarded as severely carcinogenic, are more than one order of magnitude lower with methanol than with gasoline.

INTRODUCTION

With the exhaustion, or near exhaustion of world petroleum reserves predicted within the next 25 to 30 years, it has become imperative to investigate the consequences of using non-petroleum fuels in automobiles. Substitutes receiving great attention are methanol derived from coal or wastes (municipal refuse, waste wood, and garbage from agricultural operations) and gasoline-like fuels from coal and oil shale. But liquid fuel substitutes for petroleum other than methanol are beset with serious problems. Oil shale products, for example, are estimated to be cheaper than methanol, but large scale development of shale deposits is still barred by such difficulties as the disposal and revegetation of tremendous amounts of tailings. Gasoline-from-coal production technology is known, but requires further development for getting economically attractive processes. Another very interesting synthetic fuel is hydrogen from coal, or water if nuclear energy can be used for production. But hydrogen technology for automotive application will not be available before 2000 a.d.

Since the invention of the spark-ignition, internal-combustion engine in 1876, it has been known that alcohols are potential alternative spark ignition engine fuels (1-4). Lately, in connection with control of engine exhaust emissions, interest developed due to the possibility that alcohol fueled engines could have comparatively

References p. 232.

low emissions (1, 5, 6). However, to date, two important factors have prevented the general use of alcohols: high costs compared with gasoline; and the absence of technical or economic advantages to compensate for these higher costs.

The present increased interest, particularly in methanol as an engine fuel, results from the following:

1. Methanol can be produced synthetically from coal by known technology at a competitive price (7, 8, 9).

2. Methanol can be stored and distributed as a liquid in the same way as gasoline (10, 12).

3. Methanol is environmentally more acceptable than synthetic gasoline or other alternatives because lower emission levels could be attained (10).

4. The motor fuel supply could be increased by adding methanol to unleaded gasolines in concentrations up to 15 vol. % (12).

5. The use of methanol could permit the design of engines with higher compression ratios, resulting in an improvement in fuel economy (12).

The literature contains very little information, especially on performance and exhaust emission characteristics, of an automobile equipped with an internal-combustion engine operating solely on methanol. Especially, data on emissions of hydrocarbons (HC), total aldehydes (TAH), and oxides of nitrogen (NO_x) are sparse. Therefore, our purpose is to give fundamental information on the environmental aspects of the automotive application of straight methanol. Results of this study may increase understanding of the important role which methanol could play in the future as an alternative automotive fuel.

ENGINE PERFORMANCE CHARACTERISTICS

Experimental Considerations — The experiments were conducted using water-cooled 4-cylinder, 1.6 liter Volkswagen engines operated on engine dynamometers. Fuel economy and exhaust emission measurements were made using a temperature-controlled chassis dynamometer.

Fuel-Air Mixture Preparation — The engine carburetion and intake systems were modified to allow a greater fuel flow and to produce as homogeneous a fuel-air mixture as possible. A standard carburetor was altered to improve the atomization of the fuel-air spray.

The intake manifold was heated by a constant flowrate of hot exhaust gases and, in addition, by hot engine coolant after warm-up. This was necessary because methanol has a latent heat of vaporization which is about three and a half times that of gasoline. The vaporization of liquid methanol in a stoichiometric alcohol-air mixture would result in a mixture temperature reduction of approximately 125°C when no external heating is provided. But with external heating, the mixture temperature of methanol-air mixtures could be kept between 7 and 19°C over the whole range of engine speed at full throttle. For contrast, gasoline-air mixture temperatures are approximately 10 degrees higher as shown in Fig. 1 for the mixture temperature of cylinder 1 of a water-cooled 4-cylinder engine as a function of engine speed at wide-open throttle. Fig. 1 shows that despite the extra heating of the methanol-air mixture it was impossible to get completely identical mixture temperatures with methanol and gasoline.

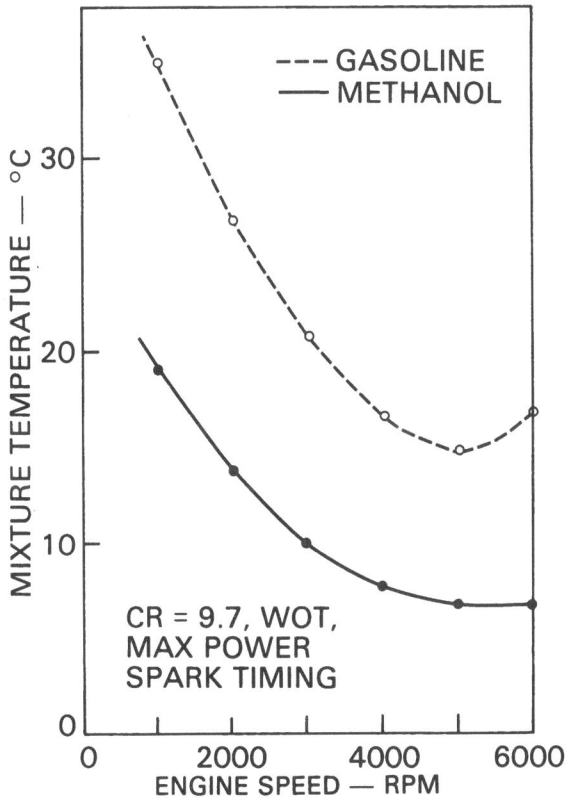

Fig. 1. Mixture temperature with methanol and gasoline.

Engine Power Output – Methanol and gasoline exhibit almost identical calorific values (in kcal/m^3) in the air-fuel mixture, related to the same temperature and overall pressure. In view of methanol's higher heat of vaporization, the intake air and thus the methanol-air mixture is cooled. This leads to a rise in effective engine power output. Fig. 2 shows the brake power outputs of the gasoline and straight methanol fueled engines as a function of engine speed. Under these, and all other operating conditions, the methanol engine had about 10% higher output than the gasoline engine (with max. power spark timing).

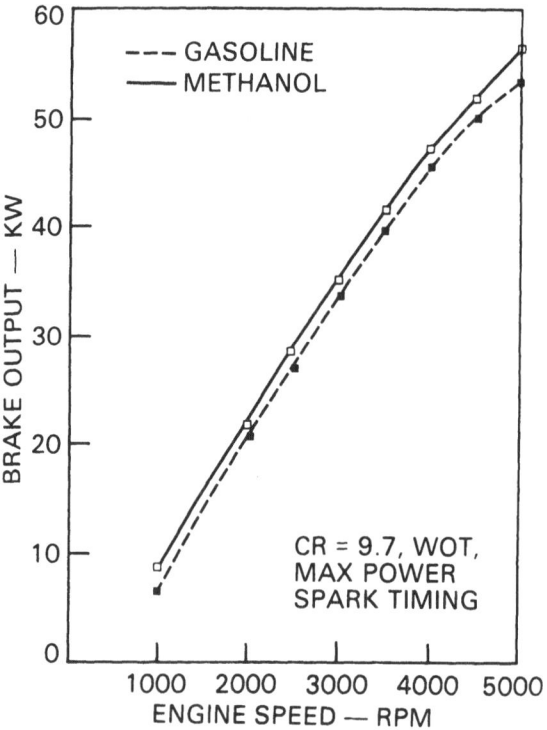

Fig. 2. Engine output with methanol and gasoline.

Engine Efficiency and Fuel Consumption — Fig. 3 shows that methanol produces a considerably higher indicated brake efficiency curve than gasoline for the same compression ratio (CR = 9.7). This is due principally to the greater volumetric efficiency which results from the high density of methanol-air mixtures (13). This high density is a consequence of the low mixture temperature generated by methanol's high latent heat of vaporization, and the greater mass of fuel per unit mass of air. Fig. 3 also indicates that an increase in compression ratio from 9.7 to 14.0 leads to a further improvement in engine efficiency. At 2,000 rpm, wide-open throttle (WOT), and stoichiometric air-fuel ratio, the gasoline engine with CR = 9.7 has a 30% brake efficiency whereas the methanol engine with CR = 14.0 reaches 36%; an increase of 20%. Hence, because of the high octane rating of straight methanol (RON $\sim$ 110, MON $\sim$ 92)(9) the compression ratio could be boosted to 14.0 to get a very economic methanol fueled power system. The influence of high compression ratio on exhaust emission characteristics will be discussed later.

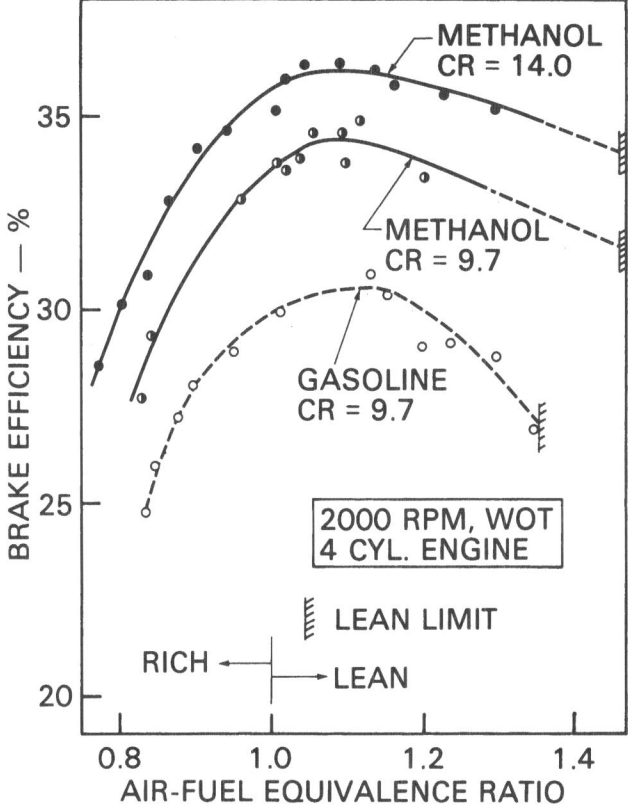

Fig. 3. Brake efficiency with methanol and gasoline.

Fig. 4 compares the fuel consumption of the methanol engine with that of the gasoline engine. Use of methanol results in higher indicated specific fuel consumption (nearly two times greater) when comparison is made on a volume or mass basis. However, when defining fuel consumption on an energy basis, Fig. 4 indicates a fuel consumption about 16% lower at 2,000 rpm and WOT than with gasoline. Similar results have been obtained at Conoco, where a V8 engine consumed from 15 to 26% less energy with methanol than with gasoline (14).

Cold Start Problems and Warm-Up Performance – When operating a vehicle on pure methanol, a cold starting aide is absolutely essential for starting the engine at temperatures below 8°C. There are several possible methods of improving cold start and warm-up, e.g. adding volatile starting additives to the methanol, using special "cold start" substances (such as butane, methyl ether, gasoline) which are sprayed into the intake air during starting, or employing a small, methanol-fueled flame preheater in the intake manifold. A conclusive and ideal solution has not been found

References p. 232.

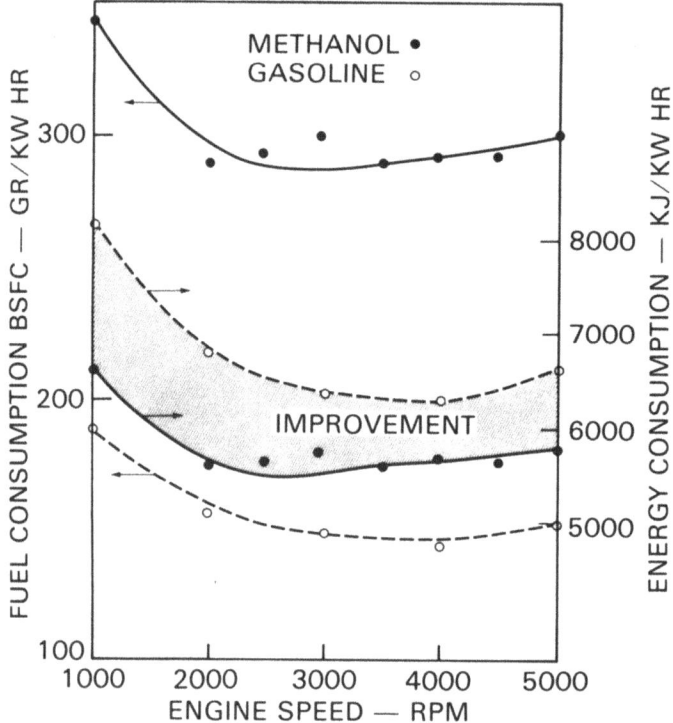

Fig. 4. Fuel and energy consumption with methanol and gasoline.

so far. Further investigations are necessary.

The warm-up performance of the methanol fueled engine depends strongly on:

1. the degree of atomization of the fuel-air spray,

2. the efficiency of the heating device which is integrated into the intake manifold, and

3. the uniformity of the fuel-air distribution to the engines's various cylinders.

Therefore, it is necessary to improve the preparation and distribution of the methanol-air mixture as much as possible. Unsatisfactory warm-up performance and bad cold start behavior are mainly due to inadequate vaporization of methanol in carburetors or fuel injection systems and in intake manifolds designed to handle gasoline. The optimal solution may lie in modified fuel injection systems, new developed methanol-carburetors, or such devices as the "Dresserator" (14, 15).

Warm-up performance of Volkswagen's straight methanol fueled test cars (AUDI 80, NSU-Ro 80, VW-Dasher, and VW-Scirocco) was measured at low ambient temperatures. As expected, the warm-up performance was very good with the heated manifold devices described. There was no pronounced hesitation when the cars were accelerated.

EXHAUST EMISSION CHARACTERISTICS

Engine Exhaust Gas Analysis — A sampling line connected to the engine exhaust system continuously fed exhaust gas to the analyzers when the engine was tested on the engine dynamometer. A second, heated sampling line was provided for collection of aqueous samples for gas chromatographic and wet chemical analyses. Instrumentation for exhaust emission measurements included a non-dispersive infrared analyzer (NDIR) for carbon monoxide, a flame ionization detector (FID) for hydrocarbons and a chemiluminescence detector for oxides of nitrogen. Total aldehydes were determined by the MBTH method (18) which provides for adding exhaust samples to a flask containing 3-methyl-2-benzothioazolone hydrazone (MBTH) solution. Furthermore, emissions of alcohol, hydrocarbons and aldehydes were measured with a Perkin Elmer Model F6 Gas Chromatograph fitted with dual columns and dual flame ionization detectors.

Oxides of Nitrogen — Exhaust emissions of oxides of nitrogen (NO_x) also differ when comparing methanol and gasoline. Fig. 5 shows the reduction in exhaust NO_x concentration for an engine speed of 2,000 rpm, WOT, and max. power spark timing. For CR = 9.7, the reduction is appreciable for rich mixtures, whereas for stoichiometric and lean mixtures the reduction is less. A quite unexpected result is a further reduction in NO_x concentration when the compression ratio of the methanol engine is increased from 9.7 to 14.0. It is not clear whether this effect is a result of the higher compression or the different combustion chamber design at the two compression ratios.

A property of methanol that may help account for the lower exhaust emission of nitric oxide is the lower temperature of burned gas in the post-flame region of the combustion chamber. As in earlier investigations (16) a two-zone model of flame propagation can be used to compute the mean burned gas temperature. The mass burned fraction, which is needed for the calculation, is obtained from a thermodynamic analysis of the engine combustion process (16). Results of the calculation of the burned gas temperature for methanol as well as for gasoline indicate a lower peak temperature (250° C lower) for methanol (13). This lower burned gas temperature influences considerably the dissociation of oxygen in the post-flame region so that less atomic oxygen is formed. Since the rate-determining reaction in the reaction mechanism for the formation of nitric oxide is

$$N_2 + O \rightarrow NO + N \tag{1}$$

the nitric oxide formation will be restrained as the result of excessively low atomic oxygen concentration.

References p. 232.

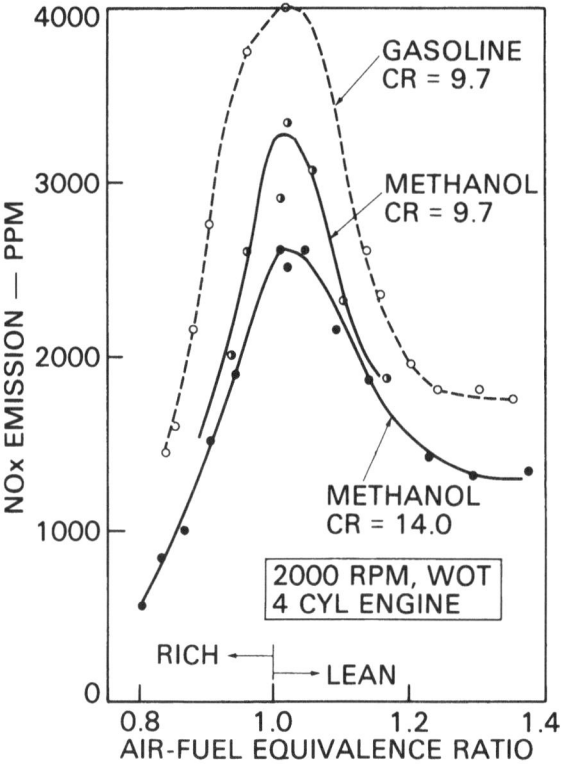

Fig. 5. Exhaust NO_x concentrations with methanol and gasoline.

It has been shown (16, 17) that a simple mathematical model which consists of a thermodynamic analysis of the engine combustion process based on a two-zone model of flame propagation, and of the Zeldovich reaction mechanism

$$N_2 + O \rightleftharpoons NO + N \tag{2}$$

$$NO + O \rightleftharpoons O_2 + N \tag{3}$$

is suitable to describe the formation of nitric oxide in a spark ignition engine fueled by gasoline. From Fig. 6 it is concluded that this mathematical model is also appropriate for methanol.

Carbon Monoxide – The higher combustion speed of methanol permits the mixture to be burned at a greater flame propagation rate. The lower proportion of carbon in the fuel molecule and more favorable dissociation properties at comparative low temperatuers lead to cleaner combustion. Carbon monoxide (CO) emissions from the methanol fueled engine correspond approximately to those from the gasoline engine. However, near stoichiometric mixture strength (air-fuel equivalence ratio

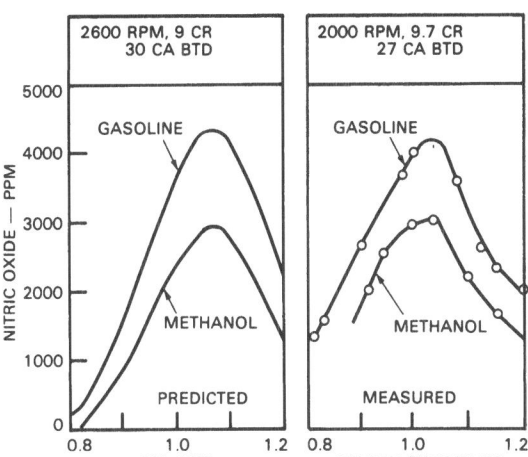

WIDE OPEN THROTTLE & MAX POWER SPARK TIMING

Fig. 6. Predicted and measured exhaust NO_x concentrations with methanol and gasoline.

$\Phi = 1.0$), and in the region $\Phi = 1.0$ to 1.1, a noticeable drop can be identified. Fig. 7 shows exhaust CO concentrations of a 4-cylinder engine (CR = 9.7) at full load and an

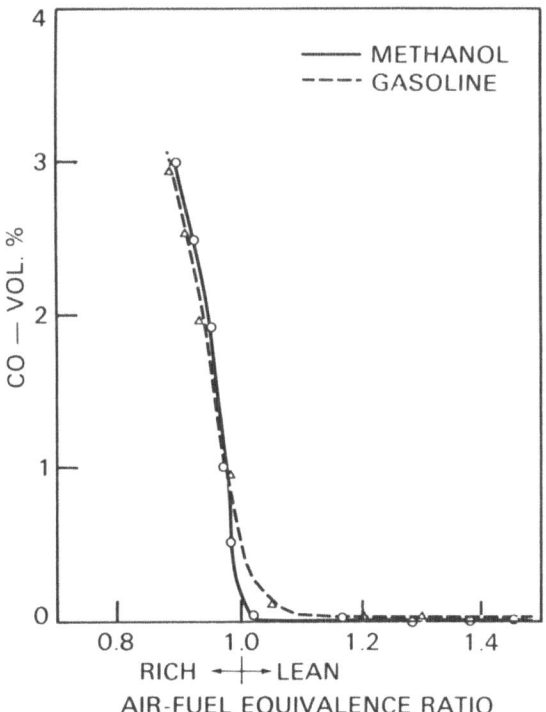

Fig. 7. Exhaust CO concentrations with methanol and gasoline.

References p. 232.

engine speed of 2,000 rpm (max. power spark timing) as a function of equivalence ratio.

Hydrocarbons – Very low levels of unburned fuel compounds (HC)* could be observed for the methanol fueled engine, though it must be borne in mind that unburned methanol and total aldehydes (formaldehyde + acetaldehyde) in the engine exhaust are not correctly registered by the FID. Comparison between FID and gas chromatographic measurements yields a factor of 1.28 which must be used to correct the FID values. However, even with this correction, the HC emissions from the methanol engine will always prove better than when running on gasoline. Fig. 8 illustrates, for CR = 9.7, 2,000 rpm, WOT, and max. power spark timing, that exhaust HC concentration with methanol is about 25% to 33% of that with gasoline. Furthermore, the number of different hydrocarbon species is significantly smaller than that for engine operation with gasoline (13).

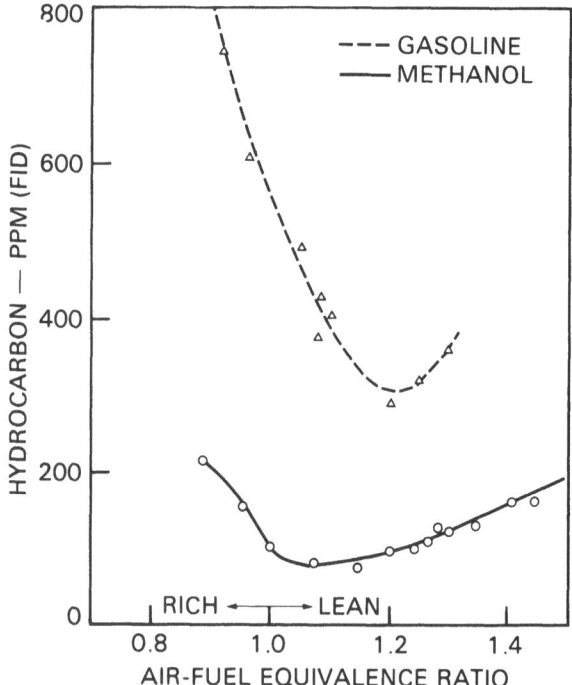

Fig. 8. Exhaust HC concentrations with methanol and gasoline.

Total Aldehydes – The total aldehyde emission from a methanol fueled engine is noticeably higher than from a gasoline fueled engine when the same compression ratio

*HC emissions of a methanol engine are: methanol, aldelydes and hydrocarbons.

(e.g. CR = 9.7) is used. Fig. 9 shows total exhaust aldehyde concentrations (TAH) measured as formaldehyde by the MBTH method. On average, the aldehyde concentration throughout the air-fuel ratio range examined is about twice as high as with the gasoline fueled engine. However, a drastic reduction in aldehydes is obtainable when the compression ratio is increased from 9.7 to 14.0. In this case, TAH from the methanol engine correspond to those of the gasoline engine. This is indeed a very important result.

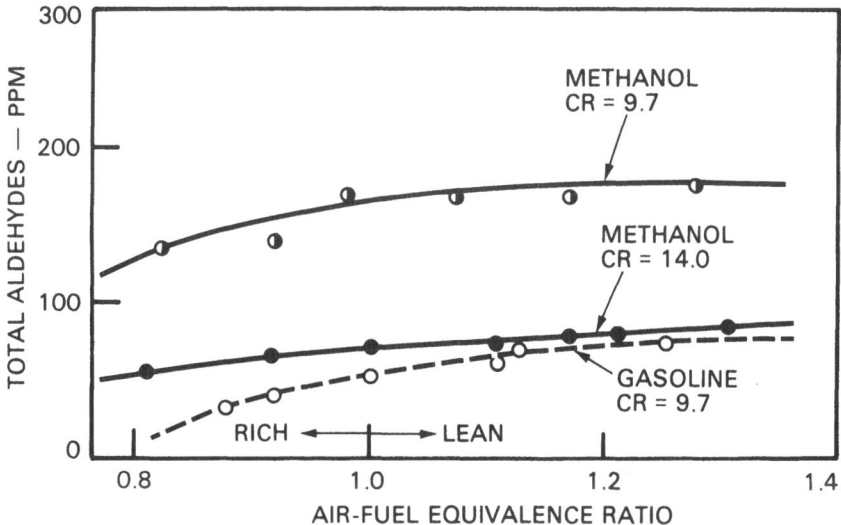

Fig. 9. Exhaust aldehyde concentrations with methanol and gasoline.

Other opportunities for a reduction in aldehydes are provided by water addition to methanol if the compression ratio cannot be increased. Simultaneously, NO_x is also reduced (Fig. 10). Up to 10% water content, HC and CO remain effectively unchanged, yet NO_x can be reduced by about 50% and aldehyde emission by about 40%. The engine power output sacrificed is about 10%. However, since the methanol fueled engine can develop more than 10% greater output than with gasoline, its output under these circumstances would be about the same as that of the gasoline fueled engine. The measurements were made with a 4-cylinder water cooled engine with a compression ratio of 9.7 at full load operation, WOT, and 3,000 rpm. Considering the water content, the air-fuel ratio was 5% lean during this investigation.

The rapid decrease of aldehydes as the water content increased is due principally to the reaction of formaldehyde and the hydroxyl radical:

$$HCHO + OH \rightleftharpoons H_2O + HCO \tag{4}$$

A source of hydroxyl radicals may be the following reaction:

References p. 232.

$$H_2O + O \rightleftharpoons OH + OH \qquad (5)$$

On the other hand, reaction (5) influences the rate-determining reaction (1) for the formation of nitric oxide. Because of a shortage of atomic oxygen caused by reaction (5), the formation of nitric oxide is restrained. Hence, a simultaneous reduction of aldehydes and nitric oxide is possible when water is added to methanol, as demonstrated by Fig. 10 (10).

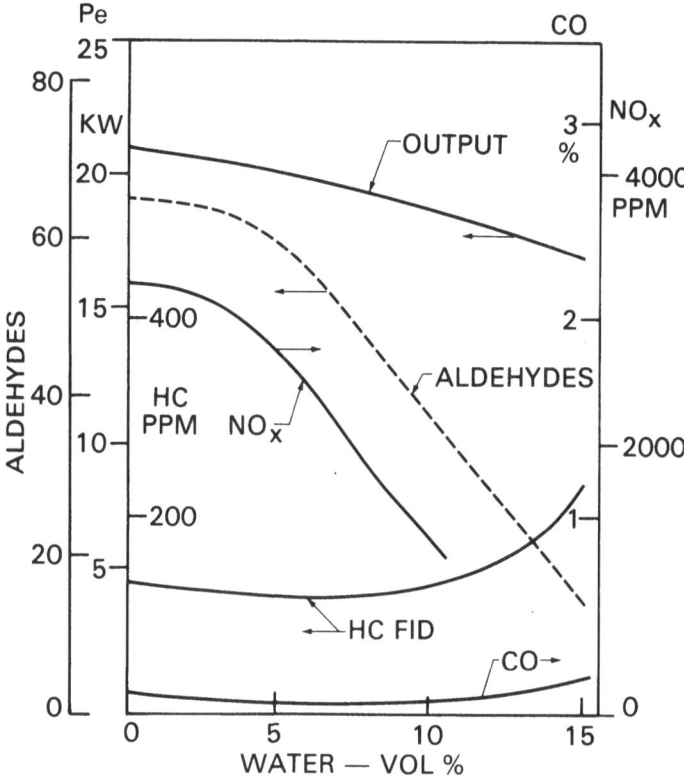

Fig. 10. Effect of water addition to methanol on exhaust emissions and engine power.

Polynuclear Aromatics – Fig. 11 shows typical results of exhaust polynuclear aromatic hydrocarbons (PNA) obtained from vehicles equipped with methanol and gasoline fueled engines. Samples were obtained during the European Test Procedure (ECE Test) and analyzed over the evaporation range from fluoranthene (FLT) to coronene (COR) (19). The PNA emissions, some of which are regarded as severely carcinogenic, are lower with methanol by more than one order of magnitude.

Particulate Emissions – The exhaust gas of methanol fueled engines contains no lead compounds. Because of the high antiknock quality of methanol, no tetra-ethyllead or other lead compounds have to be added to the fuel. Furthermore,

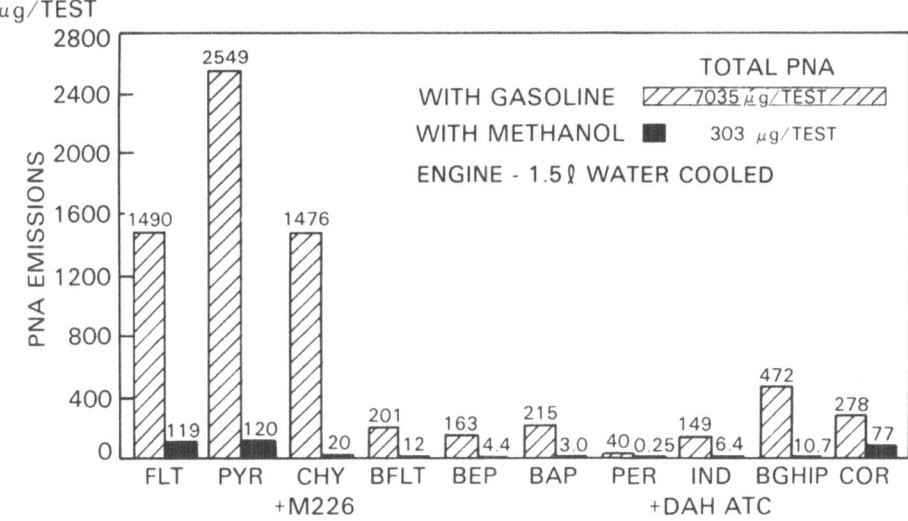

Fig. 11. Polynuclear aromatic hydrocarbon emissions with methanol and gasoline.

because of the reaction kinetics inherent in methanol combustion, no soot is formed. There are also no sulfur compounds in the exhaust.

Interaction of Emissions – Fig. 12 illustrates the interaction of the pollutants NO, HC and CO. Since the primary compound of the total oxides of nitrogen observed in gasoline and methanol fueled engine exhausts is nitric oxide, the concentration of nitric oxide represents roughly the concentration of total oxides of nitrogen. At full throttle operation, CR = 9.7, and engine speeds greater than 2,700 rpm, the concentration of nitric oxide is one-half of the corresponding concentration with gasoline fuel. The lower NO concentration at high engine speeds is due principally to a richer air-fuel mixture.

Fig. 12 shows that at WOT and especially low engine speeds, it is possible to reduce CO emissions from the engine by using methanol as fuel. Furthermore, very low levels of fuel components are observed for methanol. Taking into account that these low HC concentrations, as measured with the FID, are only 10% hydrocarbons, as demonstrated by GC techniques, it becomes clear that methanol fueled engines are environmentally more acceptable than gasoline engines.

Toxicological Aspects – All in all, the exhaust emission behavior of a methanol fueled engine is distinctly better than that of a gasoline fueled engine. Furthermore, the toxicological properties of methanol are comparable with those of traditional fuels (gasoline) and can be regarded as more acceptable than those of certain gasoline components, e.g. benzene. Nevertheless, it should be mentioned that gasoline and methanol are both toxic. Therefore, both fuels should not be ingested, inhaled, siphoned, or used as cleaner-solvents.

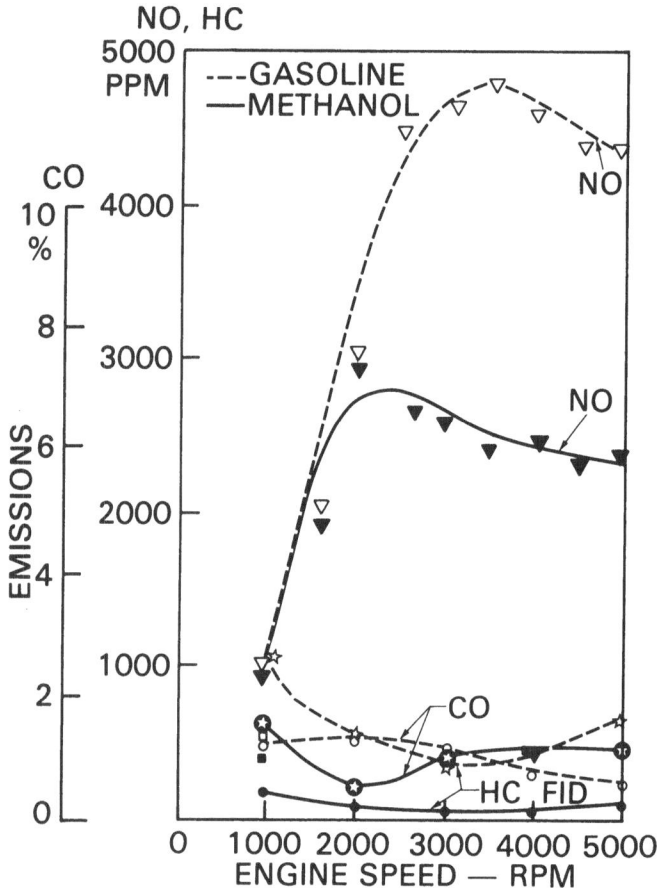

Fig. 12. Engine speed effects on exhaust emissions at full throttle with methanol and gasoline.

VEHICLE EVALUATION

Volkswagen has converted a number of standard production vehicles, including a VW Scirocco, from gasoline to methanol operation. To improve cold start performance, injection of gasoline or other volatile substances into the intake manifold during starting was required. Also, replacement of components soluble in methanol, such as plastic floats, diaphragms and gaskets, proved necessary. Since the fuel tank was not enlarged, the vehicle's operating range was about half of its range with gasoline. With the aid of production type mechanical or electronic fuel injection systems, for example injection directly into the inlet valve, mixture formation and distribution problems could be solved more readily. By the addition of fuel components which evaporate more readily, for example substances already present in crude methanol or additives such as dimethyl ether, diethyl ether, ethyl methyl ether, or methyl tertiary butyl ether (MTBE), cold start performance could be noticeably improved and emissions reduced.

Vehicle evaluation has been conducted mainly under simulated road conditions using the CVS cold-hot emissions test and the ECE Test Procedure. Furthermore, methanol fueled cars were operated on the road for testing fuel consumption, driveability and warm-up performance. As mentioned earlier, at the present state of development, energy consumption and driveability proved to be very good or at least acceptable.

Fuel Consumption – Table 1 shows the consumption of methanol during the European Test Procedure in comparison with gasoline. The energy bonus derived from methanol operation is only 4% compared with gasoline operation during the European cold-start test, but 12% for the hot-start. An improvement in engine warm-up performance when running on pure methanol is an urgent necessity. The high antiknock rating of methanol makes possible a further increase in power output and a simultaneous improvement in fuel consumption by raising the compression ratio.

Table 1 also illustrates the influence of gasoline containing 15 vol % methanol ("M 15"). In this case the European Test Procedure results (cold-start) show a 3.5% drop in fuel mass consumed. If the measured fuel consumption is converted into energy consumption, the reduction with M 15 operation during the European cold-start test is about 12%. Furthermore, Table 1 illustrates that the addition of volatile

TABLE 1

Fuel Consumption During ECE Test Procedure

Fuel	Test Condition	Fuel Consumption	
		g/Test	KJ/Test
Gasoline	Cold Start	373	16 394
	Hot Start	328	14 417
Straight Methanol (M 100)	Cold Start	801	15 757
	Hot Start	644	12 669
85% Gasoline + 15% Methanol (M 15)	Cold Start	360	14 431
	Hot Start	310	12 426
10% Gasoline + 90% Methanol (M 90)	Cold Start	716	15 725
	Hot Start	619	13 594

References p. 232.

components (such as gasoline) in concentrations around 10 vol % ("M 90") does not lead to a further improvement in energy consumption versus straight methanol ("M 100").

Vehicle Emissions – Table 2 presents exhaust emission data from test cars using straight methanol as well as M 15, operated on the ECE test cycle. In the cold-start test, the effect of methanol on emissions was a 25% reduction in CO but a 22% increase in HC (FID). In the hot-start test the CO emissions were reduced by 68% and the HC (FID) emissions by 28%. There was about a 20% reduction in nitrogen oxides in both tests. This emission behavior can probably be improved by optimizing, especially during the warm-up phase. Exhaust emissions from another straight methanol fueled car were measured during the 1975 Federal emission-test cycle (cold-start) (9). With premium gasoline the vehicle emitted HC = 4.5 g/mi, CO = 22.1 g/mi and NO_x = 5.7 g/mi. With straight methanol the effect on emissions was a 44% reduction in HC, a 58% reduction in carbon monoxide and a 67% reduction in nitrogen oxides. Hence, this vehicle met the 1975 Federal Emission Standards (HC = 1.5 g/mi, CO = 15.0 g/mi, NO_x = 3.1 g/mi) for CO and NO_x but not for HC. Taking into account that the HC emission from a methanol fueled car contains only a small percentage of hydrocarbons (mainly methane, ethane and ethylene), and that PNA emissions are very low, it becomes clear that there are really no toxicity or health problems to be expected.

TABLE 2

Vehicle Emission Test Results

ECE Test Cycle

		CO		HC (FID)	
		g/Test	Rel. Correlation to Gasoline	g/Test	Rel. Correlation to Gasoline
Gasoline	Cold Start	136	1	9.8	1
	Hot Start	112	1	7.9	1
Methanol	Cold Start	102	0.75	12.0	1.22
	Hot Start	36	0.32	5.7	0.72
M15-Mixture	Cold Start	56	0.41	8.3	0.85
	Hot Start	34	0.30	6.5	0.82

SUMMARY AND CONCLUSIONS:

On the basis of experimental and theoretical results the following conclusions can be made:

1. Methanol provides improved power, specific energy consumption, and efficiency relative to commercial gasoline.

2. A significant reduction in oxides of nitrogen emissions results from the use of methanol. The reaction mechanisms used for predicting NO_x emissions from gasoline fueled spark-ignition engines are also good for methanol combustion.

3. The carbon monoxide emissions are about the same with either methanol or gasoline. However, near stoichiometric and up to 10% lean mixtures, CO emissions are reduced when methanol is used.

4. Emissions of hydrocarbons are significantly lower with methanol than with gasoline. Special precautions have to be taken to analyze HC by FID techniques. Emissions of methanol and total aldehydes, using the same engine with the same compression ratio, are significantly higher with methanol than with gasoline. Total aldehydes in the exhaust are about twice as high with methanol as with gasoline. However, emissions of total aldehydes can be reduced substantially by increasing the compression ratio or by adding water to the pure methanol.

5. The excellent octane rating of methanol provides the possibility to increase compression ratio up to 14.0. Simultaneously, further improvement in fuel consumption and efficiency, as well as in NO_x and total aldehyde emissions could be achieved.

6. PNA emissions with methanol are lower than with gasoline.

7. Cold-start performance and driveability of methanol-fueled test vehicles are at least acceptable. Further improvements, especially in the field of fuel consumption and emissions, can be achieved by improving the warm-up characteristics of the engine and intake manifold system, particularly during the first 200 seconds after vehicle start-up.

8. Spark ignition engines, from which performance and exhaust emission data were obtained, were designed for gasoline and subsequently modified for methanol. The advantages of methanol will be maximized in engines specifically designed for methanol, and the disadvantages minimized or eliminated.

ACKNOWLEDGEMENTS

The authors would like to acknowledge the contributions of many people at the Research Department of Volkswagenwerk AG, especially, the assistance of the members of the Combustion and Reaction Kinetics Department, including Dr. A. König and Messrs. W. Geffers, P. Heidemeyer, H. Gring, H. Wischer and H. Jacob. Finally, the authors express their appreciation for the financial support by the German Ministry for Research and Technology.

References p. 232.

REFERENCES

1. R. E. Fitch and J. D. Kilgroe, "Investigation of a Substitute Fuel to Control Automotive Air Pollution," CETEC-Report No. 01800-FR, February 1970.
2. E. Fischer, "Der Motorwagen," 1926, p. 487.
3. W. Wilke, "Oele and Kohle – Erdoel und Teer," 13, 1937, p. 1030.
4. "SAE Special Publication No. 254," Society of Automotive Engineers, New York, New York, June 1964.
5. F. V. Morriss, R. Modrell, G. Atkinson and C. Bolze, "The Exhaust Content of Automobiles Burning Ethanol-Gasoline Mixtures," ACS Meeting, Preprint No. 77, September 1955.
6. E. S. Starkman, R. F. Sawyer, R. Carr, G. Johnson and L. Muzio, "J. Air Pollution Control Assoc.," 20, 1970.
7. F. H. Kant, "Feasibility Study of Alternative Automotive Fuels," Status Report prepared for Alternative Automotive Power Systems Coordinating Meeting, Ann Arbor, Michigan, October 1973.
8. J. Hellbach and W. Bernhardt, "Mögliche Alternativ-Kraftstoffe für Verbrennungsmotoren," Volkswagen Research Report F2 - 74/3, February 1974.
9. Bundesministerium für Forschung und Technologie: "Neuen Kraftstoffen auf der Spur – Alternative Kraftstoff für Kraftfahrzeuge," Gersbach & Sohn, Munchen, 1974. Translation of Part I (Methanol) available as a publication of Lawrence Livermore Laboratories (ERDA) from National Technical Information Service, U. S. Department of Commerce, Springfield, Virginia 22151.
10. H. Heitland, W. Bernhardt and W. Lee, "Comparative Results on Methanol and Gasoline Fueled Passenger Cars," Paper No. 39 presented at 2nd Symposium on Low Pollution Power Systems Development (NATO/CCSM), Dusseldorf, November 4-8, 1974.
11. H. G. Adelman, D. G. Andrews, and R. S. Devoto, "Exhaust Emissions from a Methanol-Fueled Automobile," SAE Paper 720693, 1972.
12. W. Lee and W. Geffers, "Engine Performance and Exhaust Emission Characteristics of Spark-Ignition Engines Burning Methanol and Methanol-Gasoline Blends," Paper No. 31d presented at the 90th National AIChE Meeting in Boston, Massachusetts, September 9, 1975.
13. W. Bernhardt and W. Lee, "Combustion of Methyl Alcohol in Spark Ignition Engines," Paper No. 136, 15th International Symposium on Combustion, Tokyo, Japan, August 15-31, 1974.
14. E. Faltermayer, "The Clean Synthetic Fuel That's Already Here," Fortune, Vol. XCII, No. 3, September 1975, pp. 147-154.
15. P. Bedard, "When is a Carburetor Not a Carburetor?" Car and Driver, December 1974.
16. W. E. Bernhardt, "Kinetics of Nitric Oxide Formation in Internal-Combustion Engines," Paper C149/71 presented at the Conference on Air Pollution Control in Transport Engines (Institute Mech. Engrs.), Solihull, England, November 1971.
17. W. Bernhardt, "Investigation of Nonequilibrium Processes of Nitric Oxide Reaction Occurring in the Combustion Chamber of Internal Combustion Engines," Staub-Reinhaltung der Luft (Engl. Edition) 31, No. 7, July 1971, pp. 8-13.
18. E. Sawicki, T. R. Hauser, T. W. Stanley and W. Elbert, Analytical Chem., 33, 93, 1961.
19. W. Bernhardt, W. Behrens, P. Heidemeyer, W. Geffers, "Ermittlung Polyzyklischer Automatischer Kohlenwasserstoffe im Automobilabgas in Abhängigkeit von Motorkonzept und Fahrzustand," VW-Forschungsbericht No. F2-74/75, June 1974.

DISCUSSION

A. A. Quader *(General Motors Research Laboratories)*

You indicated that a different engine was used for the high compression ratio studies. What was the difference in the combustion chamber shape between the 9.7 and 14 compression ratio engines?

Bernhardt

If you are raising the compression ratio, as a matter of fact, the combustion chamber is a little bit altered. We did not use, in the one case, one chamber and in the other case another chamber. In raising the compression ratio only the volume was altered. That's all.

N. D. Brinkman *(General Motors Research Laboratories)*

What kind of air-fuel ratio range was used with your methanol cars, and how did that compare to the gasoline operating range?

Bernhardt

We normally run our gasoline engines a little bit — maybe 2% — rich and in the case of methanol-gasoline blends, it was about 5% lean. It was also about 5% lean for pure methanol fueled cars. It's no problem to run the engine comparatively lean when you are using straight methanol.

H. C. Huffman *(Union Oil Company of California)*

In the gasoline-methanol blend study, was the storage system kept dry? If it was not, what problems did you have with separation with the methanol-gasoline blend?

Bernhardt

We took no special precautions for preventing the humidity of the air from getting into the fuel tanks. We have seen no problems so far, but we should point out that the methanol-gasoline blends used pure methanol with very little water content, less than 500 ppm in the straight methanol. The gasoline base blend had a total aromatic content around 35%. The total aromatics and water contents are very important parameters when you are working on methanol-gasoline blends.

R. M. Campau *(Ford Motor Company)*

Would you comment on whether you tried to catalytically control the aldehyde emissions in lieu of water addition?

Bernhardt

There was no reason for us, at present, to use catalysts for reducing aldehyde emissions. This would be a possibility, but we would like to avoid using a catalyst when methanol is the engine fuel.

S. Gratch *(Ford Motor Company)*

I have three questions. In your last table, the data for observed HC emission during cold start actually shows emissions 22% higher for methanol than for gasoline. I wondered how you reconciled that to your statement that HC emissions are less with methanol? The second question is on methanol toxicity. You state that it is of no concern because methanol is less toxic than benzene. At least in the United States, there are very strong restrictions on the use of benzene in fuel because of its very high toxicity. I was wondering if such restrictions do not apply in Europe? The third question relates to your comment that methanol competes favorably with gasoline in terms of cost. That is certainly not the case if you consider cost per unit of energy in the United States. Again, I wonder, is this different in Europe?

Bernhardt

You were right, HC emission during cold start is higher. We have seen that during the first 200 seconds, mixture preparation and mixture distribution is comparatively bad. We have to improve atomization in the fuel preparation system and we have to use modified carburetors which do a better job in mixture preparation. Or we have to use fuel injection systems or systems like the Dresserator for solving these problems. If you look at the emission charts during testing, you see that it takes about 200 seconds until the methanol in the inside of the intake manifold is vaporized. This is too long. You cannot beat the standards if you cannot solve this problem. This is certainly a problem we have to work on.

Now the second question. There is a lot of discussion on the toxicological aspects of methanol in Europe. At this time, it's not clear which is worse, methanol or gasoline. The German oil companies have just sponsored a toxicological study by a professor from the University of Hamburg. He came to the same conclusion our people did, you have to treat both fuels as being very similar in toxicological behavior. You have to be careful not to come in contact with methanol but you also have to be careful not to come in contact with gasoline. There's PNA's in your gasoline and the aromatic content of your gasoline can be high.

Your third question was on cost. You are right. At the moment, methanol is more expensive than gasoline. But we are looking at a time period, maybe 20 years or so from now. Technology is available for producing methanol from natural gas from the North Sea. This would be the cheapest way. The other way would be using coal as a primary source. This is a bit more expensive. In a study for the German government involving more than 15 institutions with different opinions on methanol and methanol technology, the conclusion was reached that, at least in times where we have an energy shortage as we had two years ago, methanol's price could be comparable to gasoline. For example, the price we had to pay for gasoline in Amsterdam during the Arab oil embargo was higher than the price of synthetic methanol.

COMBUSTION OF METHANOL
IN AN AUTOMOTIVE GAS TURBINE

L. W. Huellmantel, S. G. Liddle and D. C. Hammond, Jr.

General Motors Research Laboratories, Warren, Michigan

ABSTRACT

Analytical and experimental studies were carried out to assess the effects of methanol as fuel for the GT-225 experimental passenger-car gas-turbine engine. The thermodynamic analyses indicated that the engine performance with methanol would not be changed significantly from that observed with kerosene. After the engine fuel handling system was suitably modified for methanol, dynamometer engine testing confirmed that the engine performed as well on methanol as on kerosene. Using a conventional diffusion flame combustor, oxides of nitrogen emissions were reduced by about 70% for all conditions. Carbon monoxide emissions were reduced by about 25% over the normal engine operating range, but were increased by up to 165% at high engine loads when using methanol. The emissions of hydrocarbons and aldehydes were low for both fuels, but methanol operation produced somewhat more of both. Tests of the engine installed in a chassis rig in the chassis dynamometer and operated on the 1975 Federal Test Procedure Emission Schedule and Highway Fuel Economy Schedule confirmed that the energy consumption of the vehicle operating on methanol was about the same as with kerosene. At low mileage, the vehicle easily passed the 1978 Federal emission standards for HC and CO (0.25 g/km and 2.1 g/km) with either fuel, and also passed the 1977 California standard for NO_x (0.93 g/km) with methanol. No attempt was made to determine emission deterioration factors. The driveability with methanol was at least as good as with kerosene.

INTRODUCTION

The projected depletion of domestic reserves plus the uncertainty and costliness of petroleum obtained from foreign sources has caused an extensive search for suitable

References p. 259.

automotive fuels which can be derived from sources other than petroleum. One alternative fuel that has received considerable attention is methanol. Its main attractions are that it can be produced commercially from coal and that it is a liquid at normal temperatures and pressures. Some of the other alternative fuels proposed for automotive use, for example hydrogen and ammonia, require special handling such as high storage pressures or cryogenic temperatures. A significant disadvantage of methanol is its low heat of combustion; only about 20.0 MJ/kg compared to about 43.2 MJ/kg for kerosene. Table 1 presents some physical and thermodynamic properties of methanol with comparative values for kerosene.

TABLE 1

Properties of Methanol and Kerosene

	Methanol	Kerosene*
Chemical Formula	CH_3OH	$C_{12}H_{22}$
Molecular Weight	32.04	166
Specific Gravity at 289 K	0.793	0.808
Lower Heat of Combustion-Liquid (MJ/kg)	20.0	43.2
Heat of Vaporization (MJ/kg)	1.172	0.326
Boiling Temperature at One Atm (K)	338	442-562
Stoichiometric A/F by Mass	6.48	14.47

Since kerosene is a mixture of many different hydrocarbons in varying amounts, its properties are not fixed but will vary over a considerable range. The values listed represent averages taken from a number of sources.

In order to determine the effect of methanol as fuel for a gas turbine engine, an analytical and experimental program was carried out with the GT-225, an experimental regenerative passenger-car gas turbine engine. Because of its multi-fuel capability, the gas turbine engine is a logical choice for the combustion of a non-conventional fuel. The results of the study are presented in this paper.

ANALYTICAL STUDIES

Assuming that the necessary changes to the fuel supply system are made, but the rest of the GT-225 engine is left as designed for kerosene, what effect would the use of methanol fuel have on engine performance? To answer this question, an analytical study was made using a gas turbine part-load analysis computer program to calculate engine performance with both methanol and kerosene.

The first step was to calculate the properties of the combustor exhaust gas which serve as input data to the engine simulation program. Complete combustion of pure methanol in the presence of excess air was assumed. The one percent argon present in the air is treated as if it were nitrogen. The results of these calculations for kerosene and methanol are presented in Fig. 1 where the specific heat ratio of the gases is shown as a function of the equivalence ratio with combustor inlet temperature as a parameter. Commercial methanol usually contains one to three percent water which was not accounted for in these calculations.

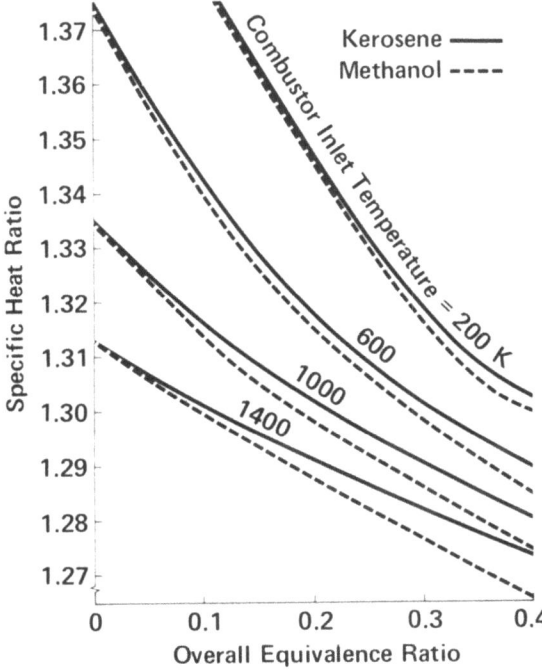

Fig. 1. Exhaust gas specific heat ratio — kerosene and methanol.

The only changes to the part-load analysis program used in this study were in those sections which calculate fuel-air ratio and exhaust gas specific heat ratio. The performance maps of the various turbomachinery components were assumed not to be affected by the change of fuel. Two additional assumptions were made. One is that the thermal conductivity of the gas is unaffected by the change of fuel, that is, the regenerator acts essentially the same for both methanol and kerosene. The second assumption is that the pressure drop in the combustion system is unaffected by the change of fuel. Since most of the exhaust gas is excess air, these assumptions are quite reasonable and should introduce little error.

References p. 259.

There are three independent variables in this program; gasifier turbine speed, power turbine speed, and, since the GT-225 engine has variable power turbine nozzles, nozzle area. Among the important dependent variables are output power, gasifier turbine inlet temperature, regenerator hot-side inlet temperature and thermal efficiency. Material limits of the turbines and the regenerator along with the surge limit imposed by the compressor, restrict the range of power turbine nozzle areas which can be used at a particular gasifier speed. For a given gasifier speed and nozzle area, there is an optimum power turbine speed which maximizes output power. All results presented in this paper are based on that power turbine speed.

The results of this analytical study can be summarized as follows:

1. For a given gasifier speed and power turbine nozzle area, the output power of the engine is nearly independent of the fuel used, but the engine thermal efficiency for methanol is lower than that for kerosene. This calculated loss in thermal efficiency ranges from about 0.4 percentage point at 50% gasifier speed to about 1.0 percentage point at 100% gasifier speed.

2. For a given gasifier speed and power turbine nozzle area, the gasifier turbine inlet temperature is lower for methanol than for kerosene. The calculated difference ranges from about 8 K at low gasifier speeds to 22 K at the engine design point. This reduced temperature is the cause of the lower efficiency with methanol.

3. For a given gasifier speed and turbine inlet temperature, the output power for methanol is greater than that for kerosene, but the thermal efficiency is slightly higher for kerosene. The maximum calculated power increase for methanol is about 5% and occurs at 100% gasifier speed.

The air flow rate at a fixed power turbine nozzle area is virtually independent of the type of fuel, varying by less than 0.03% over the entire speed range as compared to kerosene. The power required of the gasifier turbine (to drive the compressor) is therefore relatively unchanged. Since the fuel flow rate of methanol is much higher than kerosene, there is a net increase in total flow through the gasifier turbine. The power developed by the gasifier turbine is proportional to the flow through it and the inlet temperature. Since the flow is increased, the inlet temperature is reduced, as was noted previously. A similar effect is noted in the case of the power turbine. Its inlet temperature is reduced in the same ratio as the gasifier turbine and, since the flow is increased in the same manner, the product of these two factors, which is proportional to the output power, remains constant. The thermal efficiency is strongly affected by temperature and the decrease in temperature should decrease the thermal efficiency.

Fig. 2 presents some of the results of this analytical study in graphical form. Brake specific energy consumption (bsec) is shown as a function of engine power output at various gasifier speeds for both fuels. The bsec is the ratio of heat energy input (fuel flow rate times heating value of the fuel) to the engine power output, and is

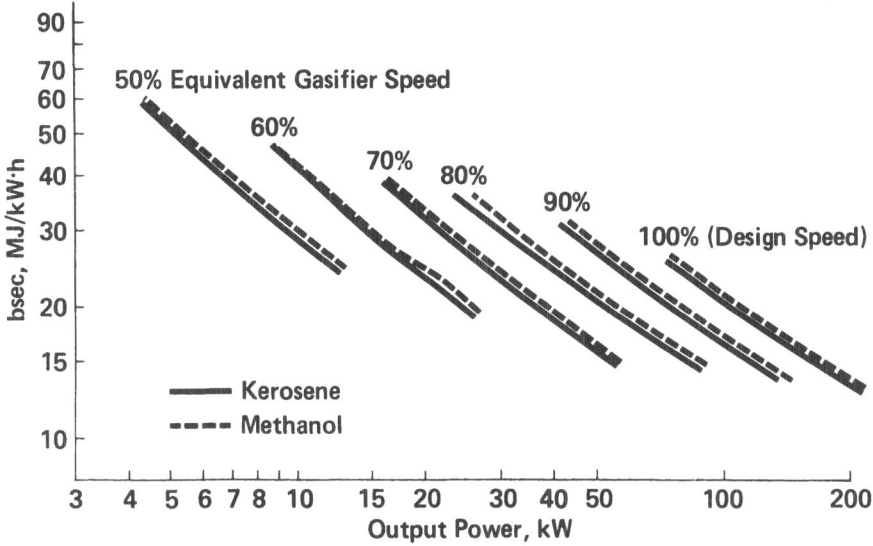

Fig. 2. Brake specific energy consumption vs. engine power — analytical results.

proportional to the reciprocal of thermal efficiency. Generally, brake specific fuel consumption (bsfc) is used to indicate engine fuel economy. However, because of the large difference in heating value between kerosene and methanol, bsfc values cannot be compared directly as can bsec values. The calculated results presented in Fig. 2 show that for a given power output, the bsec for methanol is greater (lower thermal efficiency) than that for kerosene at any gasifier speed. The difference is on the order of 5%.

EXPERIMENTAL EQUIPMENT

Engine and Control System — The GMR GT-225 experimental passenger-car gas turbine engine (1) shown in Fig. 3 was used for this fuel comparison study. The GT-225 is a dual-shaft regenerative engine incorporating variable power turbine nozzles for turbine inlet temperature control. Its design output power is 168 kW at a design gasifier speed of 44,000 rpm and turbine inlet temperature of 1283 K. An electronic control system is used to regulate fuel flow and power turbine nozzle position. The control system incorporates a speed governor which controls the gasifier speed as a function of throttle setting. The control system automatically positions the power-turbine nozzles to make the turbine inlet temperature follow a prescribed schedule which is a function of throttle position and gasifier speed. The engine also can be operated at a fixed power turbine nozzle settings by deactivating the nozzle control system.

References p. 259.

Fig. 3. GT-225 gas turbine engine.

Combustor – The same combustor was used to test both fuels and was unchanged during the test program. It is a conventional diffusion flame unit as shown in Fig. 4. Air passes through the liner via four openings: the swirler, an annular film-cooling slot between the swirler and flame-tube, a row of primary holes fitted with penetration bushings and a row of dilution holes. Air-assist fuel nozzles were used for both fuels as described below.

Fuel System Modifications for Methanol Operation – Since several of the properties of methanol differ considerably from those of kerosene, certain modifications to the engine fuel system were required to operate the GT-225 engine on methanol. The fuel nozzle passages were enlarged to permit adequate fuel flow without exceeding the pressure capabilities of the fuel control valve. The fuel nozzle used in the GT-225 employs air assist to help atomize the fuel for improved combustion.

Other fuel system modifications were the replacement of the conventional gear-type high pressure pump with an experimental swash-plate-type piston pump,

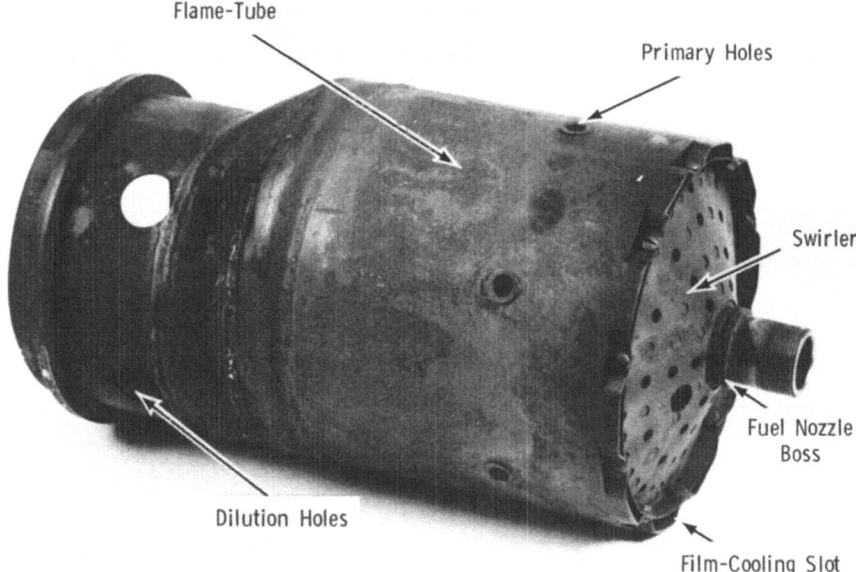

Fig. 4. Combustor used in GT-225 for kerosene-methanol comparison.

and the changing of the diaphragm in the fuel valve to a material compatible with methanol. The swash-plate piston pump and modified fuel valve were used both for the methanol and kerosene tests. Except for limited tests in which the effect of atomizing-air pressure on emissions was determined, the fuel nozzle with the enlarged passages was used only for the methanol tests while a conventional fuel nozzle was used for the kerosene tests.

Vehicle — Vehicle tests were conducted using a special rig developed exclusively for use on the chassis dynamometer. A 1969 Buick Electra 225 formed the basis for this rig. The front sheet metal was removed and the front frame modified to accommodate the engine. This arrangement allows free access to the engine during testing for observations and adjustments. In addition, the front passenger seat has been removed to allow space for extensive instrumentation and control electronics. A three-speed variable-stator torque converter transmission, and a 3.9 rear axle ratio were used in the chassis rig during this study. Fig. 5 shows this rig being tested in the chassis dynamometer.

Instrumentation — Exhaust emissions of hydrocarbons, carbon monoxide, carbon dioxide, and nitrogen oxides were measured using essentially identical facilities for both the engine-dynamometer and the chassis-dynamometer testing. The analyzers used and their concentration ranges are listed below:

Hydrocarbons — Beckman 402 Heated FID; full scale concentrations: 5000, 1000, 500, 100, 50, 10, 5, 1 ppmv as C_6.

References p. 259.

CO — Beckman 315B NDIR; full scale concentrations: 600, 200, 100 ppmv.

CO_2 — Beckman 315B NDIR; full scale concentrations: 5, 2, 1 %-v.

NO_x — Thermo Electron 10A chemiluminescent; full scale concentrations: 10000, 2500, 1000, 250, 100, 25, 10, 2.5 ppmv as NO.

Fig. 5. Chassis rig on chassis dynamometer.

The sample lines to all NDIR analyzers pass through refrigerated traps and filters. The lines to the NO_x analyzers contain only filters. The hydrocarbon analyzers are fed by separate unobstructed sample lines maintained at 450 K.

A Beckman 400 unheated FID having full scale concentrations of 1000, 100, 10, 1 ppmv as C_6 was used to measure the background hydrocarbon concentration during testing on the chassis dynamometer. The sample line to this analyzer passes through a refrigerated trap and filter.

Exhaust aldehyde concentrations were measured during steady-state operation using the MBTH (3 methyl-2-benzothiazolone hydrazone) test described by Sawicki et al. (2). The equipment used is shown schematically in Fig. 6. After first evacuating

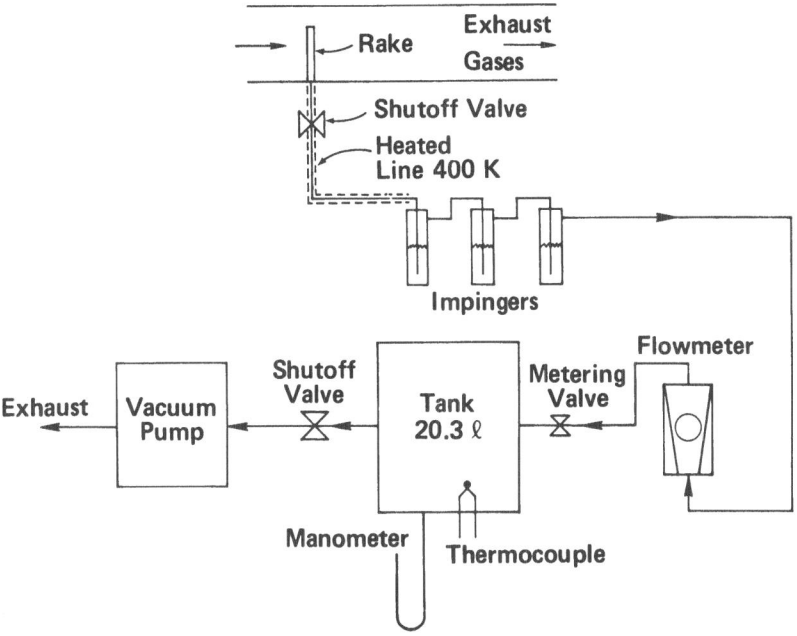

Fig. 6. Schematic of aldehyde sampling system.

the tank with the vacuum pump, the metering valve was opened slightly and a flow of exhaust gases was pulled through the three impingers in series and into the tank. The impingers contained an aqueous MBTH solution which removes water-soluble, aliphatic aldehydes from the gases. Three impingers were found necessary to insure that essentially all of the aldehydes were removed. Fig. 7 shows the sampling system installed on the GT-225 engine in the engine dynamometer facility. The volume of gases sampled was calculated from the tank volume and the temperature and pressure in it before and after sampling. Thus, an average exhaust gas aldehyde concentration was determined.

The extremely low aldehyde concentrations in the GT-225 exhaust gases (less than 1 ppmv) required that a large volume of gases be sampled. The sampling rate was limited to 500 mℓ/min to insure the removal of essentially all the aldehydes in the impingers. Therefore, sampling times on the order of one hour were necessary. These difficulties made it impossible to use this method for short-duration measurements during transient operation. The technique also is not compatible with the fuel-based emissions measurements used for GT-225 testing on the chassis dynamometer. Since a constant volume sampler is not used, the sampling rate would have to be varied directly with engine exhaust flow throughout the FTP to determine mass emissions.

Various methods of measuring engine fuel flow rate were employed during this study. For engine dynamometer tests, the test cell burette system was used when

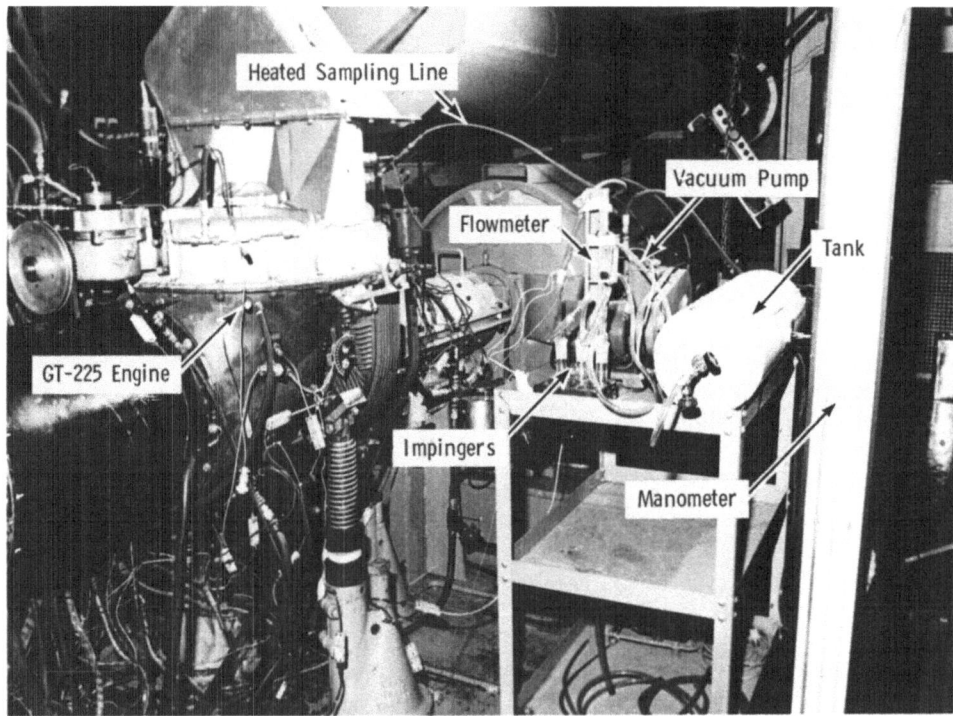

Fig. 7. GT-225 engine with aldehyde sampling system.

operating on kerosene. This is an automatic precision system which measures the time required for the engine to consume the fuel from a known volume burette. Fuel temperature is measured and appropriate density corrections are made. To avoid possible damage to this system due to incompatibility with methanol, it was not used for the methanol tests. Instead, the average methanol flow rate at each test condition was measured over a given time interval by means of a precision platform scale and stop-watch. A sufficient time interval was employed at each test point to insure fuel measurement accuracy.

For steady-speed tests with kerosene on the chassis dynamometer, a GM Proving Ground Model 5A digital fuel meter was used. This unit is a positive displacement type meter which provides direct readouts of fuel volume, fuel temperature and elapsed time. Again, to avoid possible damage to this meter due to incompatibility with methanol, the platform scale and stop-watch system was used for measuring methanol flow rate during the steady-speed chassis dynamometer tests. For the Federal Highway Fuel Economy Tests (HWFET), the scale was used to measure the total consumption of both fuels.

For the 1975 Federal Test Procedure (FTP) emission schedule, the fuel flow was measured for both fuels by means of a Model E20 Bearingless flowmeter*. Since the meter rotor has no bearing and is quite small (approximately 5 mm in diameter), it has excellent response to changes in fuel flow. Instantaneous fuel flow measured by this meter was used in the Fuel-Based Mass Emission Measurement Procedure (3) to obtain the FTP emissions and fuel economy results reported in this paper. In addition, a check of the total fuel flow during the FTP for both fuels was obtained by use of the fuel weighing technique. There was good agreement between the total accumulated fuel flow as measured by the Bearingless fuel meter and the weighing technique.

EXPERIMENTAL PROCEDURE

Engine Dynamometer Tests – Engine performance and emission measurements were made during steady-speed operation over a gasifier turbine speed range of from 50% of design (idle) to 100% of design (maximum power). Measurements were made at two temperature levels at each gasifier speed by manual adjustment of the variable power turbine nozzles. The higher temperature level at each test speed is approximately the maximum allowable steady-state operating temperature for that speed. At low and intermediate gasifier speeds, the hot-side regenerator inlet temperature is limited to approximately 1000 K. At the higher speeds, the gasifier turbine inlet temperature is limited to approximately 1280 K. At each test speed, the output shaft speed was adjusted for maximum power.

Engine performance data taken at each test point included gasifier speed, output shaft speed, output torque, fuel flow rate, air flow rate, and temperatures and pressures throughout the engine. All performance data were corrected to standard inlet conditions (288.2 K and 101.3 kPa). The emissions of hydrocarbons, carbon monoxide and nitrogen oxides were measured for at least two power turbine nozzle areas for each of six gasifier speeds ranging from 50 to 100% of design speed. Replicate measurements were made for all of these data. Aldehyde emission measurements were also made except at 90 and 100% gasifier speeds where the fuel consumed during the one hour sampling time would have been unreasonable. Only selected aldehyde measurements were repeated. Both fuels were tested at each gasifier speed within the same half-day time period as opposed to completing all tests with one fuel before changing fuels. This was done to minimize the effect of small day-to-day variations of engine performance on the relative performance of the two fuels.

Chassis Dynamometer Tests – Steady-speed and transient fuel economy and emission tests were conducted using the chassis rig operating on the chassis dynamometer. The dynamometer was set at 9.3 kW at 80 km/h and 2270 kg equivalent inertia mass, and automatic power turbine nozzle control was used for all chassis dynamometer tests. The following tests were conducted on the chassis rig with both fuels:

Bearingless Flowmeter Company, Boston, Mass.
References p. 259.

1. 1975 FTP Emission Test
2. HWFET
3. Steady-speed emissions and fuel economy tests

The steady-speed tests were run with the engine and drivetrain fully warmed up. Exhaust pollutant concentrations, fuel flow rate, engine speed and various engine temperatures were recorded while running at various fixed speeds from 20 to 96 km/h. The tests were terminated at 96 km/h because of speed limitations of the chassis dynamometer.

EXPERIMENTAL RESULTS

Engine Performance – Steady-Speed Operation – The results of the steady-speed engine dynamometer tests are summarized in Figs. 8 through 10. Fig. 8 shows equivalent output power as a function of equivalent gasifier turbine inlet temperature for the different equivalent gasifier speeds tested. Fig. 8 shows that for a given equivalent turbine inlet temperature and gasifier speed, the GT-225 engine produced more power with methanol than with kerosene. The increased power output with methanol ranges from 0% at the hotter 50% gasifier speed conditions to almost 8% at 100% gasifier speed. At all speeds other than 100%, the power difference between the two fuels tends to decrease as the turbine inlet temperature increases.

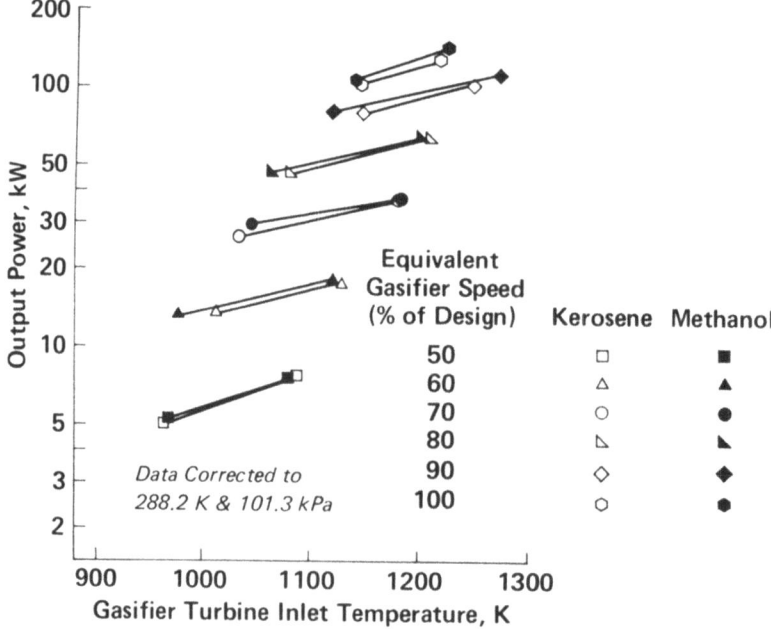

Fig. 8. Output power vs. turbine inlet temperature – steady-speed engine-dynamometer tests.

Fig. 9 compares the brake specific energy consumption (bsec) for the two fuels at the six gasifier speed settings as a function of turbine inlet temperature. The results shown indicate that there is little difference in bsec between the two fuels at a given turbine inlet temperature. In most cases, there is a tendency for the bsec for kerosene to be lower than that for methanol at the higher temperature level for a particular speed. At the lower temperature level for a given speed, the slight advantage for kerosene disappears and, in most cases, methanol has the advantage. This trend tends to agree with the relative changes in output power with changing turbine inlet temperature for the two fuels at a given gasifier speed as described previously. At 100% speed, the data indicate slightly lower bsec for methanol over the entire temperature range tested. As discussed previously, these data were taken by manually adjusting the power turbine nozzles to vary the turbine inlet temperature level. During normal engine operation, automatic nozzle control is employed, and the resulting turbine inlet temperatures approximate those at the higher temperature end of the family of curves shown. Therefore, these results indicate that under normal steady-speed engine operation (other than at 100% speed) the GT-225 engine utilizes kerosene slightly more efficiently than it does methanol. However, in most cases the difference between the two fuels is small and is on the order of the experimental accuracy of the measurements.

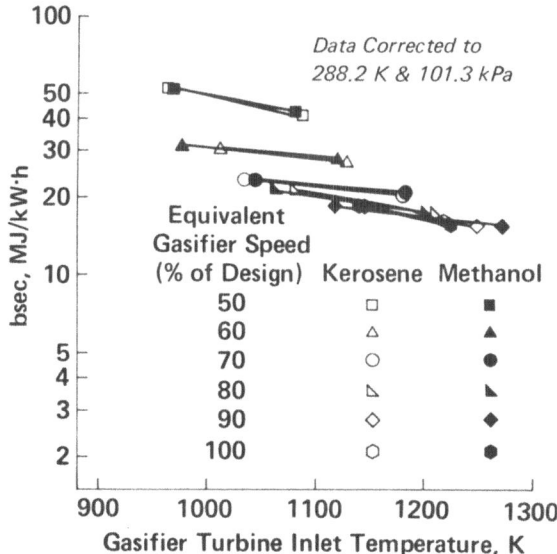

Fig. 9. Brake specific energy consumption vs. turbine inlet temperature — steady-speed engine-dynamometer tests.

Fig. 10 compares the GT-225 performance for the two fuels in another manner. It is similar to Fig. 2 and shows bsec as a function of engine power output at the six

References p. 259.

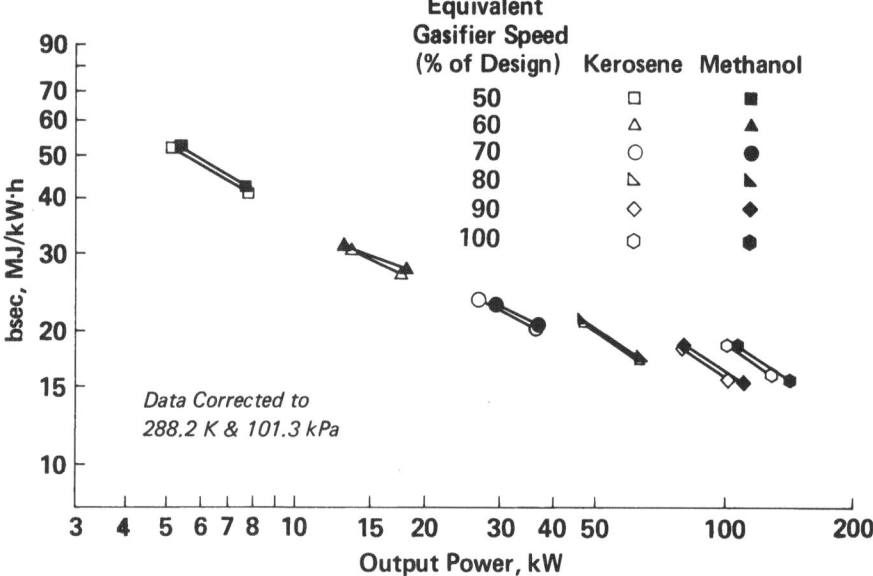

Fig. 10. Brake specific energy consumption vs. output power — steady-speed engine-dynamometer tests.

gasifier speeds tested. The right hand end of each pair of curves represents the higher temperature condition for that particular speed. These results show that the energy consumption for methanol is greater than that for kerosene at a given power output and gasifier speed. The maximum difference in bsec is about 3%.

The results of the steady-speed chassis rig performance tests on the chassis dynamometer are presented in Fig. 11. Fuel economy (corrected to 289 K) and energy economy are plotted for both fuels as a function of vehicle speed. The automatic nozzle control system was operative during these tests. The results show that the fuel economy for methanol is about 55% less than that for kerosene over the entire range tested. This large difference is due primarily to the difference in heating value of the two fuels. However, the difference is much less when compared on a unit energy input basis. On this basis, the GT-225 appears to utilize kerosene only slightly (1-3%) more efficiently than methanol over the speed range tested. These results are in agreement with the results presented previously for the engine dynamometer tests.

Engine Performance — Transient Operation — In addition to the steady-speed tests, fuel economy measurements were made for both fuels during vehicle operation on the 1975 FTP and the HWFET. The results of these tests are presented in Table 2.

The results shown in Table 2 again indicate very little difference in fuel economy for the two fuels when compared on a unit energy input basis. In both cases the

TABLE 2

Transient Operation Fuel Economy Comparison

Test	Kerosene		Methanol	
	km/ℓ	km/MJ	km/ℓ	km/MJ
1975 FTP	2.77	0.079	1.22	0.077
HWFET	5.19	0.149	2.38	0.150

differences approach the experimental accuracy of the tests. The average speed for the HWFET is about 78 km/h. The fuel and energy economies measured for both fuels during this test are slightly below those shown in Fig. 11 for 78 km/h for steady-speed operation. This difference is to be expected because of the transient nature of the HWFET.

Engine starts with methanol were similar to those for kerosene. Transient engine operation with methanol, as judged during the chassis dynamometer tests, was at least as good as with kerosene.

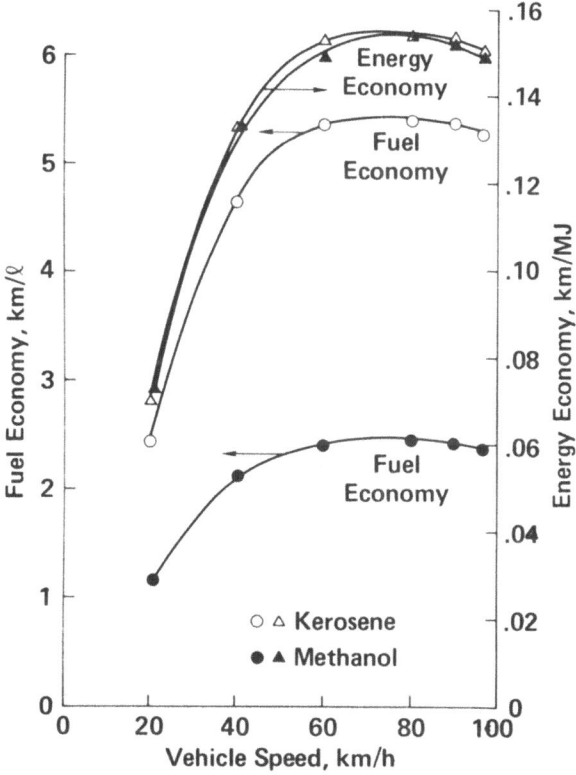

Fig. 11. Fuel and energy economy vs. vehicle speed — steady-speed chassis-dynamometer tests.

References p. 259.

Engine Emissions – Steady-Speed Operation – The most extensive evaluation of exhaust emissions was based on steady-state engine dynamometer test data. It was discovered early in the test program that the atomizing-air pressure drop had a significant effect on emissions of carbon monoxide when operating on methanol. This effect is illustrated in Fig. 12, which also depicts the effects of atomizing-air pressure drop on the oxides of nitrogen emissions. The effect on hydrocarbon emissions was slight.

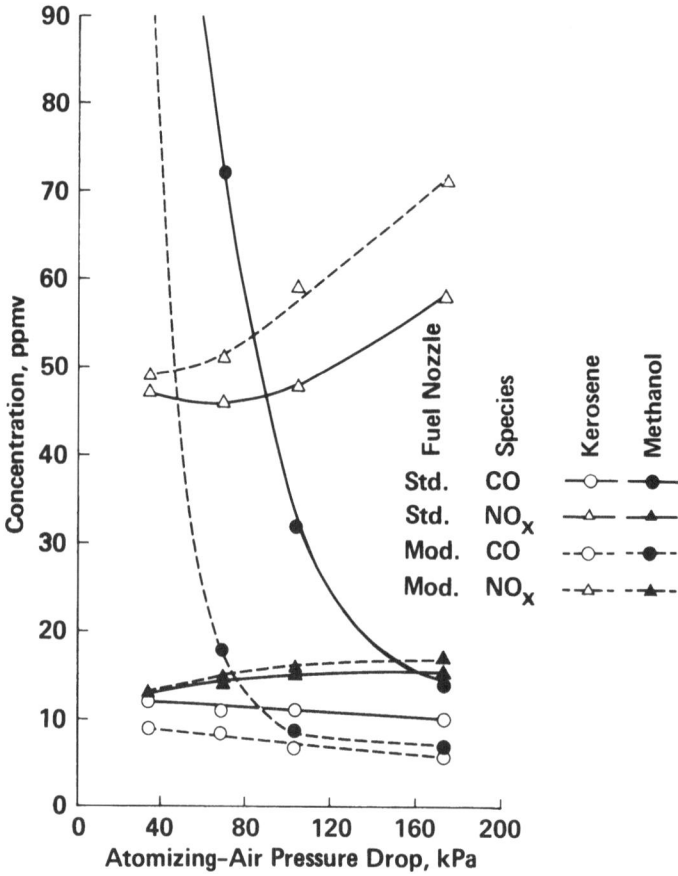

Fig. 12. Effect of atomizing-air pressure drop on CO and NO$_x$ emissions at 60% gasifier speed.

As can be seen in Fig. 12, an atomizing-air pressure drop of 69 kPa provides the best emissions compromise for kerosene with the standard nozzle and it was used in this study. The data also indicate a reduction in carbon monoxide for the modified nozzle (enlarged fuel passages) at the same atomozing-air pressure drop. Use of either nozzle with methanol required higher atomizing-air pressure drop   to reduce carbon monoxide emissions, and the modified nozzle exhibited a significant advantage.

Fortunately, nitrogen oxides emissions increased only slightly with atomizing-air pressure drop for methanol operation. Therefore, the modified fuel nozzle with atomizing-air pressure drops of 138 and 172 kPa was used for methanol operation, and significant reductions in carbon monoxide emissions were realized. Had the control of carbon monoxide emissions with atomizing-air pressure drop not been available, the emissions of carbon monoxide could have been noticeably higher. Other investigators (4, 5) have reported increased carbon monoxide emissions for methanol operation over those for #2 fuel oil. In both cases, simplex nozzles were used, and the additional control of carbon monoxide emissions from an air-assist nozzle was not available.

All steady-state emissions are subsequently reported as the mass of the species emitted per unit energy input from the fuel. This parameter is referred to as "energy specific emissions" (ESE) and is the conventional "emissions index" (EI) divided by the appropriate heating value of the fuel in use. The advantage is that the ESE for various powerplants and fuels may be compared directly.

The emissions of carbon monoxide are presented in Fig. 13. Operation with methanol produces slightly less carbon monoxide than with kerosene at low gasifier speeds (50 and 60%), but more at the higher speeds. For a fixed atomizing-air pressure drop, the fuel nozzle does not operate as well at higher methanol flow rates

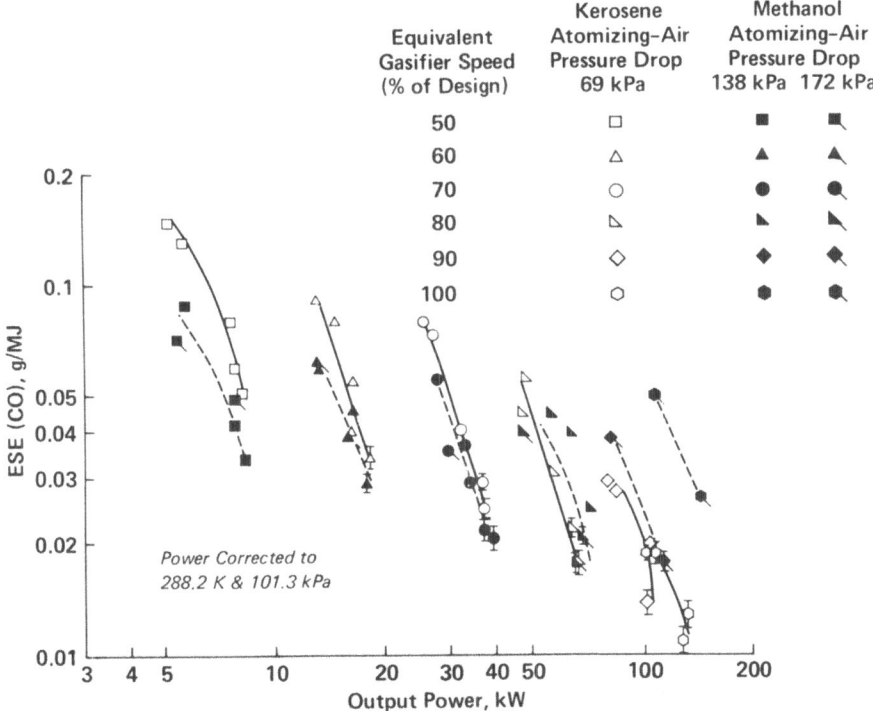

Fig. 13. Steady-speed GT-225 exhaust emissions of carbon monoxide.

References p. 259.

(higher gasifier speeds), and more carbon monoxide is expected.

The emissions of nitrogen oxides resulting from operation on methanol are approximately 65-70% lower than those produced from operation on kerosene. These data are shown in Fig. 14. The reduction in nitrogen oxide emissions results primarily from the lower combustion temperature of methanol. A portion of the reduction observed for methanol can also result from an increased post-flame quench rate and reduced nitrogen and oxygen concentrations as discussed by LaPointe and Schultz (5).

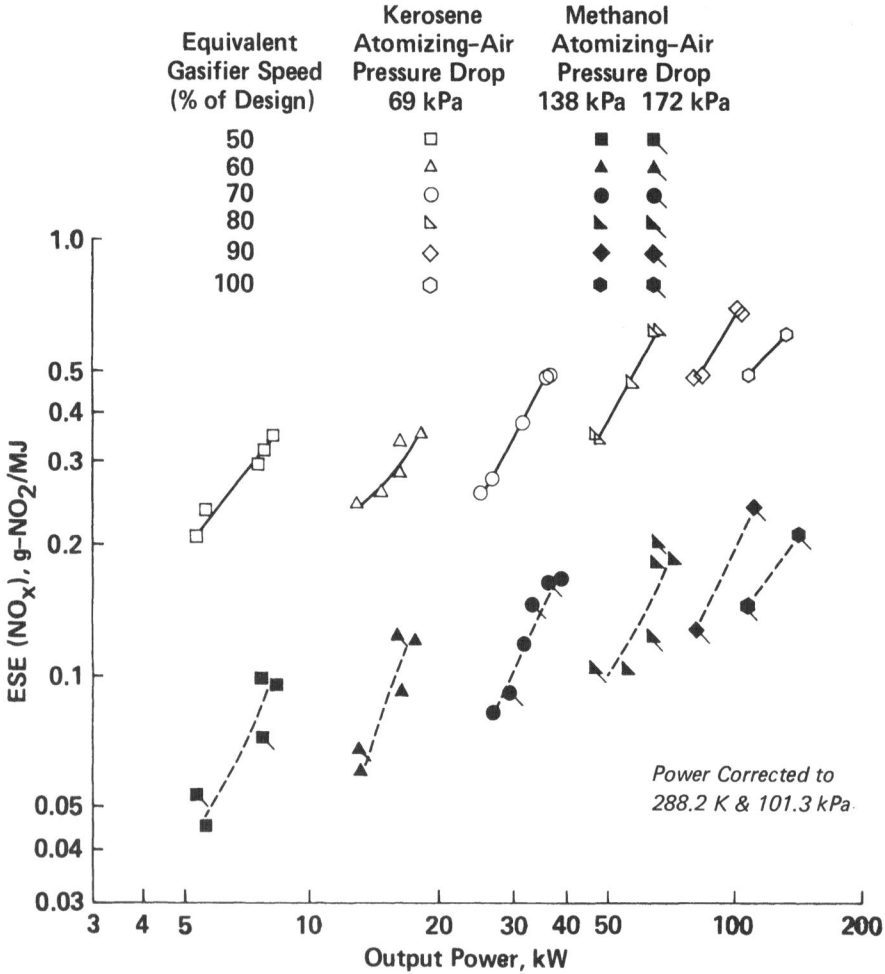

Fig. 14. Steady-speed GT-225 exhaust emissions of nitrogen oxides.

The hydrocarbon emissions of the GT-225 engine are low for all conditions evaluated (less than 0.5 ppmv - C_6). The data appear in Fig. 15. The large uncertainty limits result from the inherent inaccuracy in making such low-level measurements. The hydrocarbon emissions did not reproduce as well as the carbon monoxide and nitrogen oxide emissions. In preparing Fig. 15, the response of the FID analyzer to the exhaust hydrocarbons produced from each fuel was assumed to be equal. The FID response to methanol/nitrogen mixtures is $80 \pm 10\%$ of the response to propane/air mixtures. Therefore, depending on the exact exhaust composition, the "hydrocarbon" emissions reported for methanol fuel may be up to 30% low.

Operation on methanol produced higher hydrocarbon emissions than operation on kerosene. The emissions of hydrocarbons are thought to result primarily from quenching by film-cooling and primary air. Since the required methanol flow rate is more than twice that of kerosene for the same conditions, the concentration of

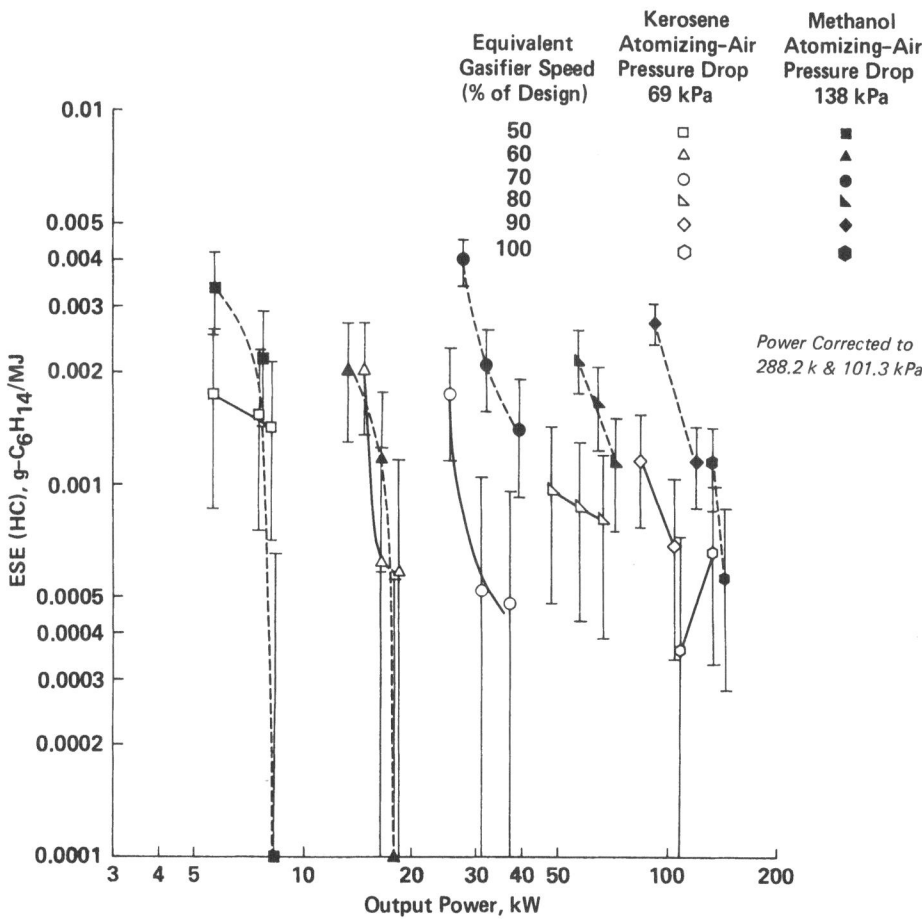

Fig. 15. Steady-speed GT-225 exhaust emissions of hydrocarbons.

References p. 259.

combustibles in any gases quenched will be higher. Thus, higher hydrocarbon emissions are to be anticipated for operation on methanol.

The exhaust emissions of aldehydes (reported as formaldehyde, HCHO) are higher for operation on methanol than for operation on kerosene. These data are shown in Fig. 16. Data scatter is high and it may be the result of any of the following: extremely low concentrations (less than 1 ppmv-HCHO) which had to be measured, instability of the engine operating point over the one hour sampling time, and differing impinger absorption efficiencies.

The higher aldehyde emissions of methanol result from the importance of formaldehyde as an intermediate in methanol oxidation (6). Any gases which are quenched are expected to have a higher formaldehyde concentration for methanol combustion than for kerosene combustion.

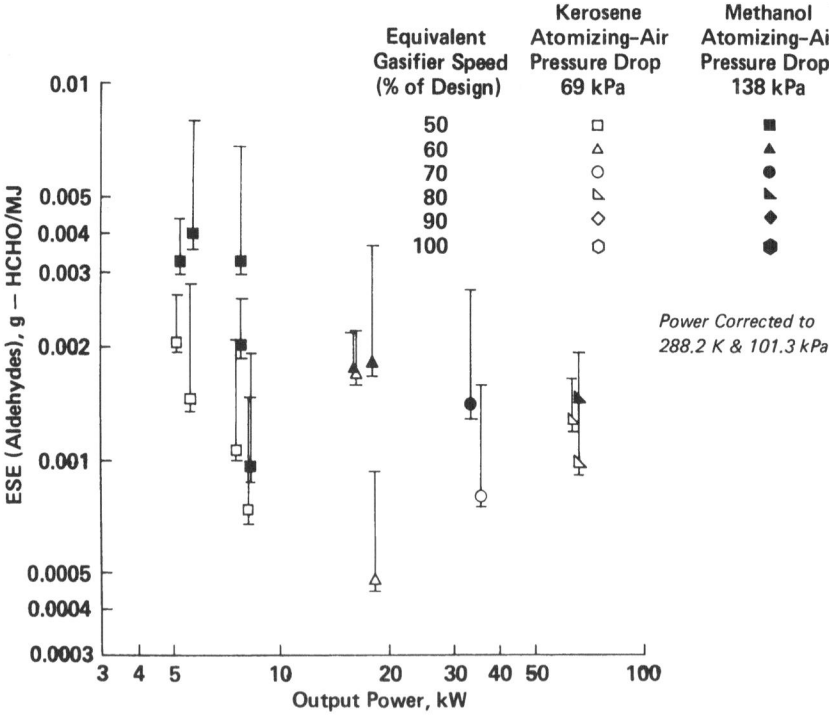

Fig. 16. Steady-speed GT-225 exhaust emissions of aldehydes.

The emissions of hydrocarbons, carbon monoxide, and nitrogen oxides also were measured for steady-state, road-load vehicle speeds ranging from 20 to 96 km/h. These data are presented in Fig. 17 through 19 both as ESE and as "distance specific emissions" (DSE), the mass of pollutant emitted per unit of distance travelled.

Exhaust emissions of aldehydes were not measured during these tests.

Operation on methanol produced consistently lower carbon monoxide emissions than operation on kerosene over the range of vehicle speeds evaluated (Fig. 17). To produce the indicated range of vehicle speeds, gasifier speeds ranging from 50 to 70% of design speed were required. The data taken on the engine dynamometer showed lower ESE of carbon monoxide for methanol operation over this range of gasifier speeds and the lower ESE at constant vehicle speed follows directly. Since the energy economies of the vehicle for the two fuels do not differ greatly (Fig. 11), reductions in carbon monoxide DSE ranging from 10 to 36% were realized.

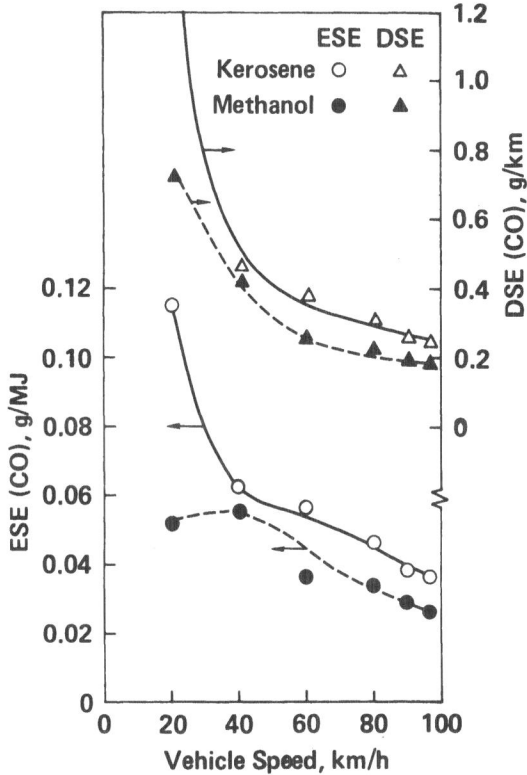

Fig. 17. Steady-speed road-load carbon monoxide emissions of the GT-225 powered vehicle.

Operation on methanol also produced considerably lower nitrogen oxides emissions than operation on kerosene. The reduced ESE shown in Fig. 18 resulted in 66 to 82% reductions in the nitrogen oxides DSE. These results are consistent with the data obtained from testing on the engine dynamometer.

The hydrocarbon emissions shown in Fig. 19 are extremely low. The hydrocarbon

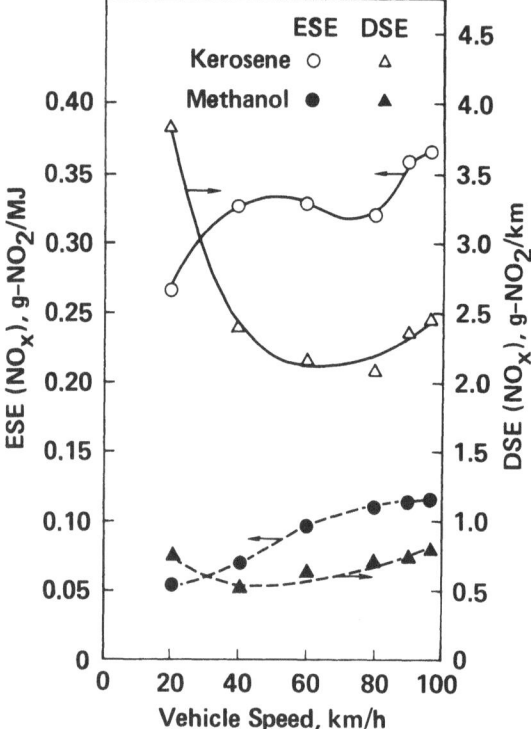

Fig. 18. Steady-speed road-load nitrogen oxides emissions of the GT-225 powered vehicle.

exhaust concentrations were consistently lower than the hydrocarbon concentration in the air ingested by the engine. These results, shown in Fig. 19, which are not corrected for inlet hydrocarbons indicate lower emissions for methanol operation than for kerosene operation, in contrast to the result obtained from the engine dynamometer tests. The inherent uncertainty in the hydrocarbon concentration measurements resulting from the extremely low levels could account for this discrepancy.

Engine Emissions – Transient Operation – Transient testing consisted of driving the vehicle over the 1975 FTP schedule on the chassis dynamometer. The resulting vehicle emissions are summarized in Table 3.

With either fuel, the vehicle emissions of hydrocarbons and carbon monoxide were less than the 1977 California and the 1978 Federal standards. The emission of nitrogen oxides was below the 1977 California standard when operating on methanol. Methanol operation over the 1975 FTP produced hydrocarbon emissions 60% higher, carbon monoxide emissions 30% lower, and nitrogen oxides emissions 73% lower than operation on kerosene. The reductions in the carbon monoxide and nitrogen oxides

TABLE 3

1975 Federal Test Procedure Emissions, g/km

	HC	CO	NO$_x$
GT-225 Kerosene	0.05	1.04	3.37
GT-225 Methanol	0.08	0.73	0.91
1977 California Std.	0.25	5.6	0.93
1978 Federal Std.	0.25	2.1	0.25

Note: HC as C_6H_{14}, NO$_x$ as NO$_2$

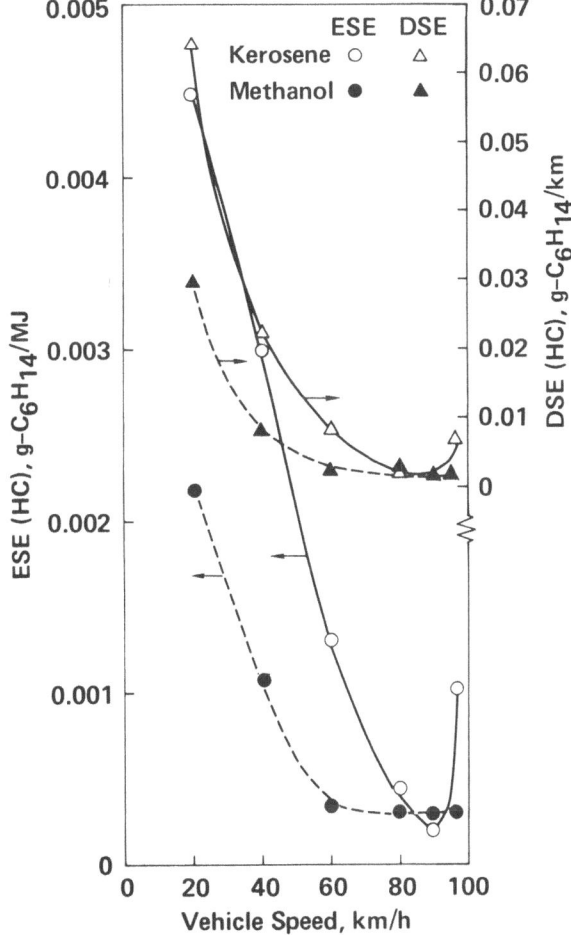

Fig. 19. Steady-speed road-load hydrocarbon emissions of the GT-225 powered vehicle.

References p. 259.

emissions were approximately the same as those experienced in the steady-state testing; transient operation does not affect these emissions significantly. The hydrocarbon emissions with methanol were somewhat greater than those with kerosene during most cycles of the FTP. This was also observed during steady-state engine dynamometer testing, but not during steady-state chassis dynamometer testing. The reported hydrocarbon emissions for the FTP consist primarily of those which occurred during the cold start and warm-up portions of the test. The observed concentrations during those periods were considerably higher, and hence, were more accurately measured than the extremely low concentrations observed during steady-state testing. The increased hydrocarbon emissions for methanol operation most likely result from its high latent heat of vaporization (over three times that of kerosene on a mass basis) which reduces the droplet evaporation rate and hence causes more fuel to escape unburned during the cold start.

SUMMARY

The results of this analytical and experimental investigation indicate that methanol is a suitable fuel for use in an automotive gas turbine engine. Its principal advantage over conventional gas turbine engine fuels, such as kerosene, is the much lower emissions of nitrogen oxides for all conditions. By controlling atomizing-air pressure drop the emissions of carbon monoxide with methanol were lower than those for kerosene over the normal engine operating range. However, methanol produced higher carbon monoxide emissions than kerosene for high power operation. The exhaust emissions of hydrocarbons and aldehydes were extremely low for both fuels.

The principal disadvantage of methanol is its low heat of combustion which requires a fuel flow rate of about 2.2 times that for kerosene at the same power. This necessitates the use of a larger capacity fuel system including such components as the fuel pump, transfer lines, control valve, supply tank and fuel nozzle. Precautions must also be taken to insure that fuel system materials are compatible with methanol.

The analytical study, which required the use of a gas turbine engine part-load computer program and separate exhaust gas property computer programs for kerosene and methanol, indicated a slight power advantage for methanol at a given turbine inlet temperature and gasifier speed. This study also indicated slightly higher specific energy consumption (lower thermal efficiency) for methanol.

The increased power output prediction was confirmed qualitatively by the experimental test results. The steady-speed tests tended to show slightly higher specific energy consumption for methanol, but the transient test results showed no difference. The increased power output with methanol is attributed to the increased mass-flow rate through the turbines due to the increased fuel-flow rate. The slightly increased specific energy consumption rate with methanol during steady-speed operation is believed to be caused by slight mismatch of the engine components which were designed for kerosene operation.

These tests also showed that the engine started as well with methanol as with kerosene although no cold start tests below ambient temperature were attempted during this investigation. Slightly higher hydrocarbon emissions were realized during cold starts which are attributed to the higher latent heat of vaporization for methanol. Vehicle driveability as judged during transient tests on the chassis dynamometer was judged to be at least as good as with kerosene fuel.

ACKNOWLEDGMENTS

Special thanks are extended to Mr. D. V. Gottschalk who aided in all of the engine-dynamometer and much of the chassis-dynamometer testing. Others who assisted in this engine test program were Messrs. J. G. Matson, C. C. Matthews, R. M. Sanford, L. P. Brown and G. R. Eschberger. Measurements of the concentrations of aldehydes were facilitated by Mr. G. J. Nebel who provided the impingers and by Dr. D. L. Hilden who provided the facilities for and assistance in the spectrophotometric analysis. In addition, the efforts of Mr. F. S. Carothers of Delco-Remy Division who assisted in modifying the fuel control valve for operation with methanol, and of personnel from AC Spark Plug Division who provided the experimental fuel pump, are greatly appreciated.

REFERENCES

1. J. S., Collman, et al. "The GT-225 – An Engine for Passenger-Car Gas-Turbine Research," SAE Paper No. 750167, February 1975.
2. E. Sawicki, et al. "The 3-Methyl-2-benzothiazolone Hydrazone Test", Anal. Chem. 33, 1, 1961, 93-96.
3. D. L. Stivender, "Development of a Fuel-Based Mass Emission Measurement Procedure," SAE Paper No. 610604, June 1971.
4. R. C. Farmer, "Methanol-A New Fuel Source?", Gas Turb. Int. 16, 3, 38-40, 1975.
5. C. W. LaPointe and W. L. Schultz, "Measurement of Nitric Oxide Formation Within a Multi-Fueled Turbine Combustor," Emissions from Continuous Combustion Systems, Plenum, New York, 1972, 211-242.
6. C. T. Bowman, "A Shock Tube Investigation of the High-Temperature Oxidation of Methanol," presented at the Fall Meeting of the Eastern Section of The Combustion Institute, Silver Springs, MD, November 1974.

DISCUSSION

J. M. Colucci *(General Motors Research Laboratories)*

There are very few gas turbine-powered vehicles driving on the streets of the United States. We know there is a significant number of gas turbine units used for peak power generation by utilities. Would you care to extrapolate your results and say whether or not methanol would also be a good fuel for those kinds of gas turbine installations?

Huellmantel

There have been tests using methanol in a power plant to generate electricity; Florida Power and Light, I think it was. Their results indicated that methanol is a good fuel. They used their present fuel system and because of the difference in heating value, they could only get about half power out of the system. They did not modify it to get the necessary extra fuel flowrate. However, their results indicated that methanol is a good fuel for utility units.

D. M. Teague *(Chrysler Corporation)*

I have some comments on using water with methanol. As it was pointed out in the previous paper, the addition of water to methanol reduces nitrogen oxide emissions very substantially. This is also true of the addition of water to methanol for use as a turbine fuel. However, water in the fuel initially, up to 10% let's say, increases the problem of sensitivity to water pick-up in storage and to corrosion of various parts of the systems, both in the spark ignition engine and presumably in the turbine.

In the case of the spark ignition engine, the addition of low boiling hydrocarbons, improves the cold starting. But if you have water in the mix, around 10% to minimize NO_x, you get into the problem of stability of the three components. And so, it is true that the addition of water is a very enticing possibility for minimizing NO_x, but it has its problems.

E. E. Spitler *(Chevron Research Co.)*

I wonder if the fuel economy measurements reported over the FTP and idle cycle with kerosene are representative of the latest technology in gas turbines?

Huellmantel

I don't think they represent the latest technology. I think we're all aware that there are some problems with gas turbine engines as far as fuel economy is concerned, especially on light loads. There are methods of improving fuel economy and they are being looked at. The most obvious method is increasing turbine temperature, and we are looking at that, as are many other people around the country.

J. P. Longwell *(MIT and Exxon)*

Were there any cold start problems in connection with gas turbines?

Huellmantel

I would not expect the cold start problem with a gas turbine to be as severe as it is with a piston engine using methanol. We have an air atomizing fuel nozzle and it does break up the fuel into a very fine mist. I think it would light off pretty well even in cold ambients. The turbine engine is noted for its good cold starting ability. Methanol may cause some problems, but I don't think as much as it does with a piston engine. Dean, what do you think?

D. C. Hammond *(General Motors Research Laboratories)*

During all the tests that we ran, we had absolutely no starting problems. Our major problem was with methanol dissolving parts of the fuel system. There was no starting problem. Our tests were run during the summer so it was fairly warm. I think that difficulty with cold start would be aggravated by lower engine temperature, but I don't think it would be insurmountable nor do I think you would have to go to any exotic preheat or atomization-type devices to solve it.

M. Lauriente *(U.S. Department of Transportation)*

Did you by any chance notice any corrosion effects on the engine or problems with any of the plastic feedlines?

Huellmantel

We noticed corrosion, not in the engine itself, but as Dean mentioned, in our fuel system. Some of the diaphragms which control fuel flow were not compatible with methanol and had to be replaced. There was some corrosion of some metal parts in our fuel control valve. Aluminum parts seem to be affected by methanol. I think you can have some problems with methanol in a fuel tank; the lining of the fuel tank can give you problems. We did all of our running out of the barrel that the methanol came in so that did not present a problem. And also you can get into some problems with the water in the methanol causing rusting of some parts.

S. Gratch *(Ford Motor Company)*

Did you notice whether the exhaust odor with the methanol in cold start was the same, better, or worse than with kerosene?

Hammond

The exhaust of the engine had absolutely no odor under any conditions whatsoever.

W. J. McLean *(Cornell University)*

I wonder if you have any idea whether the NO_x reduction is completely attributable to combustion zone temperatures or whether you had different stoichiometry with the different fuels?

Huellmantel

I did some predictions based on the adiabatic flame temperature for stoichiometric combustion for kerosene and methanol and predicted a reduction in nitric oxide emissions based on a characteristic temperature developed from the work of deSoot and came out with about 78% reduction. We measured 75% so based on that I think there is a slight effect of stoichiometry but I think the predominant effect is one of flame temperature.

ALTERNATIVE FUELS FOR AUTOMOTIVE DIESEL ENGINES

W. M. SCOTT

Ricardo Consulting Engineers, Sussex, England

ABSTRACT

With an increasing emphasis on the need to conserve energy and, more particularly, that derived from oil, alternative sources of fuel are being considered for all forms of prime mover. The diesel engine offers, currently, the most efficient means of converting hydrocarbon fuel into mechanical energy, and can play a significant part in energy conservation by its wider application in the automotive field. However, with limited known oil resources and the problem of excessive demand for the barrel of crude oil, resulting from such an expansion, there is clearly an interest in the use of alternative fuels in automotive diesel engines.

The interest in fuels other than middle distillates for compression ignition engines is not new. Experiments have been carried out on a wide range of fuels from coal dust and very viscous residual fuels through gasolines, alcohols and gases including hydrogen.

The Ricardo Laboratories have become involved in many such exercises over the years. These include both straight compression ignition, and the use of dual fuel techniques and ignition promoting additives.

It is concluded that ignition quality is the single most critical factor governing the suitability of fuels for automotive diesel engines. Storage and handling of the gaseous and lower density fuels are also a problem. These problems are demonstrated and discussed in the light of the experience gained both in the Ricardo Laboratories and by other researchers in this area.

References p. 290.

INTRODUCTION

The diesel engine, largely on account of its fuel economy, is a vital component in the transportation systems of all the developed and developing countries of the world. Possible short term and more certain long term shortages of oil based fuels, have generated growing interest in the possibility of burning alternative fuels in internal combustion engines. A combination of the diesel engine, with its high thermal efficiency, and a fuel from a less limited source is clearly of great importance.

The interest in alternative fuels for diesel or compression ignition engines is not new. Indeed, diesel oil itself might be classed as an alternative, though superior fuel to the coal tars and dust used in Rudolf Diesel's early experiments. In the last half century, the motives for the operation of diesel engines on alternative fuels have been numerous and diverse. They have ranged from the desire to make use of by-products of chemical and manufacturing processes such as methane from sewage plants and creosote from the coal gas and coke industry, to a need to guard against local shortages of diesel oil in times of national emergency.

Being in the forefront of research in the diesel engine field, Ricardo have become involved in many such exercises, ranging from the operation of airship engines on surplus hydrogen in the mid 1920's to a very recent investigation of the use of methanol in commercial vehicle diesel engines.

This paper reviews this experience and seeks to indicate the more promising sources of suitable "diesel" fuels together with some suggestions for the utilisation of those fuels which the diesel engine rejects.

SPECIAL REQUIREMENTS OF THE DIESEL ENGINE

In all reciprocating internal combustion engines, the operating cycle requires that the release of energy by combustion of the fuel should take place as near as practicable to the point of minimum cylinder volume. In the gasoline engine a homogeneous mixture is inhaled and compressed at a mixture strength that will ignite readily when a spark is introduced. The timing of the spark and therefore heat release in relation to the piston position can be closely controlled.

The diesel engine relies on the fuel being ignited by the heat of compression of the charge in the cylinder. This rules out any possibility of operating on a homogeneous fuel/air mixture inhaled during the induction period, since there are too many independent variables influencing the compression temperature and therefore time of ignition. To achieve ignition, the engine would have to be throttled to produce an ignitable mixture at part load. Throttling will lower the compression temperature as will the change of heat loss to the coolant with change of speed. Inlet temperature and fuel ignition quality would also have a significant influence on the precise time during compression when ignition occurs. Assuming that ignition were achieved, the mixture would burn in an uncontrolled fashion in the manner of the end gas detonation in a gasoline engine. It is, of course, true that the model airplane "diesel" engine operates in this way but is strictly a one load, one speed engine in spite of the variable compression ratio.

The solution to this problem is to compress the air charge and inject the fuel at high pressure into the hot air at a time which will allow the fuel to reach self-ignition conditions when ignition is required. Not only does fuel injection give control of the timing of ignition, but also forms an extremely heterogeneous mixture ranging from liquid fuel droplets to fresh air. There are thus numerous points of the correct mixture strength to ensure ignition even at very high overall air/fuel ratios, thus avoiding the need to throttle the engine. The load modulation can be controlled purely by the quantity of fuel injected, and the success of the modern diesel engine owes much to the precision of timing and metering provided by modern fuel injection equipment. The combination of operating at a high compression ratio and unthrottled at part load gives the diesel engine its high brake thermal efficiency. The economy of the diesel engine in terms of miles per gallon in automotive use is further improved by it being able to operate on a fuel of relatively high specific gravity and therefore high volumetric heat content.

The middle distillate fuels are among the cheapest end products of the refinery processes, since they require little further treatment after leaving the distillation plant. Nevertheless, the requirements of the modern automotive diesel engine are fairly precise. Of particular importance are ignition quality, volatility, viscosity, specific gravity and lubricity.

Ignition quality – Cetane number is used to indicate the self-ignition quality of diesel fuel and is the percentage of cetane in a mixture of cetane and α methyl naphthalene which gives the same ignition delay time as the fuel being rated. Low cetane fuels give long ignition delays which result in poor starting and misfire if the delay exceeds the period when the compressed air charge is above the ignition temperature. With long ignition delays, a greater proportion of the fuel charge is injected before ignition. The resulting high rates of pressure rise in the cylinder cause noise. However, combustion predominantly in the pre-mixed phase gives less black smoke since smoke is largely the product of diffusion burning when the fuel is partially in liquid form.

The ignition delay of fuel of a given cetane number is not constant, but a function of the physical conditions in the engine. In particular, delay varies with injection timing due to the variation of compression temperature over top dead center, being a minimum at about 5° early. Advancing or retarding the injection timing both result in a longer delay, the former rapidly promoting high rates of pressure rise and noise, the latter eventually resulting in misfire.

Volatility – As the fuel is injected in liquid form, evaporation is an essential prerequisite of the mixing of the fuel and air prior to combustion. Evaporation is promoted by atomization of the fuel thus presenting a high surface-to-volume ratio.

References p. 290.

However, extremely rapid evaporation and dispersal of the fuel throughout the air can be detrimental to the charge stratification essential to good ignition.

Fuel volatility affects other aspects of diesel engine operation. High volatility fuels can cause vapor lock on the low pressure side of the fuel pump. Low volatility fuels containing the heavier polycyclic aromatics are prone to slow rates of burning and the production of exhaust smoke.

Viscosity — It is difficult to separate the effects of viscosity and volatility. From the point of view of combustion, and within the range of viscosity normally associated with automotive diesel fuels, the effects of changes in viscosity are minimal. However, viscosity is of considerable importance to the satisfactory operation of fuel injection equipment. On the one hand, low viscosity can mean poor lubrication and fuel leakage, particularly at low speeds. This and the sensitivity of some metering systems using throttling can adversely affect fuel delivery characteristics over the engine speed range. On the other hand, high viscosity fuels require heating in order to be satisfactorily handled by the fuel injection pump.

Specific gravity — High specific gravity may indicate that the fuel contains heavier fractions which can be detrimental to exhaust cleanliness. However, in general, specific gravity will not affect engine operation if it remains within normal diesel fuel limits. Since the fuel is metered volumetrically, changes in specific gravity result in changes in air-fuel ratio or rating. An increase in specific gravity will raise the engine output at the expense of exhaust smoke. However, higher specific gravity means more miles per gallon, and this accounts for part of the diesel engine's superior vehicle fuel economy compared with the gasoline engine. Low specific gravity fuels are at a distinct disadvantage when tax is levied on a volumetric basis.

Lubricity — As mentioned in connection with viscosity, the fuel injection equipment is very sensitive to the lubricating properties of the fuel it delivers since in most cases the fuel is the main lubricant. The high injection pressures and high speeds result in components designed near the limits of Hertz stress.

Specifications — Typical specifications for diesel fuels for automotive engines for the U. K. and the U.S.A. are given in Table 1.

REVIEW OF POSSIBLE ALTERNATIVE FUELS

Table 2, though by no means fully comprehensive, lists the more likely alternative fuels. Automotive diesel fuel lies in the center of a whole range of fuels, many of which would not normally be considered suitable for the diesel engine, but which have been considered of interest in particular circumstances.

TABLE 1

Specifications of Diesel Fuels

Test	U.K.		U.S.A.**	
	A 1 Fuel	A 2 Fuel	1-D Fuel	2-D Fuel
Appearance	Clear and Bright	Clear and Bright	Clear and Bright	Clear and Bright
Distillation				
I.B.P., °F	370*	358*	330-390	340-400
10%, °F	410*	432*	370-430	400-460
50%, °F	522*	513*	410-480	470-540
90%, °F	675 max.	675 max.	460-520	550-610
EP, °F	688*	693*	500-560	580-660
Spec. Gravity, 60/60°	0.838*	0.865*	0.806-0.826	0.840-0.860
Total Sulfur, % wt.	0.5 max.	1.0 max.	0.05-0.20	0.2-0.5
Aromatics, % vol.	–	27.3*	8-15	27 min
Naphthenes, % vol.	–	38.2*	} Remainder	} Remainder
Olefins, % vol.	–	0.7*		
Paraffins, % vol.	–	33.8*		
Flash Point, °F	130 min	130 min	120 min	130 min
Viscosity at 100°F, CSt	1.6-6.0	1.6-6.0	1.6-2.0	2.0-3.2
Cetane Number	50 min	45 min	48-54	42-50

* *Typical figures where not included in specification.*

** *Environmental Protection Agency Specifications for Emissions Tests.*

The automotive diesel engine is designed and developed to operate primarily on liquid fuels. It uses middle distillate gas oils which represent 5-6% of the barrel of crude oil. In searching for alternative fuels it is therefore reasonable to consider first the utilisation of a greater proportion of the barrel on either side of the present middle distillate cut. A recent study by Exxon (1) has indicated that the automotive distillate cut could approach 50% of the total automotive product before becoming uneconomic. On this basis it is doubtful whether a requirement will exist for more unusual alternative fuels for diesel engines while oil is available. However, this approach would involve major modification of the refinery processes, and there could be an interest in the utilisation of alternative products of the present refinery processes. At the heavy end, these would include: other distillates, fuel oils, and residual fuels. At the light end, the available alternatives are: kerosene, motor gasoline, and liquid petroleum gas (LPG).

The remaining sources of fossil fuels are coal and natural gas. Coal yields a range of products including coke, tar and creosotes, synthetic distillates, synthetic gasolines

References p. 290.

and synthetic natural gas (SNG). Natural gas can be liquified (LNG) or converted into methanol in order to transport it more economically, and to render it more convenient for automotive use.

Finally, hydrogen should be included in the list of alternative fuels. It is perhaps the ultimate fuel for the internal combustion engine as we know it, being produced by electrolysis of water using electricity generated by nuclear energy.

TABLE 2

Some Alternative Fuels and their Sources

Raw Material	Product	Method of Production
Water	Hydrogen	Electrolysis
Natural gas	Methanol	Catalytic conversion with steam
	Liquid natural gas	Compressed and cooled natural gas
Oil	Liquid petroleum gas	Distilled and compressed
	Motor gasoline	Distilled and processed
	Kerosene	Distilled and refined
	Auto distillates	Distilled and blended
	Other distillates	Distilled
	Fuel oils	Residue from distillation
	Residual fuels	Residue from distillation
Coal	Synthetic natural gas	Pyrolysis of coal
	Synthetic gasolines	Hydrogenation of coal
	Synthetic distillates	Fischer-Tropsch or hydrogenation
	Tar creosotes	Pyrolysis and distillation
	Coke	Residue from pyrolysis

RICARDO WORK ON ALTERNATIVE FUELS

The various experiments to be described cover fairly completely the whole range of fuels listed in Table 2. Some of the investigations were carried out over 50 years ago, and by today's standards are lacking in data which would allow detailed analysis. Nevertheless, the results in all cases indicate the problems associated with the particular fuel and the limitations on its use in modern high speed automotive diesel engines.

Heavier Distillates and Fuel Oils — The application of supercharge to a single cylinder engine, fitted with the Ricardo Comet Mk III combustion system shown in

Fig. 1, rendered the engine relatively insensitive to fuel quality, more particularly, cetane number. A series of tests was therefore conducted to demonstrate the ability of this engine to digest fuels ranging from light distillate to very heavy fuels (2). All the fuels had a similar, rather low cetane number stemming from the high aromatic content. They had a wide range of viscosities at 70°F, but not at higher temperatures. Pour points were similarly wide ranging.

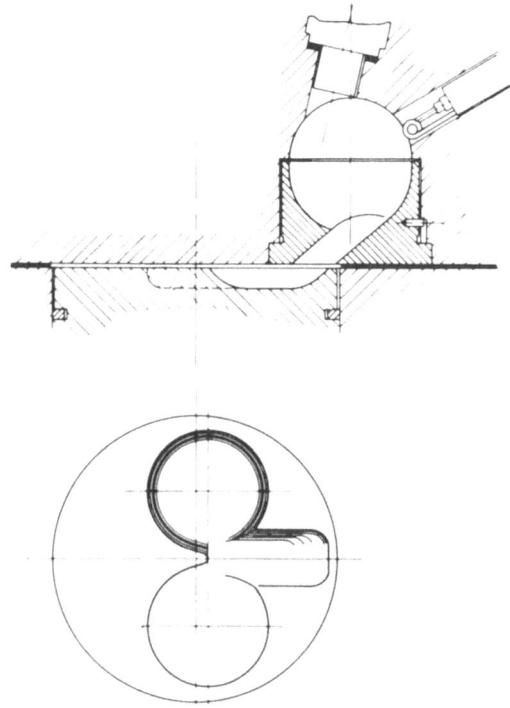

Fig. 1. Ricardo Comet MK III combustion chamber.

Tests were carried out at 500 and 1250 rpm, at various air inlet temperatures and 10 psi boost. The engine was also run naturally aspirated at 500 rpm. If we select the two conditions of naturally aspirated at 500 rpm, and boosted to 20 in. Hg at 1250 rpm we have a simulation of a turbocharge characteristic, but without exhaust back pressure. The inlet temperatures of 86 and 194°F would also be typical.

From the load range curves in Fig. 2 it is seen that at low speed, none of the fuels can match the automotive pool gas oil; the heaviest fuel gives the worst performance. At 1250 rpm the lighter fuels are matching the gas oil, but the heavier fuels still have some difficulty in competing. This inferior performance of the heavy fuels results from a slower burning rate which is apparent in the cylinder pressure diagrams taken at the time.

References p. 290.

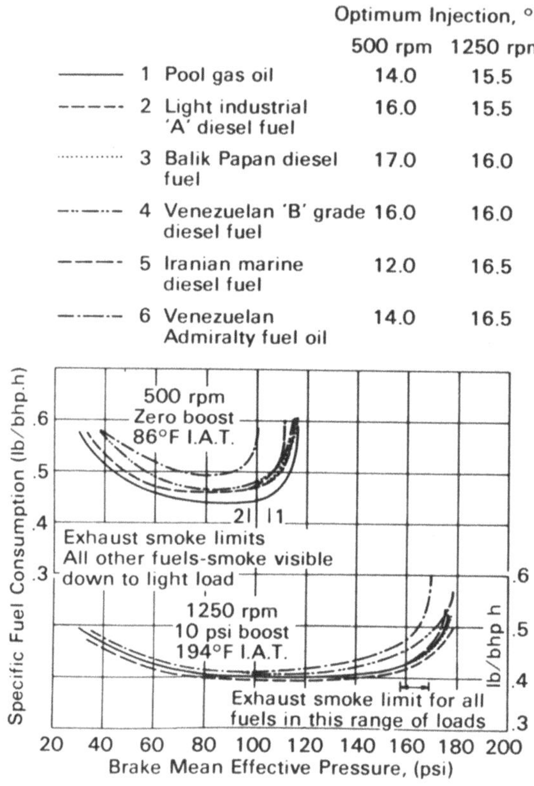

		Optimum Injection, ° E	
		500 rpm	1250 rpm
———	1 Pool gas oil	14.0	15.5
-----	2 Light industrial 'A' diesel fuel	16.0	15.5
··········	3 Balik Papan diesel fuel	17.0	16.0
—··—··	4 Venezuelan 'B' grade diesel fuel	16.0	16.0
— — —	5 Iranian marine diesel fuel	12.0	16.5
—·—·—	6 Venezuelan Admiralty fuel oil	14.0	16.5

Fig. 2. Performance of Ricardo Comet engine when operating pressure-charged on heavy fuel oils.

The improved performance of the heavier fuels at the higher speed probably results from the higher mixing rate and the high inlet air temperature. Tests at 1250 rpm, 10 psi boost, but 86°F inlet temperature showed a significant deterioration of the performance with the Venezuelan admiralty Fuel Oil.

Large, low-speed marine engines operate satisfactorily on the very heavy fuels, undoubtedly because of the greater time available for combustion. In the tests described, a quite acceptable performance was obtained using a combustion system with high swirl, and therefore mixing rate, combined with the higher temperatures associated with pressure charging. The top speed of 1250 rpm is however low compared with modern truck engines which also predominantly use direct injection (D.I.) chambers of the form shown in Fig. 3. It must therefore be concluded that modern truck engines would have greater difficulty in coping with these heavy fuels, and would not accept the heavier diesel and fuel oils.

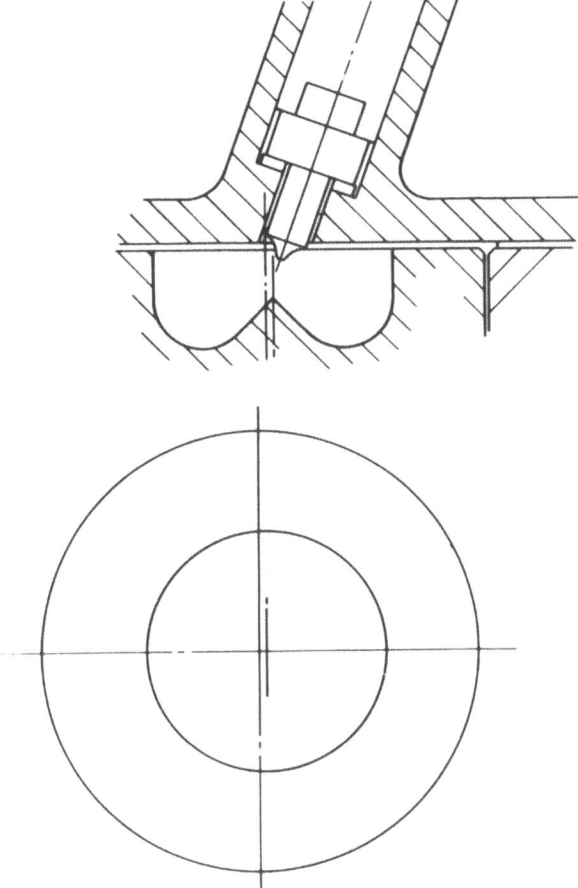

Fig. 3. Toroidal direct injection chamber.

These heavier fuels would also prove impractical since their viscosity and pour points would necessitate heating of the fuel system. A further problem associated with these fuels is nozzle fouling due to high operating temperatures. In large, low-speed engines injectors have to be cooled. Such complications are hardly acceptable in truck or automobile diesel engines.

It may be concluded from these tests, however, that the diesel engine can operate satisfactorily on fuels extending further into the heavy end than at present, but with some penalty regarding ignition delay and smoke limited performance due to slow burning.

Kerosenes – Kerosene is a distillate fuel containing some lighter and fewer heavy components than diesel fuel. As shown in Table 3, its distillation range overlaps that of gas oil, it has lower sulfur and heavier aromatic hydrocarbon contents, and it is of

References p. 290.

lower specific gravity and higher ignition quality (or cetane number) than gas oil.

TABLE 3

Physical Properties of Gas Oil and Kerosene

Test	Gas Oil	Kerosene
Appearance	Red, clear and bright	Red, clear and bright
Specific gravity, 60/60°F	0.838	0.782
Viscosity at 100°F, CSt	2.86	1.33
Flash Point, °F	172	115
Cloud point, °F	+20	- 46
Pour Point, °F	+10	- 55
Carbon Residue, % wt.	0.09	0.04
Total Sulfur, % wt.	0.76	0.02
Distillation I.B.P. °F 10% Recovery, °F 50% Recovery, °F 90% Recovery, °F E. P., °F	370 410 522 636 688	315 345 390 455 493
Recovery, % vol.	98.5	99
Residue, % vol.	1.0	1.0
Estimated net calorific value, Btu/lb	18,330	18,610
Cetane Index	55.0	56.0
Aromatics, % vol. Naphthenes, % vol. Olefins, % vol. Paraffins, % vol.	28.3 38.3 0.7 33.7	7.7 32.2 0.3 59.8

High speed automotive diesel engines have no difficulty in burning kerosene cleanly and efficiently as is seen in the comparative performance curves in Fig. 4. The engine used was of 236 cu. inch capacity in four cylinders and employed a toroidal D.I. chamber shown in Fig. 3. No changes in engine operating conditions were made when changing from gas oil to kerosene except for an increase in volumetric fuel delivery to compensate for the lower specific gravity. The numbers on the curves denote exhaust smoke level and except at the highest speed, smoke levels are improved. The improvement at 1,000 rpm is quite marked and probably indicates better fuel injection/air swirl matching with kerosene than gas oil. In D.I. engines this match is a

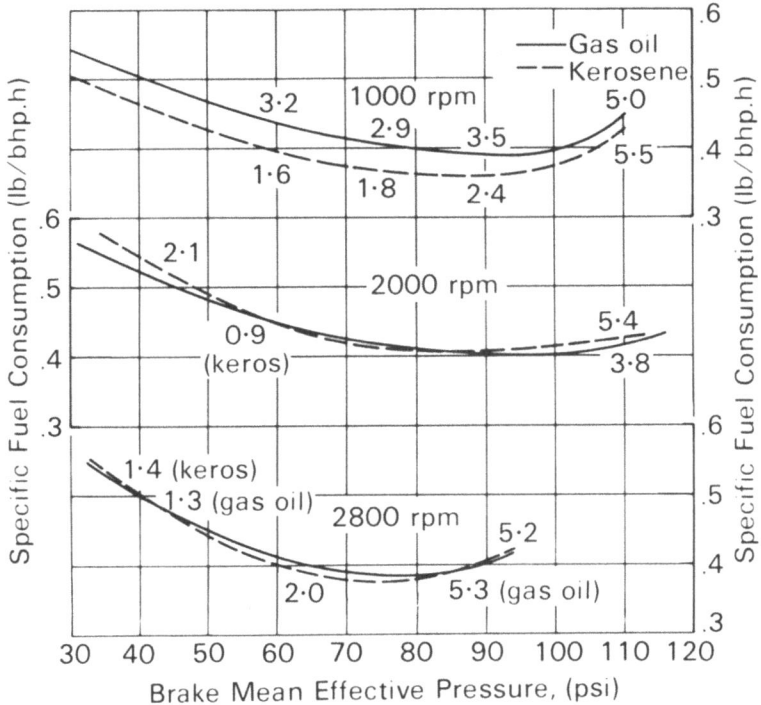

Fig. 4. Comparative performances of 4 cylinder 236 cu. inch D.I. engine on diesel fuel and kerosene.

compromise between low and high speed conditions.

The very low sulfur content of kerosene, around 0.02% is a mixed blessing. Its use in the diesel engine would go a long way to eliminating the sulfate emissions problem for which there is really no other solution so far as the diesel engine is concerned. However, the sulfur in automotive diesel fuel contributes to its extreme pressure (EP) lubricity. This enables the fuel to act as an effective lubricant for the working parts of the fuel injection pump. In particular, the cam and roller followers in distributor pumps work close to the limit of Hertz stress and the removal of the sulfur from diesel fuel, without the substitution of an alternative EP additive, can result in serious damage to these components. Fig. 5 shows typical damage caused to a cam ring which resulted from operation on kerosene

In-line fuel pumps with separate lubrication of the cams and followers do not suffer in this way, although the addition of an EP additive is still advisable.

Kerosene is of course already used in diesel engines in the colder areas of North America on account of its low pour point. Normal gas oil can give serious operating problems resulting from fuel filter and line blocking due to wax crystals which form at temperatures below 20°F. The use of dieselene based on kerosene stock eliminates

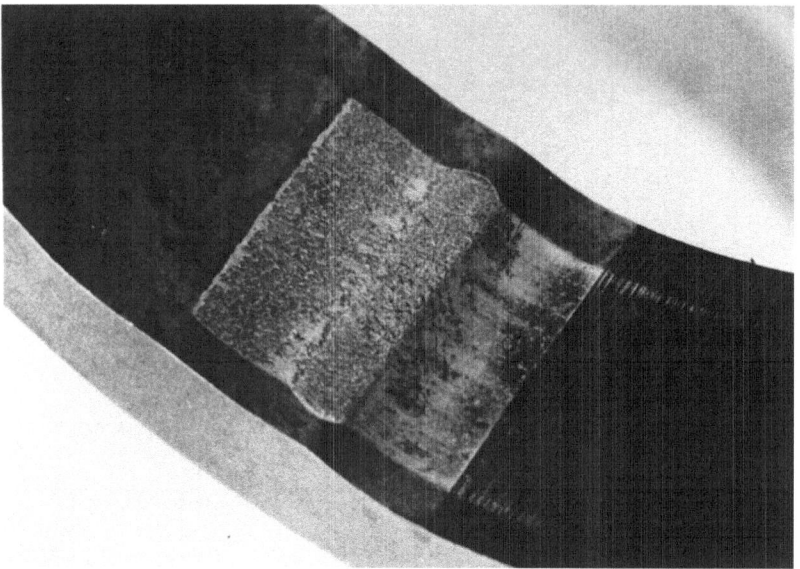

Fig. 5. Damage to fuel pump cam ring resulting from operation on kerosene.

this problem. This fuel usually includes a lubricant additive.

Kerosene is clearly a very suitable alternative to gas oil for use in high speed automotive diesel engines and requires a minimum of engine modification.

Motor Gasolines – It has been suggested that a potential problem associated with the penetration of light duty diesel engines into the passenger car field would be an imbalance in the consumption of gasoline and diesel oil requiring a change in refinery arrangements. Although Reference 1 suggests that this will not be a problem in the longer term, diesel engines operating equally well on gasoline would avoid any such problem in the short term while still providing the larger part of the fuel saving associated with the diesel engine. The fuel saving resulting from the higher specific gravity of the fuel would not, of course, be realised. This is the incentive behind the current interest in stratified charge engines.

However, prior to the present concern for the conservation of oil based fuels, the main interest in the operation of diesel engines on gasoline has come from the military demand for multi-fuel engines. These engines have been variously required to operate on fuels ranging from normal diesel fuel through kerosene to 80 octane number gasoline. This would permit the engines to be operated, in an emergency, on whatever fuel may happen to be available.

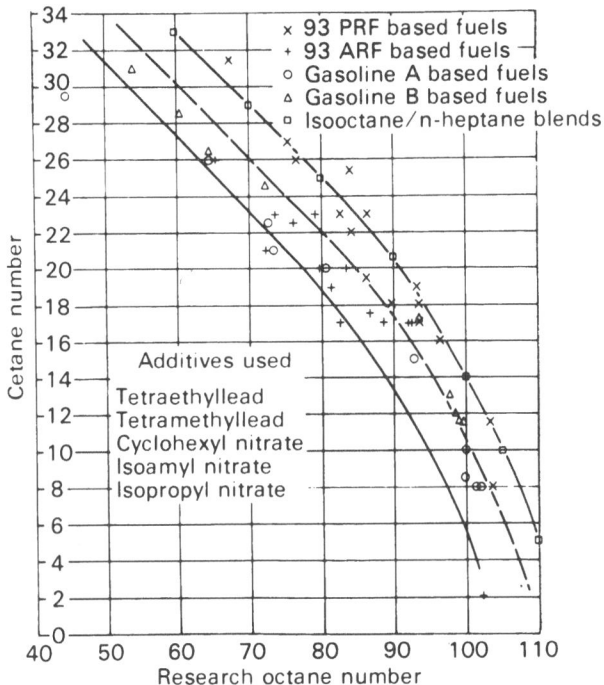

Fig. 6. Relationship between cetane and research octane numbers of various gasolines with additives.

The primary problem of operating diesel engines on gasolines stems from their inherently lower ignition quality which is further depressed by the addition of lead, which is used to raise the octane rating of gasolines. A high octane number means a low cetane number as shown by the correlation between research octane number and cetane number for a range of gasoline fuels and additives in Fig. 6 (3).

The low ignition quality of gasolines is most troublesome when a direct change of fuel is attempted because the long ignition delay causes a late start of combustion and leads to misfire at high speeds. It is necessary therefore to advance the injection timing, or preferably to take steps to reduce the delay period by heating the inlet air or raising the compression ratio. The latter is attractive because it is effective in reducing the ignition delay of the low cetane fuel while having little adverse effect on the performance of the normal diesel fuel.

Fig. 7 shows a typical comparison of gasoline and diesel fuel performance over the load and speed range carried out with a 300 cu. inch 6 cylinder D.I. truck engine (4). The compression ratio was raised from 17 to 19:1, thus allowing a compromise timing to be used for all three fuels. At the lower speeds the improvement in smoke limited performance, indicated by the J.V. (just visible) points is clearly evident. This improvement in exhaust color is attributed to a greater proportion of the fuel being burned in a fully evaporated, pre-mixed state. It is generally accepted that the smoke

References p. 290.

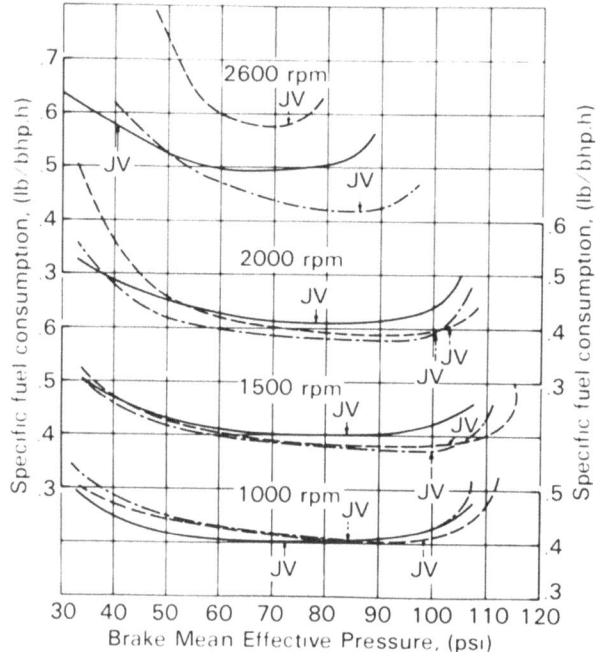

Fig. 7. Typical diesel fuel and gasoline load range curves from a 6 cylinder 300 cu. inch diesel engine.

is the result of diffusion burning in the presence of liquid fuel. The more advanced compromise timing had negligible overall efect on the gas oil performance and resulted only in an increase in noise, which was nevertheless low compared with the gasoline.

The deterioration in fuel economy at high speed and light load is due to the onset of misfiring due to the long ignition delay, particularly with the MT 80 fuel.

Fig. 8 shows typical diesel and gasoline cylinder pressure and needle lift diagrams. It can be seen that in this instance where the compression ratio was not raised, the advance of timing and long delay resulted in all the fuel being injected before ignition giving a very high rate of burning. Combustion photographs taken under these conditions (5) confirm the predominantly pre-mixed burning of the gasoline.

A further result of combustion under these conditions is the formation of pockets of end gas outside the combustion bowl, which detonate causing damage to the piston crown as shown in Fig. 9. This damage is typical of detonation damage to gasoline engine pistons and progresses to a much more advanced state of destruction than is shown if allowed to continue.

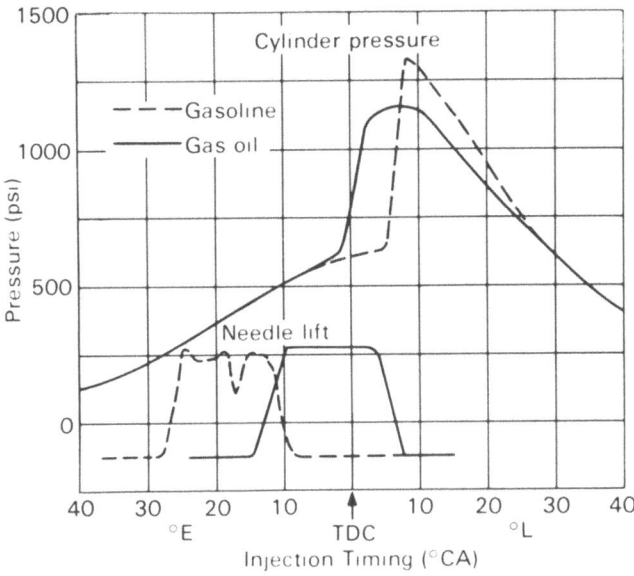

Fig. 8. Diesel and gasoline cylinder pressure diagrams.

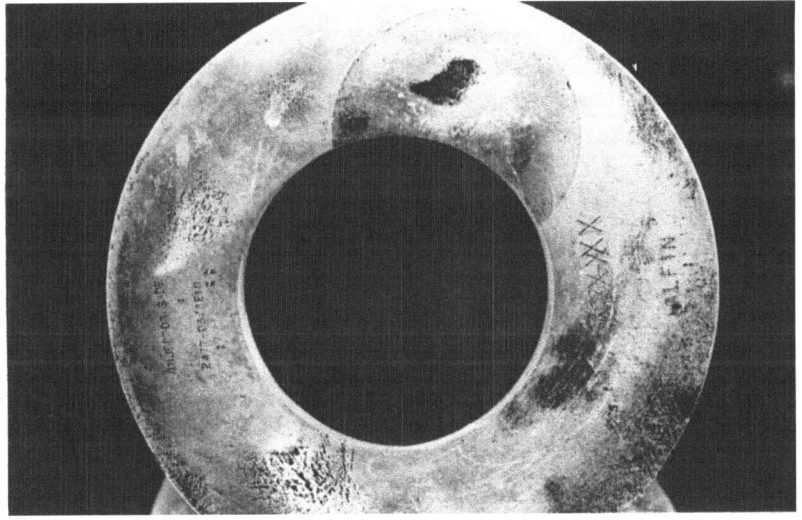

Fig. 9. Piston damage resulting from operation of a diesel engine on low cetane gasoline.

Clearly the solution to the problem of operating diesel engines on gasoline lies in reducing the ignition delay. Apart from the measures to raise the compression temperature mentioned earlier, the other alternative is the use of ignition improving additives. A number of additives have been tried and are listed in Table 4 with their effects on both octane and cetane number.

TABLE 4

Additive Effects on Cetane and Octane Numbers of Gasoline

Fuel	Cetane Number	Research Octane Number
93 PRF* Clear	19.0	93.1
93 PRF + 5% vol. DERV	21.0	89.5
93 PRF + 5% vol. Lubricating Oil	22.0	91.7
93 PRF + 3.17 gm Pb as TEL/USG	11.0	103.8
93 PRF + 3.17 gm Pb as TEL/USG + 5% DERV	13.0	100.0
93 PRF + 3.17 gm Pb as TEL/USG + 5% vol. Lubricating Oil	18.0	102.3

PRF = Primary Reference Fuel

The most effective of these in the presence of tetraethyllead (TEL) would appear to be the 5% of lubricating oil. This raises the cetane number with least depression of the octane rating and thus reduces the ignition delay while minimizing the risk of detonation damage. This has proved to be a complete solution to the damage shown in Fig. 9 while also providing necessary lubricity to the gasoline for the benefit of the fuel injection pump. It is nevertheless necessary to provide separate lubrication of the cams and followers of inline fuel pumps when operating on gasoline.

A further requirement of the fuel system is a pressurized recirculating feed to the pump gallery to avoid vapor lock which is a problem with high volatility fuels and causes incomplete pump filling.

An alternative approach to the problem of low ignition quality is dual fuel operation. This would involve a pilot charge of normal diesel fuel, to provide ignition with normal delay, injected ahead of the gasoline which might be injected or aspirated. However, the difficulty of providing a duplicate fuel system with two injectors or, alternatively, the introduction of the problems of weak mixture burning make this solution least attractive for gasoline. It is normally only used when operating on gaseous fuels as described later.

Attempts to operate divided or indirect injection chambers on gasoline are far less successful than in D.I. engines. The high rate of mixing and dispersal of the fuel coupled with high heat losses results in excessive delay, and therefore unreliable ignition under all but full load conditions when the engine is hot. The accidental refueling of diesel passenger cars with gasoline is an effective demonstration of this point. The only solution here is to provide an ignition source when the engine moves towards a stratified charge concept.

Gaseous Hydrocarbon Fuels – Gases are perhaps the most unnatural of fuels for the diesel engine. As explained earlier, compression ignition is only practical because of the high degree of charge stratification provided by liquid fuel injection. Since gases can only be introduced in gaseous form, the primary requirement of local rich mixture does not exist and it is therefore essential to resort to dual fuel operation so that the pilot charge of diesel fuel can provide the ignition sources. This system is quite commonly used in large stationary engines. The gas is introduced either by a separate gas valve in the cylinder head, or by gas carburetor, and the pilot charge of diesel oil is injected through the normal diesel injector. The subject of large dual fuel engines has been reviewed and summarized in two Ricardo reports (6, 7).

Gaseous fuels in automotive diesel engines have largely been used as supplements to achieve a cleaner exhaust or higher smoke limited rating. Aspiration of a gas, either butane or propane, for loads above 70% can achieve a higher air utilisation by reaching the air that the diesel fuel has difficulty in using. Experiences with this system used in buses in Vienna and Amsterdam have been reported (8).

Limiting the gas consumption to above 70% full load means that the actual gas consumption is small and hardly rates as an alternative fuel. On the other hand, the production of LPG is small compared with gasoline, and relatively little would be available should it have proved to be a good alternative fuel for the diesel engine.

Tests have been conducted by Ricardo (9) to assess the possibility of operating indirect injection engines on natural gas, predominantly methane. The engine used in this instance was a single cylinder research engine of 97.5 cu. inches fitted with a Comet chamber shown as State 1 in Fig. 10.

Methane has a significantly higher knock resistance than most other gaseous fuels. Nevertheless, as can be seen in Fig. 11, the maximum power was knock limited to 85 psi bmep until the throat member of the swirl chamber was completely eliminated as in State III. In this state 105 psi bmep could be reached.

Light load fuel consumption was poor due to the difficulty of burning fuel in weak mixture areas. Karim has indicated that weak mixtures of methane do not burn satisfactorily in dual fuel engines. Some throttling of the intake at part load improved the fuel consumption as shown by the dashed lines in Fig. 11, but excessive throttling would promote misfire. Unburned hydrocarbons would be a serious problem.

The modification of the chamber to operate on methane is clearly unacceptable for operation on diesel fuel only, and any possibility of rapid change-over from one to the other is unlikely.

Undoubtedly, a direct injection combustion system would be better suited to

References p. 290.

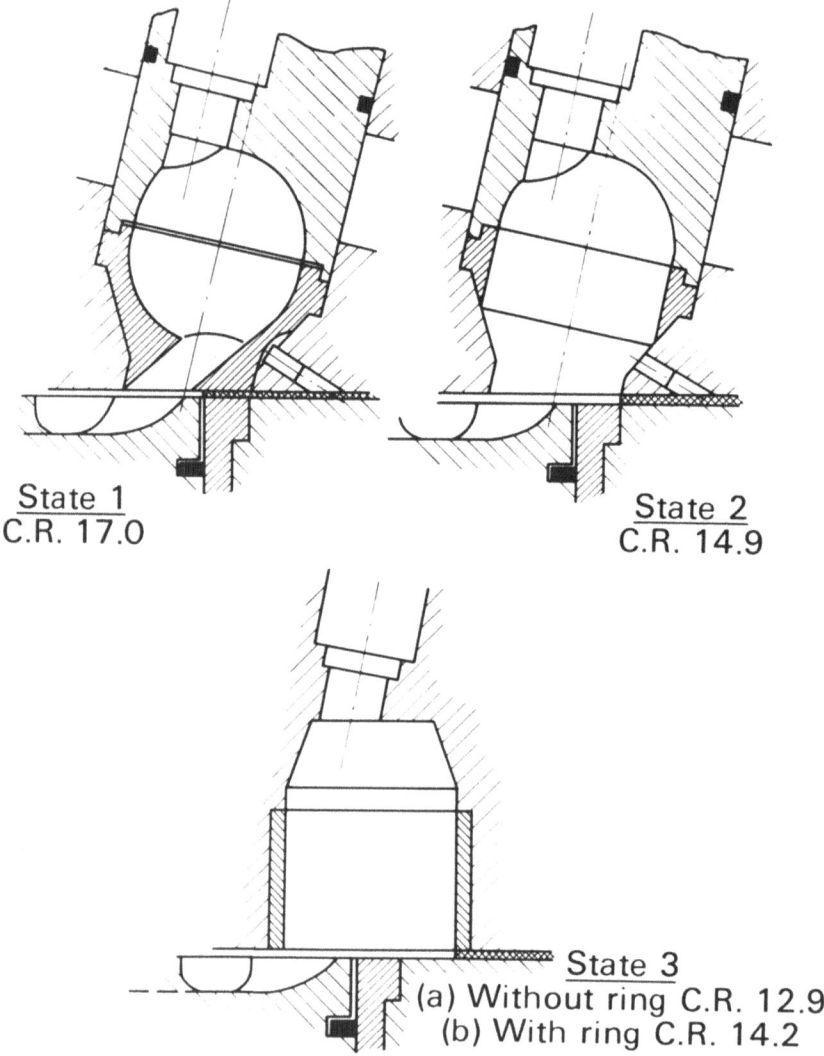

State 1
C.R. 17.0

State 2
C.R. 14.9

State 3
(a) Without ring C.R. 12.9
(b) With ring C.R. 14.2

Fig. 10. Ricardo Comet chamber variants used for operation of natural gas.

operation on the dual fuel principle. However, there are other problems connected with the use of natural gas in automotive engines. These are storage, transportation, and handling of the gas in a mobile application. Its very low boiling point means that it must either be stored at high pressure involving the use of gas cylinders, or in liquefied form in a cryogenic tank with provision for venting any gas boiled off by heat transfer into the tank. Transporting compressed or liquefied natural gas over long distances is uneconomical, and it has been shown to be preferable to convert it to methanol, which is more suitable for automotive use being a stable liquid at normal temperatures.

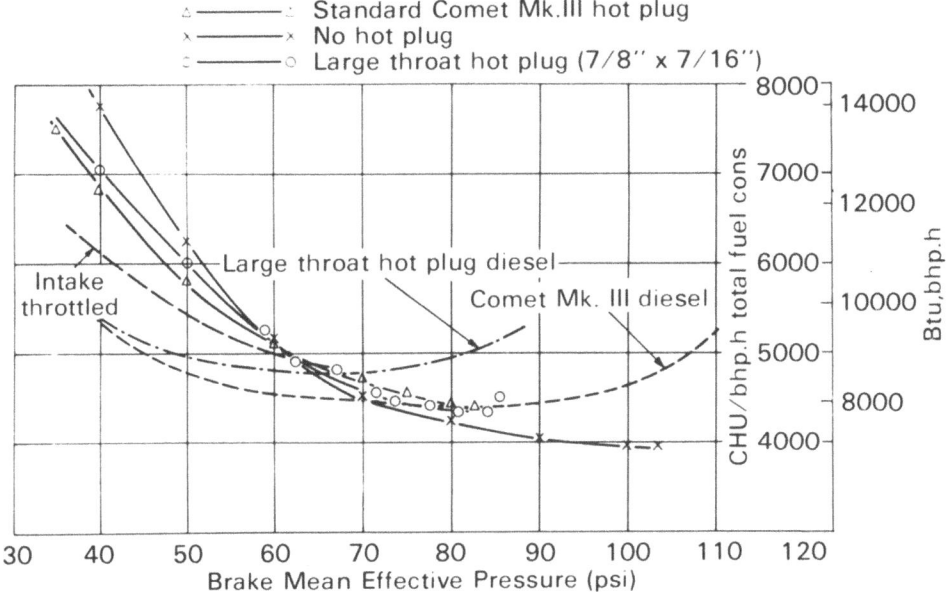

Fig. 11. Test results showing the limitations of operating I.D.I. engines on natural gas.

Methanol — It has been predicted that methanol can be produced at a price well below that of gas oil compared on an energy basis. It would therefore be an economically attractive alternative fuel for the diesel engine, and its future supply is far less limited than oil. Methanol can also be manufactured from coal and other carbonaceous materials which would give it an availability considerably beyond current estimates for oil or natural gas. Methanol is produced by passing methane over a catalyst in the presence of steam.

Ricardo have recently been involved in an investigation into the operation of automotive diesel engines on methanol, the object being to utilise as much methanol in place of diesel as possible with minimum engine modification (11).

Methanol can be blended with gasoline within limits, but it is not possible to blend methanol with most diesel fuels. Methanol has a very high research octane number (> 100). In view of its very low ignition quality it was decided that the most attractive approach would be to operate on a dual fuel principle, aspirating the methanol and using the normal diesel system to provide the pilot injection for ignition purposes. A single cylinder investigation was run, the only modification to the engine being the addition of a variable main jet carburetor, without throttle valve, to an otherwise normal automotive D.I. engine specification as in Fig. 3.

It soon became apparent that the first factor limiting the amount of methanol which could be used was the heat absorbed from the air to evaporate the methanol which has a latent heat of evaporation of 474 Btu/lb. This chilling of the charge air resulted in quenching of combustion, thus limiting the percentage of methanol which could be used, particularly at high speed. Heating the intake air was found to be impractical due to the high temperature required, so a heat exchanger was placed between the carburetor and engine to heat the mixture using engine cooling water. By this means it was possible to restore the inlet temperature to a normal diesel value of about 80°F. The presence of the methanol had a negligible effect on the index of compression and it was therefore surprising to find that, with normal diesel compression temperatures restored, the pilot fuel exhibited considerably longer ignition delays in the presence of the methanol. This was assumed to be due to a depression of the ignition quality of the diesel fuel by blending with the methanol.

Heating the air change introduced the problem of "knock" or spontaneous ignition of the methanol charge in spite of its high anti-knock qualities. Therefore further charge heating to restore normal ignition delay was not possible.

Attempts were made to restore the ignition delay by improving the ignition quality of the diesel fuel by the addition of amyl nitrate. This had the desired effect of raising the quench limit, but the resistance to knock was also reduced, particularly at low engine speed. At low quantities of diesel fuel, the knock and quench limit were very close together at full power.

It was found to be possible to operate on a methanol equivalent of 70% of fuel load at high speed (2200 rpm) using amyl nitrate, and 80% at 1000 rpm without amyl nitrate. The high speed operation was limited by quench, and the low speed by knock. Further addition of amyl nitrate at high speed caused knock to be the limiting factor. Fig. 12 summarizes the levels of methanol utilization achieved so far.

It is clear that while these levels can be achieved under steady state conditions, it is impractical to vary the amount of amyl nitrate, and it will be difficult to modulate the mixture heating accurately enough under transient conditions to avoid either quench or knock. There are clearly other systems which should be considered, including direct injection of the methanol and the variation of other engine parameters, but methanol is clearly not a natural alternative to diesel fuel.

In addition to the combustion problems it must be remembered that the low calorific value of this fuel means storage tank capacity must be about double. Concern has also been expressed about the toxicity of the fuel which can be absorbed into the bloodstream with disastrous results.

Hydrogen – It has been suggested that the ultimate fuel for internal combustion engines will be hydrogen produced by the electrolysis of water, the primary source of energy being nuclear. In the absence of hydrocarbon fuels, hydrogen would appear to be a convenient way of storing energy in a form which can be used for automotive purposes and which incidentally could be the solution to most of the current

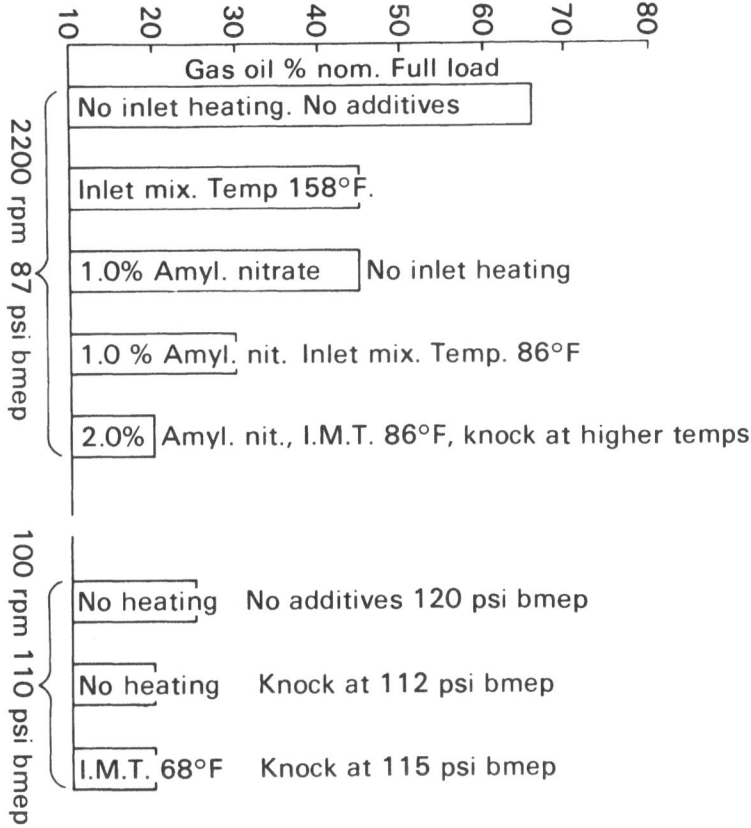

Fig. 12. Methanol dual fuel tests — % gas oil to give diesel performance at Bosch 3 smoke.

pollutants emitted by IC engines.

The diesel engine would probably not be the first choice of prime mover for the combustion of hydrogen. However, in 1926, Ricardo carried out some tests on behalf of the Air Ministry (12) to investigate the possibility of burning the hydrogen which has to be valved from air ships to compensate for the loss of weight as the diesel fuel was consumed. This offered the possibility of extending the range of the air ship by conserving the diesel fuel.

The engine used for these tests was a single cylinder research unit of 5½ in. bore x 7½ in. stroke employing a single sleeve valve and an open combustion chamber in the head as shown in Fig. 13. This chamber used a high level of swirl, a single fuel spray, and the chamber wall was insulated with a steel sleeve. The latter was found to give smooth combustion when operating on diesel fuel. The hydrogen was introduced into the induction air at a choke tube fitted in the induction system and at a pressure of 0.018 psi.

References p. 290.

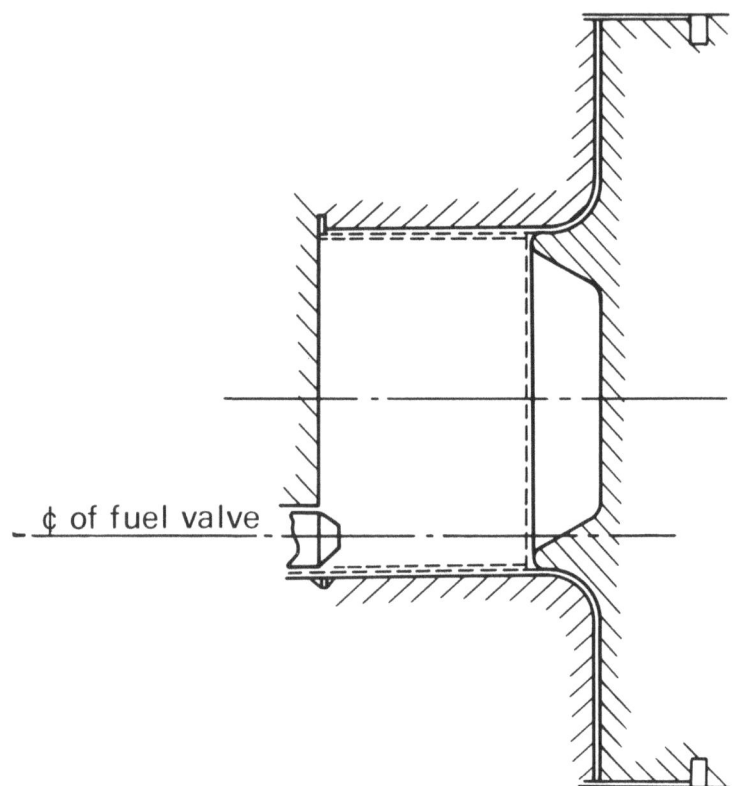

Fig. 13. Vortex chamber used for hydrogen tests.

The tests involved increasing the supply of hydrogen at various levels of diesel fuel delivery until the combustion became intolerably "bumpy" or rough due to pre-ignition of the hydrogen from the heat insulated sleeve fitted in the combustion chamber. Removal of this sleeve delayed the onset of pre-ignition and resulted in the performance shown in Fig. 14. Engine roughness was monitored by maximum cylinder pressure and limited to 800 psi.

The valving ratio of hydrogen (1 lb H_2 for 13 lb oil) can be matched at about 80 psi bmep. Below this load the engine can easily consume all the hydrogen which has to be valved.

Fuels Derived from Coal — It is possible to synthesize a whole range of fuels using coal as the basic source. Processes involving hydrogenation are available for the production of gasoline and distillate fuels. There is also the Fischer-Tropsch process for producing diesel fuel. In general, the production of liquid fuels from coal has not proved economic when the price of the end product is compared with that from oil. However the projected long term availability of coal extends so far beyond that of oil

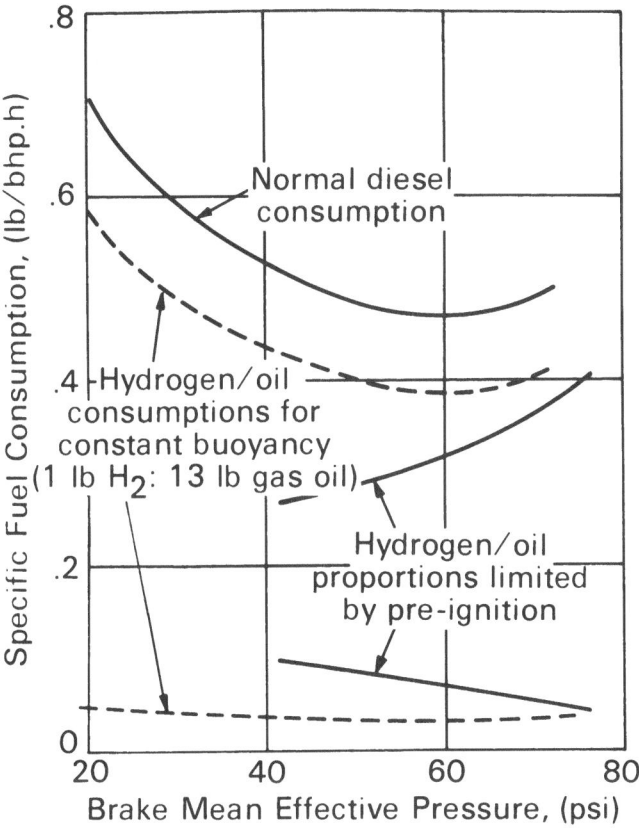

Fig. 14. Hydrogen test results showing limit of hydrogen consumption imposed by pre-ignition.

that interest in its use as a source of automotive fuel must increase. Synthetic diesel oil played an important part in the German war effort from 1936 to 1945, and it is interesting to note that the South African company SASOL is currently producing diesel fuel from coal on a commercial basis in competition with petroleum fuels.

Tipler (12) has reviewed the prospects of operating diesel engines on coal and its derivatives and it is not proposed to consider the alternatives in detail here. Of course, where a fuel of equivalent quality if produced there is no problem in substitution, and operational difficulties would be limited to certain side effects due to minor departures from the characteristics of oil based fuels.

The operation of diesel engines on coal tars and creosotes is not uncommon, and papers have been written on the subject (13, 14). Ricardo have carried out tests on a range of creosotes from various sources (15). These tests were confined to an investigation of the ignition quality of the fuels when blended in various proportions with gas oil. Each creosote was mixed with gas oil at 25% and 50% concentration, and rated against mixtures of high and low ignition quality reference fuels giving a range

References p. 290.

of cetane numbers. The basis of comparison was the throttled induction pressure at which misfire occurred. These tests were carried out in a Ricardo Whirlpool IDI engine (Fig. 15) and a Gardner engine with a D.I. chamber as shown in Fig. 16. Operating conditions were 1,000 rpm, 3 30 psi bmep.

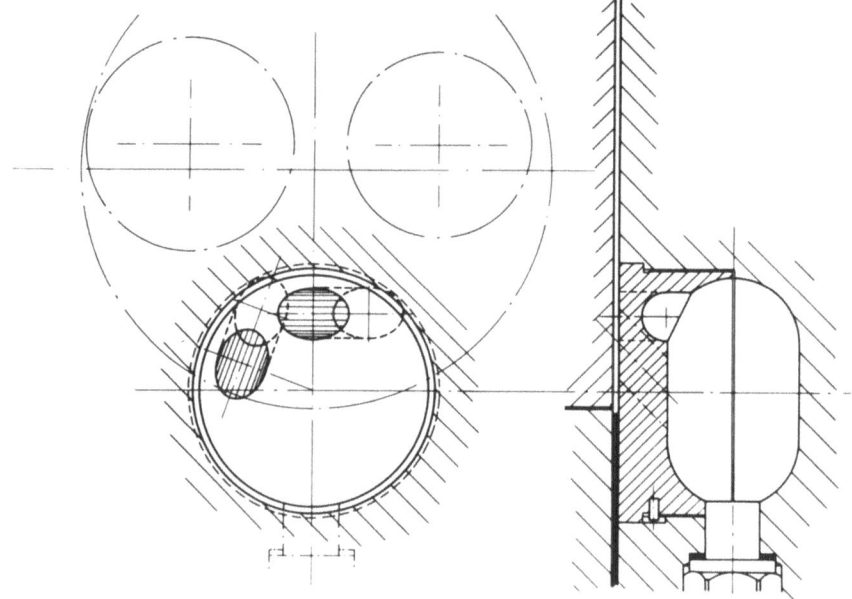

Fig. 15. Whirlpool chamber.

Fig. 17 shows the results obtained with cetane number plotted against induction pressure. Good agreement is evident between the two engines, the difference in induction pressure only indicates that the Whirlpool chamber could operate on a lower cetane fuel.

The cetane ratings of the blends were broadly as follows:

Blend	Cetane Number
100% Gas oil	50
25% Creosote	36-41
50% Creosote	24-33
100% Creosote	0-16 by straight line extrapolation

The characteristics of the individual fuels are somewhat academic, but it is clear that ignition quality is the major problem with creosote having extremely low cetane numbers based on these tests. Clearly these fuels are hardly worth considering as alternative fuels for automotive diesel engines even on a 50/50 basis mixed with gas oil.

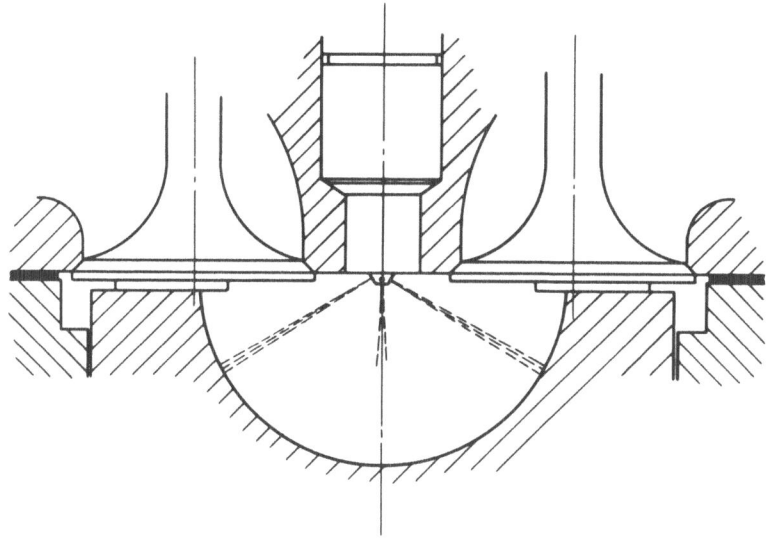

Fig. 16. Gardner D.I. chamber.

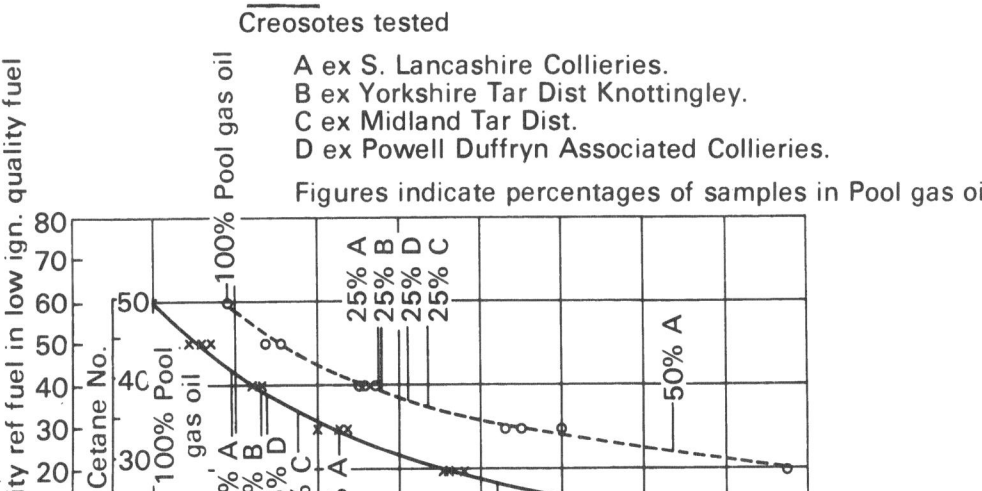

Fig. 17. Cetane rating of creosote/gas oil blends.

The only way in which these fuels could be used is to convert them into lighter distillate type fuels by hydrogenation, if the quantities available and process cost make it economically worthwhile.

DISCUSSION

In considering alternative fuels for automotive diesel engines, it is assumed that in the shorter term it may be necessary to switch existing engines to other fuels because, for some reason, the traditional fuel is in short supply or has become uneconomic. The interest in the long term is to retain the high thermal efficiency of the diesel when having to utilize alternative energy sources. It must be remembered in considering alternative fuels, that energy consumption is not confined to the conversion of the fuel into work in the engine. Depending on source of the fuel, the energy used to produce the fuel in usable form may be considerable. Energy used in transportation of the fuel may also be excessive. The preparation and supply of alternative fuels for use in diesel engines may not therefore prove the most economic overall even with the diesel's high efficiency. Modification of the engine could also prove difficult or uneconomic.

This paper contains a cross section of the many experiments carried out in the Ricardo Laboratories which cover the whole range of possible alternative hydrocarbon fuels. The problems encountered and complexity of measures taken to use some of the less likely fuels is a measure of their suitability as alternatives to gas oil, and offers a means of rating them in order of suitability.

First—Kerosenes and Some Other Distillates – These fuels can be substituted directly for gas oil with no change in combustion requirements. Maximum fuel settings must be adjusted to compensate for the varied volumetric calorific values, and provision must be made to ensure adequate fuel pump lubrication.

Second—Gasolines and Heavier Fuel Oils – Because of significantly lower ignition quality, injection timing and/or compression ratio will have to be changed. Heated inlet air or additives may also be needed.

These fuels also show a preference for combustion system i.e. direct injection (D.I.) for gasoline, indirect injection (I.D.I.) for heavy fuel oils.

Third—Methanol and Gases Including Hydrogen – On account of the very low ignition quality and, in the case of gases, the difficulty of achieving locally rich mixtures for ignition, it is necessary, in compression ignition engines, to use the dual fuel approach. This means either duplicate fuel injection systems or a means of introducing the fuel with the air. Aspirated fuel/air mixture introduces the problems of "knock", pre-ignition, and unburned hydrocarbons due to weak mixture areas in the engine cylinder. Knock and pre-ignition have the overall effect of limiting the proportion of the alternative fuel which can be burned. In order to maximize utilization, the metering and control of engine operating parameters become complex, especially under transient conditions or where a wide speed range is required.

Fourth—Coal Tars and Residual Fuels — The very low ignition quality and/or other physical properties preclude their being used as automotive fuels unless blended with gas oil or other light distillates. Even so, to maintain adequate ignition quality and burning rate, the diluted fuel will be predominantly gas oil and the substitution rate will be too low for these fuels to be entertained as real alternatives. The only possibility of their use, undiluted, lies in the provision of heated fuel systems and probably a means of ignition.

General Considerations — It is clear that the automotive diesel engine will only continue in its present form while fuels in the first two categories, whether derived from oil or coal, exist; and while they can be produced at a competitive price and energy consumption. Except in some special applications, dual fuel operation to utilize methanol or gas is unlikely to be an acceptable system for road vehicles due to its complexity.

It is likely, of course, that all the fuels in the first two categories which have their origin in oil will become unavailable simultaneously, and there is a need therefore to conserve them. To this end it is suggested that the combination of a wider cut distillate fuel embracing these two categories and a diesel engine, perhaps with assisted ignition, would be of value. Current activity in the field of the stratified charge engine could be directed to this end.

In the third category, methanol is probably the most attractive automotive fuel, being a stable liquid at normal temperatures. It is clearly not an ideal diesel fuel, although there are still some virtually unexplored areas. The long term availability of methanol which does not rely entirely on fossil sources makes it deserving of a special engine developed to maximise its efficient use. Its high knock resistance suggests an engine of high compression ratio and spark ignition.

Hydrogen, although placed in category 3 is not really considered a suitable alternative fuel for diesel engines owing to the relatively small substitution rate.

The suggestion of a 'wide cut distillate' engine and a 'methanol' engine highlights an area of little knowledge — that of spark ignited fuel sprays. It is suggested that a study of this area would be profitable to all engaged on advanced combustion systems for any fuel which cannot easily be compression ignited.

CONCLUSIONS

1 — Real alternative fuels for the automotive diesel engine are limited to those having a good self ignition quality. They include kerosene and the heavier distillates whether from oil or coal.

References p. 290.

2 — The modification and complication of adopting the diesel engine to run on lower ignition quality fuels may not prove economic or practical except in the larger truck engines.

3 — There would appear to be a case to widen the diesel fuel cut of the barrel, and develop a diesel engine to operate on this fuel in order to conserve oil-based fuels in the automotive field.

4 — For the more difficult fuels, particularly methanol, special engines should be developed to ensure their most efficient use, since it is unlikely that the diesel will cope satisfactorily with their low ignition quality.

5 — Spark ignition of fuel sprays may play an important part in future alternative fuel engines by helping to perpetuate the high compression ratio, unthrottled operation of the diesel engine. Work should be directed to this area.

REFERENCES

1. *F. H. Kant, A. R. Cunningham and M. H. Farmer, "Effects of Changing the Proportions of Automotive Distillate and Gasoline Produced by Petroleum Refining," EPA 460/3-74-018 July 1974.*
2. *"Note on Comparative Tests with Various Fuels under Supercharge Conditions at 500 and 1200 rpm on the E18/1 Engine with Comet Mk III Combustion System," Ricardo Report Number DP.145.*
3. *"The Effects of Additives on the Octane Number and Cetane Number of Motor Gasoline," Ricardo Report Number DP.1489.*
4. *"Multi Fuel Project, Progress Report Number 1," Ricardo Report Number DP.4303.*
5. *W. M. Scott, "Looking in on Diesel Combustion," SAE Paper No. 690002 (SP345), January 1968.*
6. *"Operation of Internal Combustion Engines on Gases," Ricardo Report Number DP.9292.*
7. *"Future Use of Gas in Internal Combustion Engines," Ricardo Report Number DP.13548.*
8. *"Visit to Municipal Bus Depot in Vienna to Discuss LPG, May 1970," Ricardo Report Number DP.12370.*
9. *"Report on Dual Fuel Tests on E16-2 (4¾" x 5½" Comet Mk III)," Ricardo Report Number G. O. 1820.*
10. *G. A. Karim and S. A. Klat, "The Knock and Auto Ignition Characteristics of Some Gaseous Fuels and Their Mixtures," Inst. of Fuels Journal, March 1966.*
11. *"Further Investigations of Methanol in a Dual Fuel Engine," Ricardo Report Number SN.19949.*
12. *W. Tipler, "Prospects for the Operation of Diesel Engines on Coal or Its Derivatives," Paper C18/75. Conference on Power Plants and Future Fuels, Automobile Division, Inst. of Mech. Engrs.*

DISCUSSION

W. T. Lyn *(Cummins Engine Co.)*

My impression from listening today is that methanol is a very good fuel for everything except the diesel. Other engines can use methanol and we can save the good fuel for the diesel. I would like to make a plea that, if we are going to express efficiency realistically, we use megajoules per kilowatt hour instead of BTU's per horsepower hour. Finally, my question, did you have any emission data for your testing with the various fuels in diesel engines?

Scott

We don't have any emission data.

P.C.T. de Boer *(Cornell University)*

I take exception to your comment that the diesel engine is not suitable for many other fuels including hydrogen and I'd like to remark that when you run with hydrogen you could inject the hydrogen rather than induce it as you did with your engine. That is a slight modification. You may also need some modification, such as spark ignition, because it won't compression ignite, but it is still possible with slight modification to make diesel engines suitable for some of these fuels at least for hydrogen.

Scott

Yes, I think you're probably correct regarding hydrogen. But if you're going to have the spark ignite it, I guess you have to throttle, and once you throttle I don't consider it to be a diesel engine.

J. M. Colucci *(General Motors Research Laboratories)*

I recall many stories about Prof. Diesel operating one of his original engines on powdered coal. Do you have any comments regarding the use of powdered coal as a diesel engine fuel?

Scott

No, we've never used powdered coal as a diesel fuel. My friend Bill Tippler of Perkins has written a very good paper reviewing the situation regarding solid fuels in diesel engines. If you want to get caught up on history, you can read his paper. I think my worry about running on coal is that we know we have enough problems with diesel engine durability when we run with EGR. If we're going to put rather larger particles through the engine I think we'd be in trouble.

N. A. Schilke *(General Motors Research Laboratories)*

I wonder if you have general observations on noise-related characteristics of the fuels you tested?

Scott

In general, of course, the low ignition quality fuels give much higher noise levels. Anything you can do to increase the cetane number of the fuel makes it easier to control the combustion noise.

F. L. Dryer *(Princeton University)*

I would appreciate your comments regarding the use of water in diesel fuels. As you know, Mike Khan at CAV Ltd. has demonstrated some remarkable reductions in soot and NO_x using this technique.

Scott

We've done work on water injection mainly in the intake of the engine. This is not a very satisfactory way to add water. However, it reduces NO_x quite successfully without losing engine performance. But undoubtedly from the results that I've seen and you've probably seen, putting water in either alongside the fuel or with the fuel appears to be much more effective. I don't know that I can comment very much more than that because I don't know much more than you do about what CAV has done. But the main problem of course is what do you do about carrying this water around and stopping it from freezing and so on.

S. G. Liddle *(General Motors Research Laboratories)*

Have you done any work with vegetable oils, peanut oil or cottonseed oil,

M. A. Elliott *(Energy Consultant)*

The Department of Agriculture in the olden times, back in the late 40's, had a laboratory in which they had some vegetable oils like peanut oil and cottonseed oil, and they did some diesel ignition quality testing.

Scott

My understanding is when they ran on corn oil the exhaust smelled like newly-baked bread.

R. W. Hurn *(Energy Research and Development Administration)*

Some years ago I did use peanut oil in a diesel engine. In this case, though, the exhaust smelled absolutely horrible.

SESSION III – SUMMARY

J.B. HEYWOOD

Massachusetts Institute of Technology

I guess I have a few minutes to try and sum up and perhaps pick out some of the key elements of this engines-fuels interactions area that we've been reviewing this morning. I think in this area it is important to remember that the emissions requirements and especially fuel economy requirements are currently forcing the changes in engines, not the alternative fuels. So there really is a problem in getting at the interface of the alternative fuels characteristics of 5, 10, 15 years from now with the engines of that same time frame. In the papers that we went through today, I picked out various parts of this interface problem.

Dick Hurn showed that, at present, there are difficulties getting suitable fuels of sufficient quality to make detailed characterizations worthwhile. It's a short-term problem and I'm sure we'll move through that quite quickly. However, very careful characterization is required in evaluating these alternative fuels in conventional types of engines. We're looking for small differences in fuel characteristics and small differences in performance, and we really don't know whether or not they will be important.

We had several papers looking at methanol. One of the heartening characteristics of several of these was that, to me, it wasn't just experimental data. There was a good attempt at analysis trying to fit the experimental data with methanol into the kind of frameworks and scaling laws that we had developed for use with conventional fuels like gasoline or kerosene. It seems to me that it's very important in this wide open area that we do more than just take experimental data because it is hard to guess the operating conditions and precise engine details of the future. What we really need is a broad data base and some assurance that we understand the trends qualitatively and that we understand those trends quantitatively. As the future becomes more precise, we can scale from the data that we already have to the appropriate operating conditions with the new fuel's characteristics.

Joe Harrington showed that the thermodynamic analysis of engine cycles is a field that is certainly not dead at all and with the new types of instrumentation and data analysis systems that we now have available, we can do a number of things that in the past were very laborious. We can do these very much more quickly.

Dr. Bernhardt from Volkswagen, I think gave us a different perspective on methanol. The sort of applications that could be important in Europe are a little different from those that might happen in the States. It's clear we have a much better idea of the advantages of using methanol in spark ignition engines and some of the disadvantages regarding the engine modifications required. There's been a lot of progress in the last couple of years regarding methanol, and it's clear that we can explain many of the trends in methanol quantitatively with the same kind of methodology that we worked out for gasoline engines.

It's good that we didn't neglect continuous flow combustion systems. We tended to assume that we can burn a range of fuels in Brayton cycle or Stirling cycle engines without very much impact on emissions or performance. That may or may not be true. There may be some special characteristics of fuels like methanol, as we heard in the GMR paper, that allow you to get lower emissions, especially NO_x, without going to some very advanced combustion concepts. Going the other way, if we try to use wide range distillate fuels in these types of combustors there may be problems that we haven't anticipated. We've got to look at them.

Our last paper by Murray Scott emphasized to me that today's well established engines and fuels have been very carefully optimized for each other. And if you try to make changes in one, the fuel, you're going to have to make changes in the other, the engine. Obviously, that should be the case. We've had a long time to optimize the diesel and the gasoline engines and their appropriate fuels. One point that emerged that I feel should receive more emphasis, is a look at a spark-ignited diesel engine. If it has a fuel tolerance it can accept a much less refined fuel. I think we should look at that rather more carefully than I think we have.

I'd like to end up with one concern over the type of R&D that we've been doing in this engines-fuels interaction area, that I have been concerned about for the last couple of years. However, I'm somewhat less concerned after the kind of papers that we've been through this morning. If we look back on our work on engine-fuel interactions, it really started a couple of years ago with the fuel crisis. It seemed to me that many of our problems resulted in R&D programs with the wrong focus. They had a very short time frame, and they did not appropriately relate changes in engine technology to the appropriate time frame for these alternative fuels. I think our papers today show that we're getting to a much more balanced approach to this problem, and I really feel encouraged. I hope that trend will continue. I think a year or two ago fools rushed in where angels fear to tread. I hope we're not fools, and I know we're not angels, and if I keep scrambling that metaphor a little bit more, wise men will move steadily on a broad front in this area. For the reasons that I imply, there are many options in the future, both in the engine technology and the fuel characteristics and availability. We can't predict what they are. There are many possible future combinations that we need to look at. What we need are broad base programs that recognize that careful characterization is important. You've got to build up a data base and an understanding so that as that future becomes much more precise, we can relate what we're doing now to that future with much less effort. And I think, on that note, I'll close.

SESSION IV

LONG–TERM FUTURE FUELS
AND AN OVERALL ASSESSMENT

Session Chairman
S. S. PENNER

University of California at San Diego
LaJolla, California

HYDROGEN AS A RECIPROCATING ENGINE FUEL

W. J. McLEAN, P. C. T. de BOER, H. S. HOMAN and J. J. FAGELSON

Cornell University, Ithaca, New York

ABSTRACT

An analytical and experimental investigation of hydrogen as a reciprocating engine fuel has been carried out. The analytical portion involved application of a thermodynamic model to predict trends in power, efficiency and emissions with hydrogen fuel. The experimental investigation was carried out using a single cylinder CFR engine modified to run on hydrogen. Both aspects of the investigation focused particular attention on the unthrottled direct injection hydrogen engine because of its suitability in terms of power, efficiency, and control of flashback, preignition and emissions. Model calculations and experimental results are described and compared.

The analytical model was of the thermodynamic type with the burning rate specified by a semi-empirical turbulent flame speed correlation. The hydrogen was assumed to be injected and fully mixed prior to ignition. The model was adiabatic and the burnt gases were assumed to remain unmixed in order to estimate the temperature gradient across the cylinder. Nitric oxide (NO) emissions were computed by applying the finite rate extended Zeldovich mechanism to the combustion products. Reactions involving N_2O were found to be of no importance in nitric oxide production even under the leanest operating conditions. Model calculations demonstrated the feasibility of operating a direct injected, hydrogen fueled reciprocating engine unthrottled over a wide range of conditions. The lean region with equivalence ratio less than 0.6 gives a particularly attractive combination of high efficiency and low NO emissions. Due to their relatively high flame temperatures and fast burning rates, hydrogen-air mixtures give rise to pronounced NO decomposition during the expansion stroke for mixtures richer than an equivalence ratio of 0.8. As a result, the specific NO emissions show a maximum near an equivalence ratio of 0.8 and a sharp decrease for richer mixtures. The specific emissions are also relatively insensitive to compression ratio.

References p. 317.

The experimental investigation was carried out using both the high swirl prechambered CFR cetane rating head and the standard CFR octane rating Otto cycle head. Tests with fuel injection near top center piston position on the prechambered head demonstrated the feasibility of quality regulated operation on hydrogen. At high compression ratios (greater than 17), spark plug ignition was accompanied by large and rapid pressure fluctuations. On the other hand, glow plug ignition gave rise to very smooth combustion. There was also evidence that late injection did not allow sufficient mixing of fuel and air in the engine, and hence, resulted in incomplete combustion.

Experiments with the Otto cycle head indicated that increasing the duration of fuel injection, and advancing the beginning of injection to an early part of the compression stroke resulted in improved combustion efficiency and increased power and thermal efficiency. However, near stoichiometric mixtures, engine operation under these conditions exhibited preignition tendencies at high compression ratios. The experiments indicate the necessity for careful control of fuel-air mixing in the directly injected engine. Measured nitric oxide levels generally are in satisfactory agreement with the model calculations. Again, lean operation leads to high efficiency with very low emissions.

It is concluded that the unthrottled reciprocating engine, with direct cylinder injection of hydrogen fuel is a feasible and attractive mode for hydrogen usage. Significant fuel economy and emissions benefits, obtained by ultra lean operation, are possible with hydrogen.

INTRODUCTION

In considering prospects for future automotive fuels, much of the attention is currently focused on fuels derived from abundant U.S. fossil resources. Near-term moderation of the demand for imported petroleum will require not only exploitation of these resources, but also a vigorous conservation effort. However, in the longer term even these fossil reserves, primarily coal and shale oil, will be depleted, and a substantial fraction of the energy demand will have to be satisfied by nuclear, solar, or geothermal resources. In this long term scenario, the need for portable fuels of high energy density, especially for use in the transportation sector, will have to be satisfied by fuels generated from readily available substances.

Hydrogen, generated thermally or electrolytically from water, has been suggested as such a long term fuel. The concept of a "hydrogen economy" has received considerable attention during the past several years (1,2,3,4) and will not be further discussed here. The principal drawback to the application of hydrogen to motor vehicless is that presently available storage systems are heavy and occupy a large volume. Both metal hydride and cryogenic storage systems have been suggested, and both are under active further investigation and development.

There have been numerous previous experimental investigations of hydrogen as a reciprocating engine fuel, and a recent review of much of this literature is available (5). In brief, it has been found possible to take advantage of hydrogen's wide flammability limits to operate engines under highly efficient lean unthrottled conditions. Care must be taken to prevent preignition and knock when using an easily ignited and fast burning fuel such as hydrogen. Exhaust emissions of oxides of nitrogen can be controlled by lean operation, but under full power conditions these emissions may exceed those found with gasoline fuel. The reader is referred to the cited review for more details and for specific references.

The present paper considers the use of hydrogen in spark ignited reciprocating engines, and focuses on engine performance (power and efficiency) and emissions over a wide range of operating conditions. The study is both experimental and analytical with the principal objective being a quantitative determination of hydrogen's potential for low emissions and high fuel economy.

HYDROGEN COMBUSTION CHARACTERISTICS

Some indication of the performance of an engine operating with an alternative fuel can be obtained by considering the fuel's thermodynamic properties and combustion characteristics. In this section, various properties of hydrogen are considered especially as they compare with similar properties of typical hydrocarbon fuels. A summary of relevant properties for hydrogen is given in Table 1, along with reference values for isooctane.

The oxidation of hydrogen is governed primarily by relatively fast, nearly thermally neutral, bimolecular branching chain reactions. In contrast, hydrocarbon oxidation involves thermal chain mechanisms due to the slower endothermic reactions associated with fuel break-down. As a result of this fundamental difference in oxidation mechanisms, hydrogen exhibits combustion characteristics considerably different from those of typical hydrocarbons. For example, hydrogen's extremely broad flammability limits, relatively high flame speed, and low ignition energy are direct results of the nonthermal nature of the chain branching process.

For a reciprocating engine which inducts a charge of given volume at given manifold pressure and temperature, the power output at a given equivalence ratio is proportional to the energy content (heat of combustion) per unit volume of inducted mixture. On this basis, the data in Table 1 indicate that stoichiometric hydrogen air mixtures are expected to produce about 15% less power than stoichiometric isooctane air mixtures, provided cycle efficiencies are nearly the same. Direct cylinder injection of hydrogen could be used to increase the charge density and thereby increase the power output.

The relative performance of an Otto cycle engine using various fuels can be estimated by comparing engine performance on the adiabatic ideal fuel-air cycle, where temperature and composition dependent thermochemical properties are used, and residual fractions and throttling losses are accounted for, but combustion is

References p. 317.

TABLE 1

Thermodynamic and Combustion Properties of Hydrogen and Isooctane

Property	Hydrogen	Isooctane
Heat of Combustion, kJ/kg	1.20×10^5	4.44×10^4
Flammability Limits, Vol. % in Air or (ϕ)		
Lean	4.0 (0.10)*	1.0 (0.6)
Lean	9.0 (0.23)**	
Rich	75.0 (7.1)	6.0 (3.8)
Stoichiometric Mixture Properties		
Air-fuel ratio (mass)	34.6	15.2
Air-fuel ratio (volume)	2.38	59.7
Volumetric energy content at STP, J/cm^3	3.18	3.73
Approximate Laminar Flame Speed at $\phi = 1$, P = 0.1 MPa, cm/sec	180	40
Minimum Ignition Energy at $\phi = 1$, P = 0.1 MPa, mJ	0.02	1.0

* upward propagation of a non-coherent flame

** upward and downward propagation of a coherent flame

instantaneous at the top center piston position.

Power and efficiency on ideal fuel-air cycle analyses are given in Fig. 1 for a hydrogen or isooctane fueled engine with displacement equivalent to the single cylinder CFR engine. Actual CFR valve timings were used rather than the top and bottom center timings usually used in fuel-air cycle analysis. This does not affect efficiency results, but does give slightly lower power output. Unthrottled operation at 0.1 MPa manifold pressure is assumed for hydrogen fuel due to its wide flammability limits, and the power and efficiency are therefore given as functions of the equivalence ratio. Two hydrogen cases are considered. One of these assumes induction of a premixed unthrottled hydrogen-air charge, while the other assumes induction of air only (unthrottled), with the hydrogen added by direct cylinder fuel injection just prior to ignition. For isooctane, equivalence ratios of 0.6, 0.8 and 1.0 are considered, and power regulation is assumed to be achieved by throttling of the intake mixture, so that power and efficiency are shown as functions of the intake manifold pressure.

As mentioned already, premixed hydrogen produces about 15% less power than isooctane when both fuels are used in stoichiometric proportions and at atmospheric manifold pressure. Direct cylinder injection of the hydrogen greatly increases the charge density so that the power output is increased about 40% when compared with premixed stoichiometric mixtures. With respect to thermal efficiency, the unthrottled hydrogen cases exhibit increasing efficiency at lower power due to favorable thermodynamic

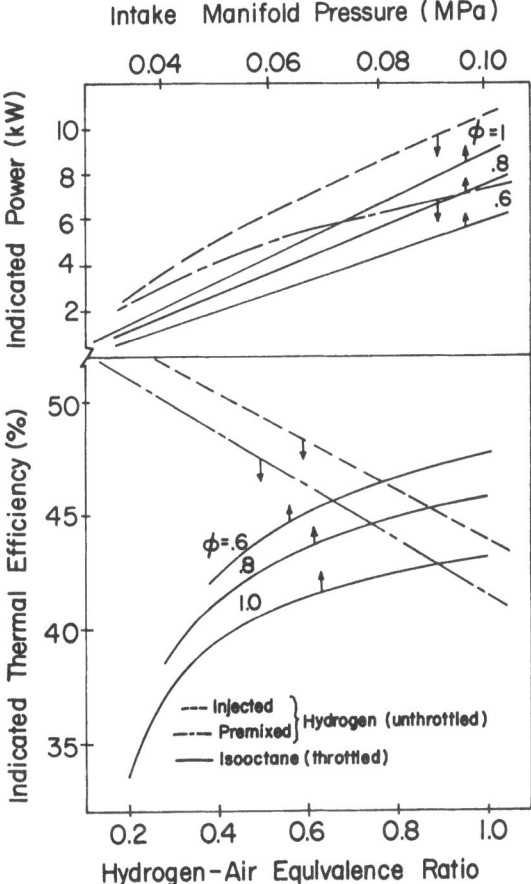

Fig. 1. Ideal fuel-air cycle power and efficiency comparisons for hydrogen and isooctane; engine displacement 611 cm^3, compression ratio 8. Hydrogen is unthrottled with atmospheric manifold pressure and is either premixed or injected directly into the cylinder. Isooctane is throttled.

effects (higher specific heat ratio, less dissociation). Similar results are obtained with leaner isooctane cases, but here losses due to the throttling required for power regulation lead to reduced efficiencies for light loads (low manifold pressures).

The results shown in Fig. 1 may be used to compare hydrogen and isooctane as reciprocating engine fuels. Comparisons may be made either by comparing efficiencies at equal power output or by comparing power outputs at equal thermal efficiency.

ENGINE MODELING METHODOLOGY

While the ideal fuel-air cycle results discussed above give some indication of hydrogen's potential as a fuel, more detailed cycle analysis is required if model calculations are to give results in reasonable agreement with experimental trends. Finite combustion rates and heat transfer are the principal processes which must be accounted for if the engine model is to give good efficiency and power estimates. Temperature gradients across the cylinder due to progressive burning must also be accounted for if the highly temperature sensitive NO_x emissions are to be calculated.

In the present studies, principal interest was in the identification of significant trends in power, efficiency and emissions. Therefore, it was not considered necessary to introduce the complexity of a detailed heat transfer analysis. However, finite combustion rates were accounted for. The computed power and efficiency are then expected to be greater than measured since heat losses reduce both power and efficiency. Most NO_x formation occurs in the high temperature bulk gases away from the cylinder wall thermal boundary layer, so that neglect of heat transfer should not introduce a serious error here. Bulk gas temperature reduction due to wall heat transfer occurs only when the bulk gases isentropically expand to maintain continuity with the cooler, higher density boundary layer gases. For example, if heat transfer effects cause a 10% reduction in cylinder pressure, then the corresponding bulk gas temperature reduction will be only about 2%.

The finite burning rate has been accounted for by assuming that a turbulent flame having a speed S_T with respect to the unburned gases propagates spherically outward from the ignition source. The flame speed S_T is assumed to be proportional to the laminar flame speed S_L, with the proportionality factor involving the engine speed, so that burning rates increase with speed as it is usually observed in experiments. The laminar flame speed S_L, with the proportionality factor involving the engine speed, so that burning rates increase with speed as is usually observed in experiments. The laminar flame speed has been taken from the Semenov formulation, and involves It should be noted that S_T is not constant, but depends upon the burned and unburned gas properties. In a fixed coordinate system then, the flame accelerates across the cylinder due to the combined effects of burnt gas expansion and increasing flame speed S_T.

The value of the proportionality factor in the flame speed was fixed by choosing a constant which gave good agreement between computed and measured best torque spark advance. This spark advance is a direct indication of burning rate, since faster burning mixtures require less advance for maximum power output. The comparison between measured and computed spark advances determined in this manner is shown in Fig. 2 where spark advance is given as a function of equivalence ratio. The model parameters were initially chosen by comparing computed advances with those measured by Stebar and Parks (7), and good agreement was obtained except for richer mixtures, where engine knock necessitated experimental spark timings retarded with respect to optimum. More recent spark advance data from our own experiments are also shown in Fig. 2. These experiments and the procedure for obtaining the optimum timings are

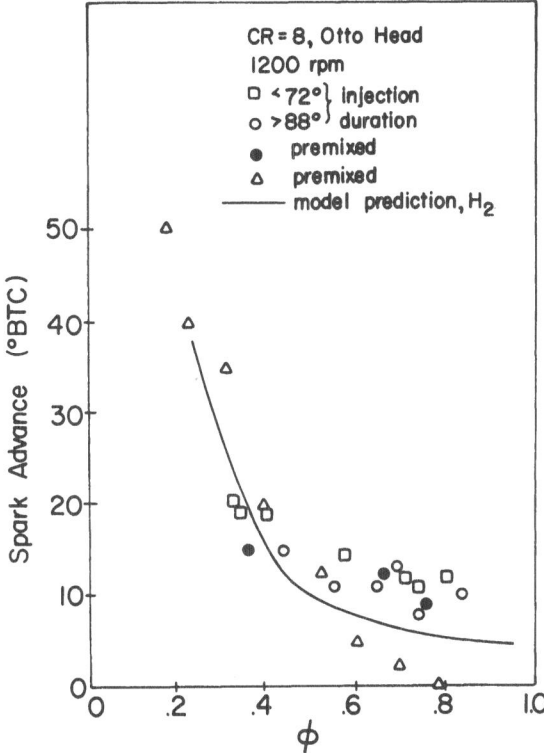

Fig. 2. Comparison of measured and predicted optimum spark timings with hydrogen fuel in a CFR engine with either a premixed charge or direct cylinder fuel injection. The △ indicate the data of Stebar and Parks (7).

described in the following section. The agreement between measured and computed advance is considered adequate, generally being within five degrees crank angle.

Further evidence for the adequacy of the burning rate model is shown in Fig. 3 where a comparison is given between computed and measured combustion durations, again as functions of equivalence ratio. The combustion duration is defined here as the time from spark plug discharge to attainment of 99% of charge mass burned. The predicted combustion durations were obtained by scaling available model calculations to conditions corresponding to the experimental conditions with a premixed charge. The experimental combustion durations were obtained from cylinder pressure traces as the time between spark discharge and attainment of maximum cylinder pressure. Although combustion is generally not complete at maximum cylinder pressure, maximum pressure and complete combustion are very nearly coincident with a rapid burning fuel such as hydrogen.

Fig. 3 shows that the model gives a trend in combustion duration with equivalence ratio which is quite similar to the measured trend, with the predicted duration generally

References p. 317.

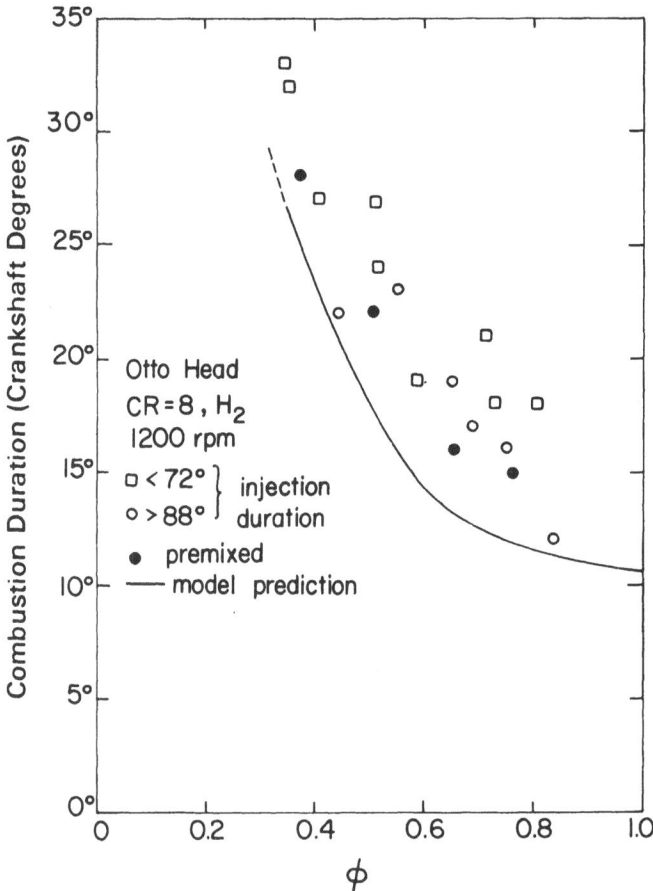

Fig. 3. Comparison of measured and predicted combustion durations in a hydrogen fueled CFR engine with either a premixed charge or direct cylinder fuel injection.

being a few degrees shorter than the measured duration. It can also be noted that the measured durations for the cases with direct cylinder fuel injection tend to be longer than the durations in the premixed cases, in contrast to the faster burning which might be expected due to the higher cylinder pressures resulting from direct injection. This slower burning rate in the injected cases is evidently due to incomplete fuel-air mixing prior to combustion, a topic which is given extensive discussion in a later section. It is notable with regard to this mixing situation that the shorter injection durations tend to result in long combustion durations (Fig. 3). This trend also correlates with the poorer mixing which is obtained with the shorter durations (cf Results and Discussion).

The NO_x emissions were evaluated by applying the finite rate, extended Zeldovich reaction mechanism to the hot combustion products. The burnt gas temperature gradient due to progressive burning was accounted for by considering the products in several discrete zones, each of uniform temperature. Nitrogen atoms were assumed to be in steady state, and oxygen atoms in equilibrium with oxygen molecules. Further details of the model calculations are given elsewhere (6,8).

EXPERIMENTAL DESCRIPTION

A detailed experimental investigation was carried out using a Waukesha CFR (Cooperative Fuel Research) single cylinder engine. The engine was run both with the Diesel head originally designed for determining cetane number of a fuel, and with the Otto head originally designed for determining octane number. Most of the results obtained with the Diesel head have been reported (5), and the present paper is concerned mainly with Otto head results. The output torque of the engine was measured using a GE type TLC D.C. cradle dynamometer, which could be operated either as a motor or a generator, so that both friction and brake power could be measured.

The modifications made for adapting the engine to operation on hydrogen included the installation of a fuel injector into the cylinder wall (see Fig. 4). The injector was actuated hydraulically, using the standard Bosch fuel injection pump (9). While direct cylinder injection as provided by this injector was used for most of the work, a separate series of experiments was carried out in which hydrogen and air were premixed in the intake manifold. For this purpose, a 33 cm long copper tube with a large number of small holes was inserted into the intake pipe close to the intake valve, and the hydrogen was admitted to this tube through a needle valve. For both injection and premixing, the air flow rate was measured with a Fischer and Porter B6N-25 rotameter and with a Merriam laminar flowmeter, while the hydrogen flow rate was measured with a Fischer and Porter 1/8 -20 Tri-Flat rotameter.

A Sundstrand 601 B piezoelectric pressure transducer was mounted on a modified spark plug in order to measure cylinder pressure. A Tektronix rotational function generator provided a timing mark for every ten degrees of crankshaft rotation, and the standard "injection indicator points" provided an electrical signal showing the beginning and the ending of the injection pulse. All signals were continually displayed on a Tektronix oscilloscope.

Sketches of the NO sampling system and of the NO chemiluminescent detector are shown in Figs. 5 and 6, respectively. The reservoir in the exhaust line was added to check whether unsteadiness of the exhaust flow velocity and the exhaust NO concentration influenced the results. Sample probes were located upstream as well as downstream of the reservoir. It was found that these probes gave essentially identical results. The entire sampling line was heated to 350K to avoid condensation of water vapor, since condensed water is known to absorb NO_2. At the very lean fuel-air mixtures used in part of the investigation, an appreciable fraction of the total emitted oxides of nitrogen consists of NO_2, and preservation of NO_2 in the sampling line is therefore crucial for obtaining accurate measurements of total NO_x. Of course, not condensing the water in the exhaust

References p. 317.

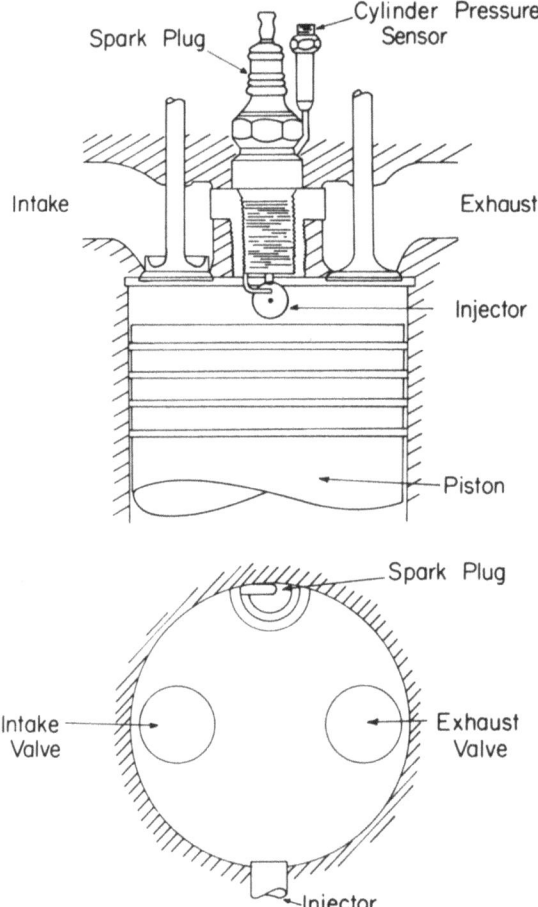

Fig. 4. CFR engine with Otto cycle cylinder head, modified for operation on hydrogen by addition of a hydrogen injector.

gas means that the NO detector was exposed to a wet rather than a dry sample. There is evidence that the presence of water vapor in the sample does not significantly influence the determination of NO by the chemiluminescence method (10).

When making comparison runs using gasoline under rich conditions, it was found that the NO_x to NO converter greatly reduced the NO concentration. Such a reduction also took place with hydrogen fuel when hydrocarbons were introduced artificially into the exhaust gas. When the converter was bypassed, the NO readings were restored to acceptable values. Therefore, the converter bypass was used in all measurements involving rich gasoline-air mixtures.

The chemiluminescence NO detector was constructed using well-known principles (10), and was calibrated before each test series by using a standard mixture of 1075 ppm NO in dry nitrogen.

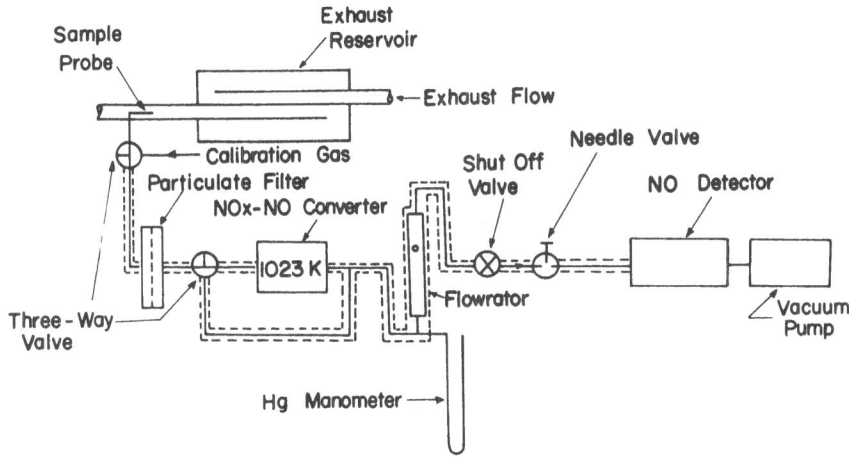

Fig. 5. Exhaust sampling system for determination of oxides of nitrogen.

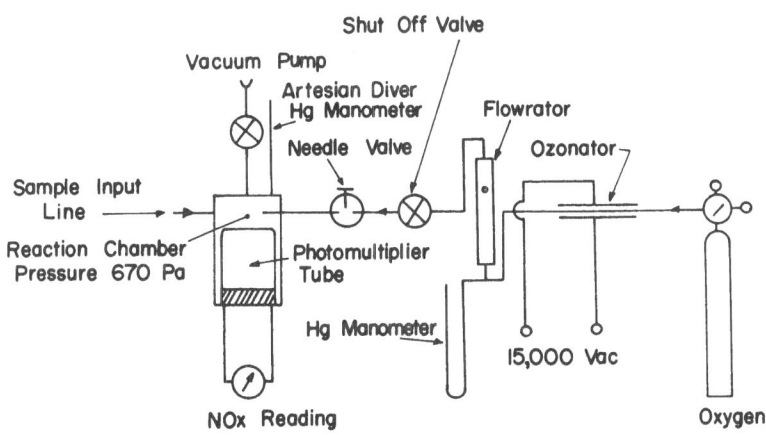

Fig. 6. Chemiluminescence nitric oxide analyzer.

References p. 317.

To determine completeness of combustion, a polarographic oxygen analyzer was connected to the exhaust line (Fig. 7). The oxygen sampling system was completely separate from the NO sampling system. The analyzer was a Beckman model 742, which measures the electrical current developed in an amperometric sensor in contact with the sample. The sensor consists of a gas-permeable teflon membrane, a gold anode, a silver cathode, and a potassium chloride electrolyte. It was calibrated using room air. Before reaching the analyzer, the gas sample was dried by cooling it to room temperature, removing the condensed water, and passing it through "Dehydrite" drying agent.

In all of the work reported here, spark advance was set to obtain best torque. Because the dependence of torque on spark advance is very small near best torque, the criterion used for setting spark advance was based on the cylinder pressure trace. The pressure rise due to combustion was always clearly identifiable, and the spark was advanced until the halfway point of this rise coincided with top center. This procedure gave timings equivalent to those obtained by the usual procedure involving maximization of torque, but gave more reproducible results. The pressure rise with hydrogen was generally quite fast so that this halfway point was easily located.

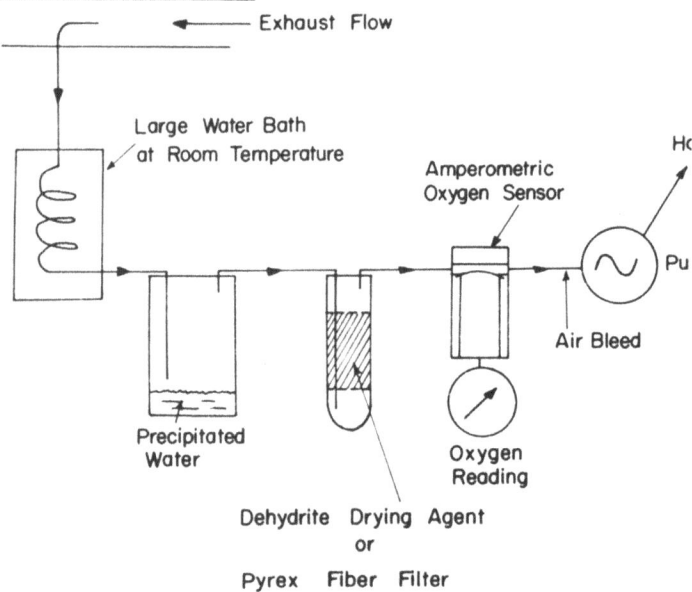

Fig. 7. Exhaust sampling system for determination of oxygen.

RESULTS AND DISCUSSION

Experimental results for engine power, efficiency, and NO_x emissions along with comparisons of model calculations are presented and discussed in this section.

Comments on hydrogen peroxide emissions are included. Evidence for poor fuel-air mixing in some direct injection tests is also discussed, as are observations concerning limitations due to rapid pressure rise, preignition and knock.

Power, Efficiency and Emissions — The measured and computed indicated mean effective pressures are shown as functions of the hydrogen-air equivalence ratio (ϕ) in Fig. 8. Several data points with carbureted gasoline (Indolene) are also shown. The predicted mean effective pressure shown in Fig. 8 was computed for a premixed charge at one atmosphere intake pressure, and it is higher than the measured premixed values principally due to the neglect of heat transfer in the model calculations. Both model computations and experimental data exhibit the trend of increasing power with richer mixtures characteristic of unthrottled quality regulated operation.

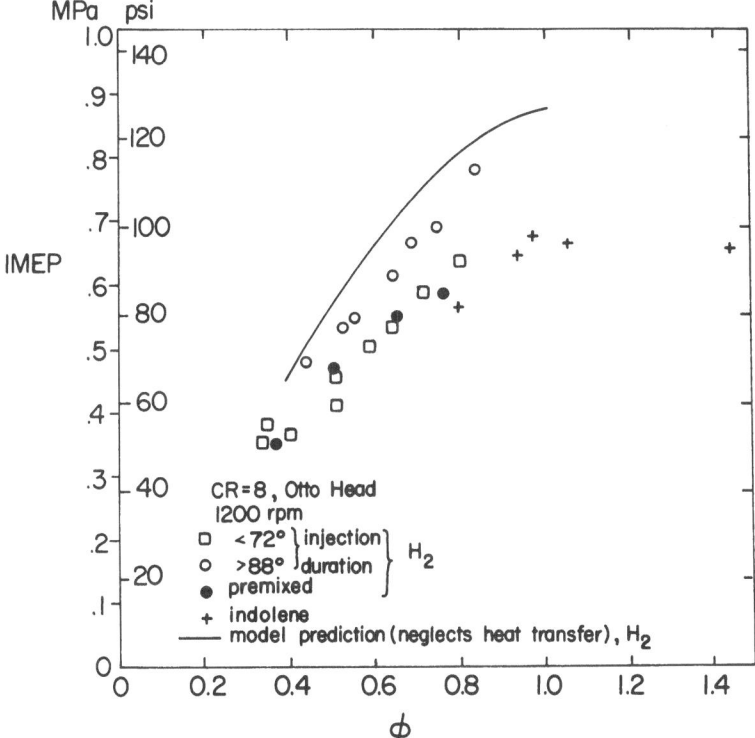

Fig. 8. Indicated mean effective pressure for hydrogen fueled CFR engine, with either a premixed charge or direct cylinder fuel injection. The model prediction is for a premixed charge.

Direct cylinder injection with the longer injection durations gave significantly greater power than either premixed hydrogen or gasoline, in agreement with the trends expected from the ideal fuel-air cycle analysis. Direct injection with the shorter durations led to limitations in fuel-air mixing and incomplete combustion, thereby

References p. 317.

resulting in lower than expected engine power. This mixing problem is discussed in detail in the following section.

The computed and measured indicated thermal efficiencies are shown as functions of equivalence ratio in Fig. 9 for the same conditions as the data in Fig. 8. The maximum computed thermal efficiency for hydrogen is reached at an equivalence ratio near 0.4. Although mixtures leaner than this can be employed, the engine efficiency is reduced due to the extended combustion duration in these lean, low flame temperature mixtures. Measured indicated thermal efficiencies for the premixed hydrogen cases were in the 25-35% range, and were about the same as the indicated efficiency for the lean gasoline cases. Direct cylinder injection of hydrogen resulted in lower thermal efficiencies due to losses associated with incomplete combustion. Again, the correlation between the shorter combustion durations and increased loss of efficiency is observed.

The computed efficiencies are higher than the measured values principally due to heat transfer effects. The differences are somewhat larger than normally attributed to heat transfer alone. It may be that throttling losses occurred either in the relatively small CFR intake manifold or across the intake valve during the intake process. A comparison of measured and computed air flows would indicate the magnitude of such throttling losses, but this was not done due to lack of computed air-flow data.

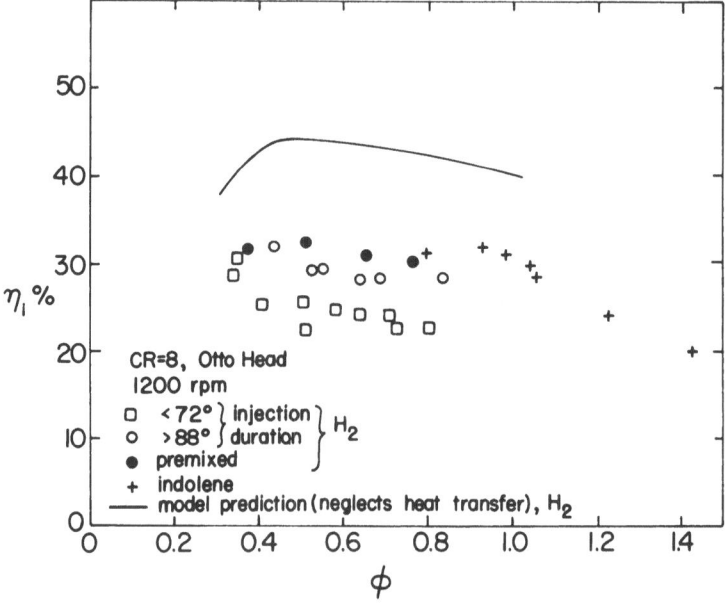

Fig. 9. Indicated thermal efficiency for hydrogen fueled CFR engine with either a premixed charge or direct cylinder fuel injection. The model prediction is for a premixed charge.

The results for nitric oxide emissions as a function of equivalence ratio are presented in Figs. 10 and 11. In Fig. 10 the mole fraction of NO is given as calculated using the previously described model, and as measured in premixed CFR engine tests in our laboratory as well as at General Motors Research Laboratories (7). The agreement between measured and computed NO concentrations is quite good, although the measured values seem to be better correlated by a curve corresponding to somewhat richer mixtures than the predicted values. In a previous discussion (6), it was suggested that such a shift might be produced by heat transfer effects which are neglected in the model calculations.

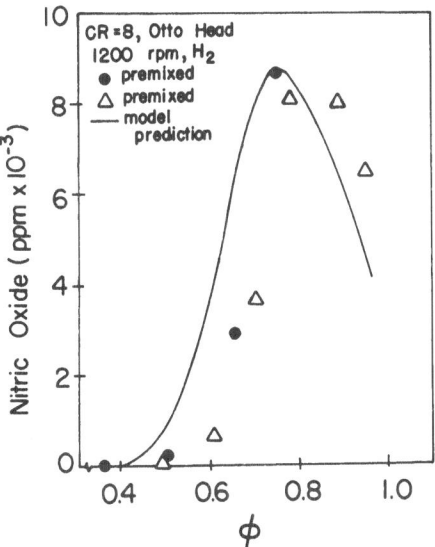

Fig. 10. Comparison of measured and computed exhaust concentrations of oxides of nitrogen (as NO) for a hydrogen fueled CFR engine with a premixed charge. The △ indicate the data of Stebar and Parks (7).

The computed and measured emissions on a specific emissions (mass of emissions per unit energy output) basis are shown in Fig. 11. Measured emissions for direct cylinder injection of hydrogen, for premixed hydrogen, and for carbureted gasoline and methanol are also plotted. In contrast to Fig. 6, the predicted specific emissions curve for premixed hydrogen falls well below the maximum measured value due to the appearance of the computed power in the denominator of the computed specific emissions. That is, the model does adequately predict the NO_x concentrations, but results in a low value for the specific emissions because it gives too high an engine power (Fig. 8).

The NO_x emissions for direct injection with the longer duration give about the same trend as the premixed cases, while the shorter injection duration and

References p. 317.

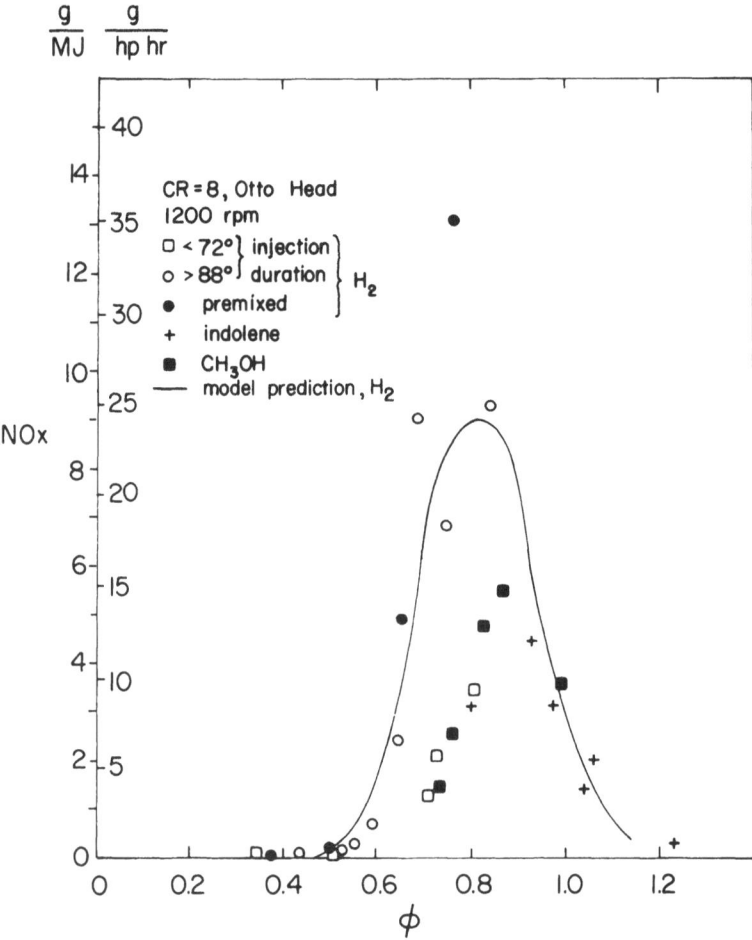

Fig. 11. Specific emissions of total oxides of nitrogen as NO for a CFR engine with hydrogen, gasoline, or methanol fuel. Hydrogen is either premixed or injected directly into the cylinder, gasoline and methanol are carbureted. The model prediction is for a premixed charge.

accompanying incomplete combustion lead to lower NO_x emissions. NO_x emissions with either gasoline or methanol are significantly lower than with hydrogen for lean and stoichiometric mixtures. This is in agreement with the results of Stebar and Parks (7) who also found significantly higher NO_x emissions for hydrogen when compared with emissions for isooctane at the same equivalence ratio in the same engine. Earlier suggestions (11), that hydrogen would somehow result in very low NO_x emissions compared to hydrocarbon fuels are thus found to be incorrect. Of course, very low NO_x emissions can be obtained with H_2 by operating under lean conditions ($\phi <$ 0.6).

Hydrogen peroxide (H_2O_2) has also been observed in exhaust from hydrogen fueled engines (12, 13). The engine operating conditions were not determined in these tests, but water condensed from the exhaust products was found to contain 220 ppm (12) and 800 ppm (13) H_2O_2. Since the water is approximately 30% of the exhaust volume, the concentration of H_2O_2 in the exhaust would presumably be about one-third of that measured in the condensate.

Hydrogen peroxide formation would require the presence of hydroperoxy radicals (HO_2), which are normally present only in very small amounts during high temperature hydrogen oxidation. To produce the required HO_2, one must hypothesize low temperature regions (such as wall quench layers or other flame quench zones) and hydrogen atom availability, so that the termolecular HO_2 formation reaction may progress:

$$H + O_2 + M \rightarrow HO_2 + M \qquad \Delta H^O_{298} = -197 \frac{kJ}{mole}$$

Once the HO_2 is formed then H_2O_2 may be formed by recombination:

$$HO_2 + HO_2 + M \rightarrow H_2O_2 + O_2 + M \qquad \Delta H^O_{298} = -178 \frac{kJ}{mole}$$

Further analysis and experiments are necessary to establish the H_2O_2 emissions potential with hydrogen fueled engines. If H_2O_2 is produced in quantities large enough to be undesirable in the atmosphere, then hydrogen vehicles might require thermal or catalytic reactors to decompose H_2O_2 to H_2O and O_2.

Mixing Limitations in Direct Cylinder Injection — The data obtained from the oxygen analyzer were used to determine the completeness of combustion. Typical results for percentage of oxygen in the exhaust as a function of equivalence ratio are shown in Fig. 12. The line marked "no combustion" simply indicates the dilution of the oxygen concentration arising from adding hydrogen to air. The corresponding curve for gasoline (Indolene) is not drawn, but is close to the upper boundary of the figure (21% O_2). The curves marked "complete combustion" indicate the remaining oxygen after complete combustion, when all water is removed from the combustion products. The experimental results obtained with direct cylinder injection are seen to fall between these two lines, indicating that combustion was only partially complete. Comparison experiments with premixed hydrogen yielded almost complete combustion as did experiments with carbureted gasoline (Fig. 12). Further experimentation with hydrogen injection showed that the fraction of fuel burned (as deduced from the exhaust oxygen measurement) increased with duration of injection as shown in Fig. 13. There also was a tendency toward more complete combustion as the timing of injection was advanced; however, this tendency was not very pronounced, and the

References p. 317.

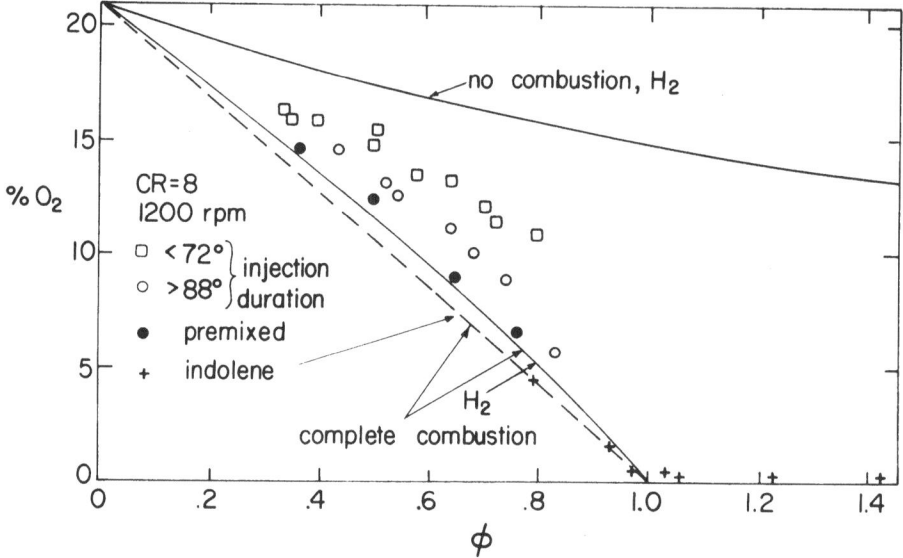

Fig. 12. Measured oxygen content in exhaust sample, dry basis. Complete combustion and no combustion lines calculated from mass balances.

main dependence of fraction burned was on injection duration. For injection durations of 88° and longer, the fraction of oxygen burned was close to 0.8, while for injection durations of less than 72°, the fraction burned was between 0.4 and 0.7.

It is believed that the lack of complete combustion arises from incomplete mixing of hydrogen and air. A similar finding was reported by Oehmichen (14), who made an extensive investigation on a hydrogen engine with direct cylinder fuel injection. He found that radial injection of hydrogen led to poor thermal efficiencies. After changing the orientation of the hydrogen injector so as to inject the hydrogen tangentially, Oehmichen obtained very high indicated thermal efficiencies (e.g. 54% at a compression ratio of 15.4, an equivalence ratio of 0.4, and an engine speed of 1500rpm). In view of this result, our plans for further work include changing the direction of injection from radial to tangential, and further experimentation with the timing of injection. Clearly, the mixing problem must be solved in order for direct cylinder injection to be attractive. Oehmichen's work indicates that this is possible, and that indicated efficiencies of the order of 50% can be achieved under suitable conditions.

Ignition Phenomena — It has been well established that hydrogen engines are susceptible to preignition of the combustible mixture by reactions at hot surfaces. These reactions may occur at the surfaces of either free floating particles or cylinder wall deposits. Preignition leads to rough running, and to backfiring into the carburetor. It can be avoided by keeping the engine cylinder very clean, and by eliminating any hot spots that may exist at the exhaust valve or the spark plug.

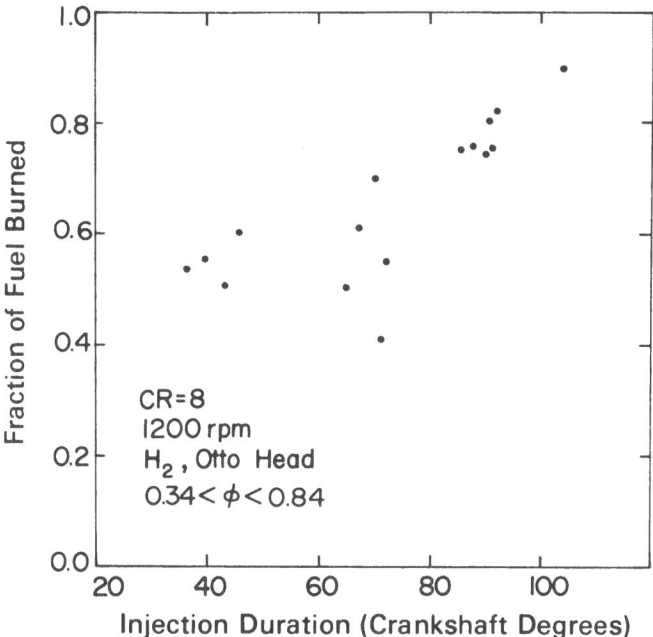

Fig. 13. Fraction of hydrogen burned in CFR tests with direct cylinder fuel injection as a function of the injection duration, τ, in crankshaft degrees. The fraction burned is deduced from measurements of exhaust oxygen content.

Another way to avoid preignition is to mix the hydrogen and air as late as possible during the engine cycle, either by fuel injection or by the hydrogen induction technique proposed by Swain and Adt (15). However, these measures by themselves do not guarantee elimination of preignition, as we found in our work on the CFR Otto head engine. For example, at a compression ratio of 8 and an engine speed of 1200 rpm, we could increase the equivalence ratio (ϕ) up to about 0.8. The pressure-time record obtained under these conditions was quite smooth. Increasing ϕ much beyond 0.8 led to preignition, as evidenced by a gradual advance of the pressure rise due to combustion to a time close to beginning of injection. This advance took place over a period of about 15 seconds. Correspondingly, upon decreasing ϕ it took about 15 seconds before preignition ceased. We interpret this as a gradual heating or cooling, respectively, of the preignition source.

True combustion knock is generally accepted as being caused by autoignition of the end-gas mixture, before arrival of the flame front, and after the end mixture has been compressed and heated as a result of combustion elsewhere in the cylinder. Oehmichen (14) determined the limiting values of compression ratio, equivalence ratio and engine speed for knock-free operation. He showed that knock-free operation at high compression ratios was possible for all equivalence ratios, provided the engine

References p. 317.

speed was sufficiently high. For example, he reported knock-free operation at a compression ratio of 10 over the full range of ϕ, provided the engine speed was greater than 750 rpm. Oehmichen furthermore reported that in his experiments, neither preignition nor auto-ignition occurred under any conditions.

True compression ignition apparently has never been observed in hydrogen engines. By definition, such ignition is caused by auto-ignition of the mixture upon compression by the piston, independent of surface-ignition.

In a recent investigation carried out jointly with the Jet Propulsion Laboratory, we used compression ratios as high as 29. We still failed to obtain compression ignition. For this investigation, the CFR-Diesel head engine was modified by placing an insert in the combustion prechamber. Pressure and temperature of the mixture after compression by a ratio of 29 are such that explosion certainly would occur if the mixture were left undisturbed. However, in the engine the mixture soon is expanded again. The ignition lag times apparently are too long compared with the time available.

Excellent combustion characteristics can be obtained by using glow plug ignition, as was found in the investigation mentioned in the preceding paragraph. By mounting a Champion AG 40 cartridge type glow plug close to the fuel injector, it was possible to achieve burning of the hydrogen very soon after injection. Combustion started only about 10 crank angle degrees after beginning of injection. This type of operation is promising for the conversion of Diesel engines to operation on hydrogen.

CONCLUSIONS

Analytical and experimental studies have shown hydrogen to be a suitable fuel for spark ignited reciprocating engines. As a result of the wide flammability limits for hydrogen-air mixtures, unthrottled quality regulation of power is possible with hydrogen. Such operation makes significant efficiency gains possible compared with throttled operation on isooctane.

Although the low density of hydrogen limits engine power output with a premixed charge, direct cylinder fuel injection may be used to significantly increase the charge density and hence the power output. However, with direct cylinder injection, care must be taken to insure that incomplete combustion due to inadequate fuel-air mixing is avoided. In the current experiments, late injection directed radially inward in the cylinder resulted in incomplete combustion, while injection earlier in the cycle resulted in nearly complete combustion.

Nitric oxide emissions from hydrogen fueled engines are governed by the same thermochemical processes which determine these emissions when hydrocarbon fuels are used. For near stoichiometric mixtures, NO_x emissions with hydrogen are considerably higher than with hydrocarbon fuels. However, the lean operation possible with hydrogen enables operation in regimes of very low NO_x emissions. Also hydrogen's rapid burning velocity indicates a high tolerance for EGR control of NO_x without fuel economy penalty. Thus, the flexibility of hydrogen fuel due to its combustion characteristics permits tailoring of the engine to minimize NO_x emissions.

ACKNOWLEDGEMENT

The studies described here were supported by the U. S. Department of Transportation, Office of University Research under Contract No. DOT-OS-30113.

REFERENCES

1. *D. P. Gregory, D. Y. C. Ng and G. M. Long, "The Hydrogen Economy," in The Electrochemistry of Cleaner Environments, J. O. Bockris (ed.), Plenum Press, 1972.*

2. *S. Linke (ed.), "Proceedings of the Cornell International Symposium and Workshop on the Hydrogen Economy," Cornell University, Ithaca, New York, 1975.*

3. *T. N. Veziroglu, "Proceedings of the Hydrogen Economy Miami Energy (THEME) Conference," University of Miami, Coral Gables, Florida, 1974.*

4. *T. N. Veziroglu (ed.), "Hydrogen Energy Fundamentals – Symposium Proceedings," University of Miami, Coral Gables, Florida, 1975.*

5. *P. C. T. de Boer, W. J. McLean and H. S. Homan, "Performance and Emissions of Hydrogen Fueled Internal Combustion Engines," International Journal of Hydrogen Energy, Vol. 1, 1976.*

6. *J. J. Fagelson, W. J. McLean and P. C. T. de Boer, "Analysis of Hydrogen as a Reciprocating Engine Fuel," Preprints 20, ACS Division of Petroleum Chemistry, 1975, pp. 114-126.*

7. *R. F. Stebar and F. B. Parks, "Emission Control with Lean Operation Using Hydrogen-Supplemented Fuel," SAE Paper No. 740187, 1974.*

8. *J. J. Fagelson, "Hydrogen as a Reciprocating Engine Fuel: An Analytical Study," M. S. Thesis, Cornell University, 1975.*

9. *P. C. T. de Boer, W. J. McLean, J. J. Fagelson and H. S. Homan, "An Analytical and Experimental Study of the Performance and Emissions of a Hydrogen Fueled Reciprocating Engine," Proceedings 9th Intersociety Energy Conversion Engineering Conference, 1974, pp. 479-486.*

10. *H. Niki, A. Warnick and R. R. Lord, "An Ozone-NO Chemiluminescence Method for NO Analysis in Piston and Turbine Engines," Transactions, Society of Automotive Engineers, Paper No. 710072, 1971.*

11. *R. G. Murray and R. J. Schoeppel, "Emission and Performance Characteristics of an Air-Breathing Hydrogen-Fueled Internal Combustion Engine," Proceedings 1971 Intersociety Energy Conversion Engineering Conference, 1971, pp. 47-51.*

12. *E. J. Griffith, "Hydrogen Fuel," Nature 248, 1974, p. 458.*

13. *J. J. Reilly, Private communication quoting results of tests at Brookhaven National Laboratory.*

14. *M. Oehmichen, "Wasserstoff als Motortreibmittel," Verein Deutsche Ingenieur, Deutsche Kraftfahrtforschung, Heft 68, 1942.*

15. *M. R. Swain and R. R. Adt, "The Hydrogen-Air Fueled Automobile," Proceedings of the 7th Intersociety Energy Conversion Engineering Conference, 1972, p. 1382.*

DISCUSSION

R. K. Pefley *(Santa Clara University)*

I presume that your model would differentiate between the premixed and the injected cases. However, your plot showed only a single trace. Can you explain?

McLean

We do not see much difference in thermal efficiency between direct injection and premixed cases. However, we do see a large power difference.

Pefley

What about the NO_x?

McLean

The NO_x emissions tend to be somewhat higher with the direct injection, because of higher combustion temperatures.

F. A. Williams *(University of California, San Diego)*

What kind of flame speed correlation did you use?

McLean

We assumed that the turbulent flame speed correlates with the laminar flame speed. We obtain the stoichiometry effects by using a Semenov form for the laminar flame speed. We then fit the scaling constant by observing combustion duration and optimum spark advance timing.

P. T. Vickers *(General Motors Research Laboratories)*

Speaking of combustion duration, Bill, you used a different definition of combustion duration than we use. We use the time between the point at which significant chemical heat release began, until 99% of the chemical heat release has been accounted for. This means that there would be some time between the time when the spark plug is fired and the time at which you see a heat release indication on your pressure trace. It further means, or implies, that there may be some chemical heat release beyond peak pressure because of the crank mechanism. With hydrogen, are those times so small that you feel comfortable using your definition?

McLean

A brief answer is yes. A longer answer is that when you calculate the cylinder pressure traces with this type of flame speed approach, you do actually see an ignition delay time which is attributable not to a true ignition delay, but to the fact that the first portion of the charge to burn and the first portion of the volume to burn cause very little pressure rise. You do get somewhat of a lag if you define that as the time between ignition and first significant pressure rise.

HYDRONITROGENS AS FUTURE AUTOMOTIVE FUELS

E. W. SCHMIDT

Rocket Research Corporation, Redmond, Washington

ABSTRACT

A group of synthetic fuels called hydronitrogens has been evaluated as automotive fuels for the time period beyond the year 2000. Hydronitrogen fuels are composed of hydrogen and nitrogen. As such, they can be synthesized from air and water without the use of fossil fuels. The main hydronitrogen fuels are hydrazine, N_2H_4, and ammonia, NH_3. Ammonia by itself has already been extensively tested by other investigators and was found to have poor combustion properties. No work has been reported to date on hydrazine combustion in internal combustion engines, but hydrazine burning velocity in air is expected to be higher than that of ammonia or hydrocarbons. This will result in more rapid and more complete combustion. When completely burned, and after removing eventually formed nitrogen oxides, hydronitrogen fuels would be non-polluting to the environment.

So far, other authors have considered ammonia or hydrazine for automotive fuels as pure substances only. However, the full advantages of hydronitrogen fuels can best be achieved in binary or ternary mixtures of hydrazine with ammonia and/or water, which have freezing points as low as -65°F. The selection criteria for hydronitrogen fuel mixtures will be discussed.

The paper summarizes the preparation of hydronitrogen fuels, production statistics, adaptability and performance in automotive engines, handling, safety and materials compatibility. The results of an evaluation matrix of hydronitrogens versus other non-conventional fuels for the time period beyond the year 2000, are discussed. The results of preliminary tests at Rocket Research Corporation with hydrazine and hydrazine mixtures in a single-cylinder internal combustion engine will be presented.

INTRODUCTION

The term "hydronitrogen" fuels has been coined in analogy to the more familiar category of hydrocarbon fuels, to which the currently used gasoline and distillate fuels belong. Hydronitrogen denotes the fact that in these fuels the carbon has been replaced by nitrogen. Nitrogen is the main constituent of the air which we breath and it is more abundant than fossil fuels, currently our main source of carbon.

Of all theoretically possible hydronitrogens only two are stable enough to be considered as automotive fuels: ammonia and hydrazine. The other hydronitrogen compounds are very unstable and are not suitable as fuels. By analogy to the hydrocarbon family (methane), it is not the lowest member of the hydronitrogen family (ammonia) which offers the greatest utility as an automotive fuel. Instead, molecules with more than one carbon or nitrogen atom, gasoline or hydrazine, are more promising fuels. In particular, they are more easily handled because they are liquids at ambient temperature.

Ammonia and hydrazine have already been proposed as automotive fuels (1) (2). Extensive experimental testing of ammonia as a fuel in internal combustion engines has been conducted (2) (3) (4) (5), but to date no experimental tests have been reported for hydrazine. Based on its background in the use of hydrazine and hydrazine mixtures as rocket propellants, Rocket Research Corporation (RRC) has conducted an experimental and analytical evaluation of hydrazine-based hydronitrogen fuels as automotive fuels. Some of the tests were only of a preliminary nature, but the results indicate that additional work is warranted.

One of the earliest tests used a Cox model airplane engine which was operated both in an air-breathing and in a monopropellant mode without the use of ambient air for combustion. It was thought at the time that this feature would make hydrazine suitable as a fuel for lunar roving vehicles. The ability of hydrazine to not only burn in air just as gasoline does, but also to decompose in a controlled mode in the absence of air makes it a very versatile fuel for a variety of applications. Future automobiles using hydronitrogen fuels may also benefit from this possibility.

Currently the only published application of the earthbound propulsive use of hydrazine is a remotely piloted research airplane called MINI-SNIFFER (6). This research airplane can reach an altitude of 30.4 km and will be instrumental in the analysis of ozone and pollutant profiles of the upper atmosphere. Hydrazine is also used as a fuel in the CONCORDE SST and the Space Shuttle auxiliary power units.

WHY HYDRONITROGEN FUELS?

The proposal to use hydronitrogen fuels instead of hydrocarbon fuels may sound absurd at a time when proven reserves of coal in the U. S. appear to be sufficient to satisfy energy and synthetic fossil fuel requirements for several hundred years.

References p. 339.

However, the question remains if it would be environmentally desirable to mine and burn all fossil fuels just for their energy value. Very soon not only petroleum, but also coal may have to be reserved as a feedstock for the chemical processing industry where alternates are not as easily found as in the primary energy sector. The availability of comparatively cheap electric power from nuclear fusion or solar power plants would change the energy picture in favor of synthetic nonfossil fuels, which can be prepared from water and air when such electric energy is available.

Furthermore, we should not restrict our point of view to the United States which is fortunate enough to have such a large supply of coal. Instead, the fossil fuel shortage may impact other countries sooner than the United States. There may be important economic and political advantages in the long run in developing nonfossil fuel technology in the U. S. and exporting this technology to other countries where the synthetic fuel age is likely to start earlier.

PREDICTED TRANSPORTATION FUEL DEMAND

Forecasts of transportation fuel production and consumption in the U. S. and worldwide are shown in Fig. 1. If the U. S. transportation petroleum demand were allowed to continue to increase at its current rate, it would exceed the total world production in the year 2020, obviously an impossible situation. It has been predicted to exceed the U. S. domestic oil production within a few years (7).

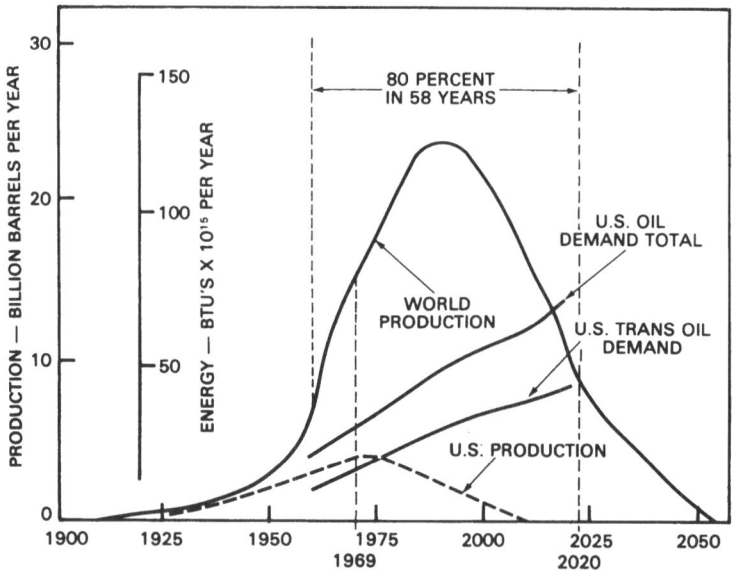

Fig. 1. World supply and U. S. demand for oil.

Another forecast on nonfossil fuels (8) has greatly stimulated our interest in hydronitrogen fuels (Fig. 2). Even though the time scale on this chart may be too

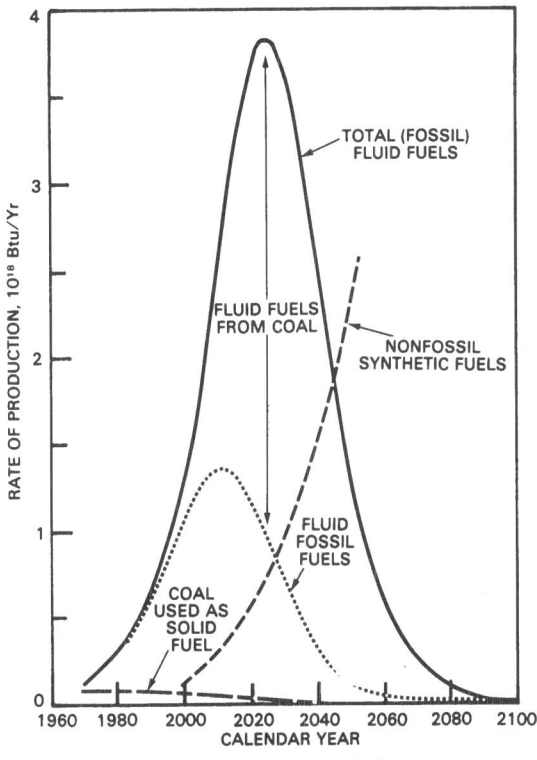

SOURCE: ATOMIC ENERGY COMMISION, TID 26136

Fig. 2. Production of fossil and nonfossil fuels (total world estimate).

compressed (in particular with regard to the duration of fluid fuels from coal) it serves to illustrate that beyond the fluid-fuels-from-coal age there is bound to be a transportation economy based on non-fossil synthetic fuels. It may not be too early to prepare for this time now in an effort to extend fossil fuel resources.

PHYSICAL PROPERTIES OF HYDRONITROGEN FUELS

Table 1 summarizes some of the physical properties of hydronitrogen fuels in comparison to other nonconventional fuels and n-octane. n-Octane has been chosen as a baseline for comparison and as a representative for gasoline. We are well aware that actual physical properties of gasoline, such as boiling range and density, vary significantly from those shown in the table. It is common practice to evaluate fuels based on their heating value. Thus, in terms of BTU/lb, hydrogen obviously has the highest heat of combustion and this is one reason that much work has been conducted on hydrogen in automobile engines. However, hydrogen in either the liquefied, compressed, or hydride state has a very low density. The storage of hydrogen in an automobile would require a very bulky storage tank which would occupy a large portion of the space which is usually allocated for passengers and baggage.

References p. 339.

TABLE 1

Physical Properties of Nonconventional Fuels

	Methanol	Ethanol	Liquid Ammonia	Hydrazine	Liquid Hydrogen	For Comparison: n-Octane (Gasoline)
Melting point, °F	-144	-174	-107.9	34.7	-434.6	-70.2
Boiling point, °F	148.5	173.1	-28.0	237.6	-423.0	258.1
Heat of formation at 77°F kcal/mole	-57.036	-66.36	-10.97(G)	+12.054	-1.887(L)	-59.74
Lower heat of combustion Btu/lb	8,570	11,507	7,534	7,169	49,920	19,089
Upper heat of combustion Btu/lb	9,752	12,740	9,201	8,348	59,312	20,592
Heat of vaporization at normal boiling point Btu/lb	473.6	360.2	588.2	583.0	191.8	129.35
Liquid density at 77°F g/cm^3	0.7821	0.7894	0.6819	1.0037	0.0708*	0.69849
lb/cu. ft.	48.82	49.28	42.57	62.659	4,420	43.60
Vapor pressure at 77°F psia	2.31	0.85	145.4	0.274	–	0.29

At -424°F

In view of the fact that an automobile is a very space-limited vehicle, energy storage density is an important evaluation criterion. Designing a more bulky car for carrying hydrogen will inevitably increase aerodynamic losses which have not been included in many performance comparisons. A more realistic way of evaluating nonconventional fuels is therefore in terms of the volumetric heating value as shown in Fig. 3. The unit used for comparison here is thousands BTU/cubic foot instead of BTU/pound. As seen on this chart, gasoline is indeed the most desirable fuel in terms of energy density. However, right next to it and even higher than the widely proposed methanol is hydrazine, our key hydronitrogen fuel. Unfortunately, the liquid range of hydrazine is very similar to that of water, and the freezing point of 35°F imposes some restrictions on the geographical areas where pure hydrazine could be used as a fuel.

Freezing point considerations and other operating parameters make it advisable to add an antifreeze to hydrazine in order to use it as an automotive fuel. The choice of antifreeze additives is limited to compounds which are miscible with hydrazine and which will not result in unacceptable pollution problems. The most effective freezing

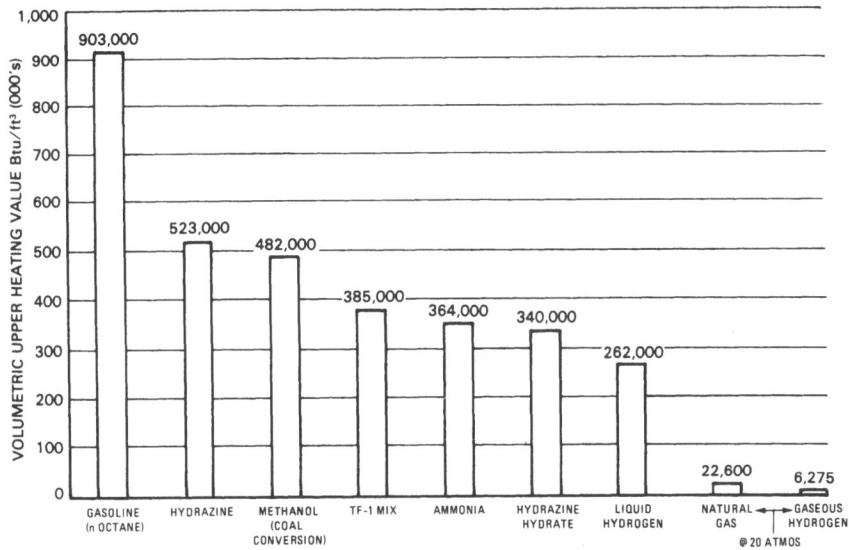

Fig. 3. Comparison of nonconventional fuels.

point depressing additives for hydrazine are water and/or ammonia. Both additives lower not only the freezing point (desirable), but also the volumetric heating value (undesirable). Depending on the freezing point required, binary and ternary mixtures of hydrazine with ammonia and/or water have been considered and tested at RRC.

The freezing point diagram of the ternary system hydrazine/ammonia/water is shown in Fig. 4. The lowest freezing point required for numerous military applications is $-65°F$. Consequently, a ternary fuel designated TF-1 composed of 64% hydrazine, 10% ammonia, and 26% water has been selected for initial testing. TF-1 is also shown in Fig. 4 and its heating value, based on equal volume, is 20% less than that of methanol. However, there is a tradeoff between performance and freezing point, and more energetic mixtures can be composed if a freezing point higher than $-70°F$ can be tolerated.

TF-1 and similar higher freezing mixtures were selected in the area of maximum density in the ternary system hydrazine/ammonia/water, an unexpected but desirable minor advantage in the fuel selection. It would be desirable to show a similar ternary diagram for the octane numbers of these synthetic fuel mixtures. However, these data are not available. Preliminary indications are that pure hydrazine/air mixtures have a very low octane number and tend to knock. This tendency can be averted by adding water and/or ammonia. Ammonia is known to have a very high octane number. Thus, along the N_2H_4/NH_3 side of the diagram we have an entire gamut of octane numbers from very low to very high. Along the N_2H_4/H_2O side of the diagram, the octane number increases until the composition reaches the limit of noncombustible mixtures

References p. 339.

in the water corner of the ternary system. Hydronitrogen fuels can thus be blended to obtain any desired intermediate octane number. Again, there is a tradeoff between octane number and energy content of the mixture.

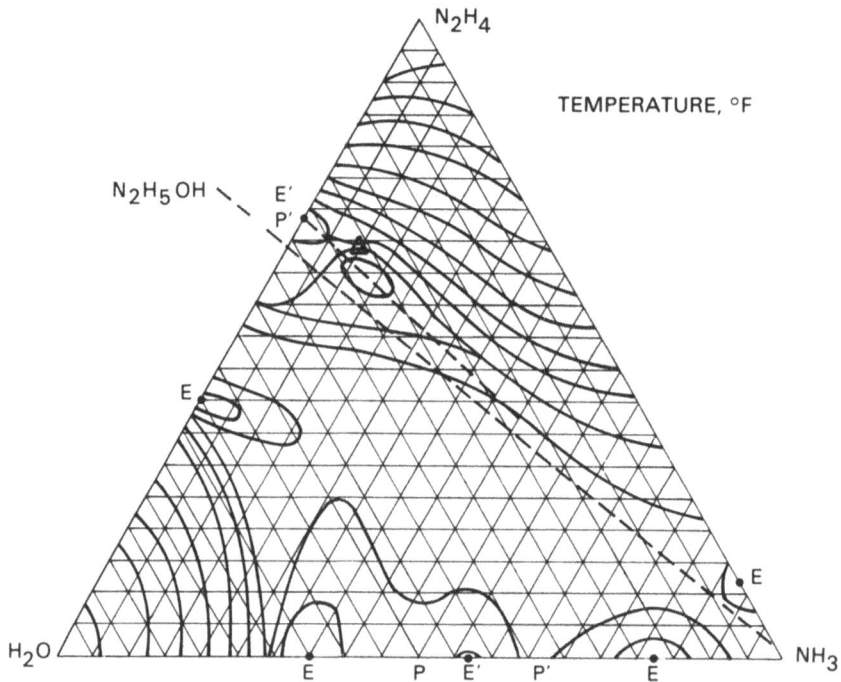

Fig. 4. Freezing point diagram of the ternary system hydrazine/ammonia/water.

When evaluating hydronitrogen fuels and, in particular, hydrazine as a hydrogen carrier and energy source, it is important to realize that the storage density of hydrogen in hydrazine N_2H_4 is actually almost twice as high as in liquefied hydrogen itself: 0.12 g/cm^3 as compared to 0.07 g/cm^3 in liquid hydrogen. Furthermore, because hydrazine has a positive enthalpy of formation, the heat of combustion is higher than that of a mixture of 12.5% hydrogen and 87.5% nitrogen. Nitrogen does therefore not simply act as a diluent of hydrogen when hydrazine is considered as a hydrogen carrier.

PREPARATION OF HYDRONITROGEN FUELS

Ammonia is currently produced in bulk quantities and its main use is as a fertilizer. The 1973 U. S. production was 15.5 million short tons and the 1972 world production was 63.5 million short tons. Additional capacity is coming on stream every year, but the natural gas or naphtha feedstock supply for some of the ammonia synthesis plants is becoming uncertain. Obviously it would be better to convert these to a liquid hydrocarbon fuel instead of ammonia if one wanted to synthesize transportation fuels. The same is true for hydrazine which is currently derived from ammonia which in turn is made from hydrocarbon feedstock.

The future implementation of hydronitrogen fuels as energy carriers hinges therefore on the development of a synthesis which will make hydrazine production independent of ammonia as an intermediate. The most promising development in this direction is a chemical reaction called nitrogen fixation. Nitrogen fixation occurs naturally in bacteria hosted in the roots of certain plants (e.g., leguminosae) and plays an important role in maintaining the nitrogen balance in agricultural crop land. Attempts to reproduce this reaction "in vitro" in the laboratory were not successful until a few years ago. The main objective of these tests was to find a new method for producing ammonia. Surprisingly, hydrazine instead of ammonia was obtained by some investigators and was initially considered as an undesirable by-product. With a little more specific research in this area it may well be possible to develop a synthesis of hydrazine literally based on water and air as feedstock, consuming electrical energy to perform a recycling reduction of one of the reactants. So far this nitrogen fixation reaction has only been demonstrated in the test tube scale (9) and is still many years away from industrial utilization. It may be assumed that electric power required in this process will be available from nuclear fusion reactors after the turn of the century.

Currently, the annual production of hydrazine in the U. S. is 23 million pounds, but only a small fraction of this is used as anhydrous hydrazine. The majority is processed to other intermediates for herbicides or foam blowing agents without concentrating the dilute product solution. The current U. S. hydrazine production capacity is insufficient to make any contribution to hydrazine as a transportation fuel. As a matter of fact, it is insufficient even to meet current demand in a rapidly growing market of new applications. The implementation of a hydronitrogen fuel based transportation economy will therefore require the build-up of a huge synthetic fuel industry.

HANDLING OF HYDRONITROGEN FUELS

Because the public will be exposed to nonconventional fuels in the future in much the same way as it is now to gasoline, the potential hazard of gasoline substitutes has to be thoroughly evaluated. On the other hand, throughout this evaluation one must keep in mind that in spite of its widespread use, gasoline is a hazardous chemical, in particular with regard to flammability. Even though the hydronitrogen fuels are admittedly more aggressive with respect to inhalation and skin irritation, it is believed that with advanced technology and adequate instruction of the public these fuels can be handled safely. Ammonia and hydrazine have a long standing handling history in agricultural and industrial usage. Although accidents involving ammonia or hydrazine have occurred occasionally, their number in relation to the amount of chemicals

References p. 339.

handled is small in comparison to the large number of injuries or fatalities due to gasoline mishaps. This comparison must not be made on a number basis alone, but must always be based on the amount of chemicals being handled. Once the transportation system has switched to hydronitrogen fuels, and people have been properly educated, we predict an actual decrease in the number of fuel-related accidents. In particular, the more obvious odor of ammonia will warn the operator of fuel leakage. The numerous accidents happening with gasoline siphoned from cars for other purposes (cleaning agent, fire starter) will no longer occur.

A thorough safety evaluation will include toxicity (inhalation, ingestion, skin contact) and flammability of substitute fuels. For evaluating inhalation hazard during handling, it is common practice to use the Threshold Limit Values (TLV's) established by the American Conference of Governmental Industrial Hygienists, and the substance's vapor pressure. The TLV data are established for the working population for an eight hour/day, 40 hour/week occupational exposure. The data may not be directly applied to the general public because of the different population involved. On the other hand, the public would not be exposed to fuel vapors as frequently as a service station attendant. A hazard comparison based on TLV alone is misleading because one compound may be less volatile than another. Instead, the so-called hazard index, formed by dividing the vapor pressure by the TLV, should be used for a hazard evaluation.

The hazard indices of gasoline and some other fuels are illustrated in Fig. 5. Admittedly, hydrazine and ammonia are more toxic than gasoline, but this is one of the risks one has to take in finding an alternative fuel for the time when gasoline may no longer be available. TF-1 as a mixture of hydrazine, water, and ammonia will have a hazard index between that of hydrazine hydrate and ammonia, most likely in the order of 20 mm Hg/ppm. Yet, this is only one order of magnitude higher than benzene and other aromatic fuels. Synthetic hydrocarbons prepared by direct coal hydrogenation are expected to have a higher aromatic content than petroleum-derived gasoline and their hazard index could be close to that of benzene.

The inhalation safety evaluation of nonconventional fuels should not be restricted to the fuel itself, but should also include the exhaust products. In spite of efforts to eliminate carbon monoxide emissions from internal combustion engines, this danger will persist for many more years. Motor vehicle exhaust is a leading cause of death by poisoning. In 1970, 803 deaths out of a total of 1,620 gas poisonings were due to motor vehicle exhaust gases while the vehicle was standing. In contradistinction to gasoline, hydronitrogen fuels burn cleanly to nitrogen and water (traces of NO_x possible), and completely eliminate the carbon monoxide poisoning problem. This advantage is certainly also shared with hydrogen as a motor fuel.

Another significant environmental advantage of hydrazine vapor which may escape into the air or rinse waters is that it is not very persistent and will not accumulate. It will readily autoxidize to nitrogen and water when exposed to sunlight and air, or air and water. Dilute solutions of hydrazine are decontaminated in a trickle flow reactor over activated charcoal.

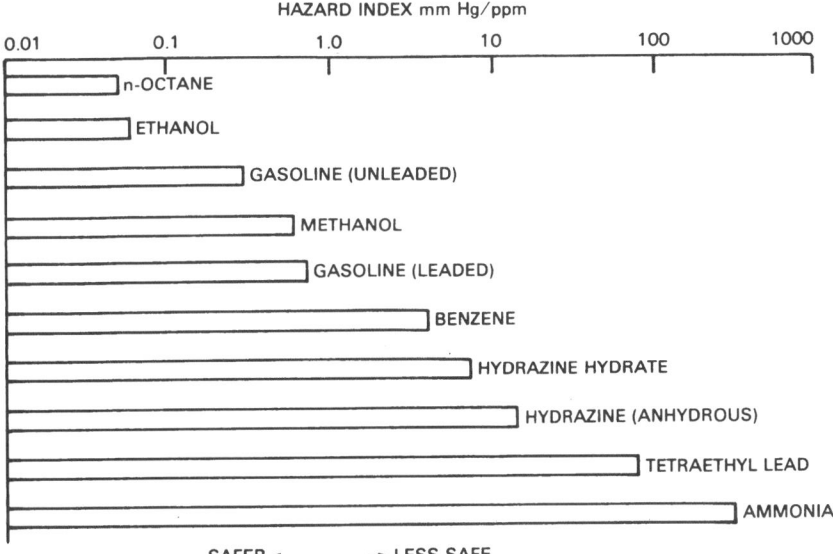

Fig. 5. Hazard index of some fuels and related chemicals.

The chance of inadvertent ingestion of motor fuel by drinking is significantly smaller for a fuel which has a distinct warning odor such as ammonia in TF-1.

The skin and eye hazard of hydronitrogen fuels is higher than that of gasoline, but less severe than that of battery acid which every automobile mechanic is accustomed to handling. The hazard involved in handling TF-1 is comparable to that of handling household ammonia (10% aqueous solution of ammonia).

Flammability and fire hazard of current and anticipated automotive fuels is an important evaluation criterion. The evaluation of fire hazard of fuels is based on physical properties which can be determined in laboratory tests and can be expressed in numbers, such as flash point and limits of flammability. These numbers are important not only for safety evaluation, but also for the design of an ignition system for an internal combustion engine burning these fuels. The flash points of gasoline and nonconventional fuels are shown in Fig. 6. The more volatile constituent of TF-1, the 10% by weight ammonia, will determine the flash point of hydronitrogen fuel mixtures. However, one should not only look at the flash point, but also at the flammable range and ignition energy required. The flammable range of ammonia (Table 2), ranges from 15 to 28% by volume. Ammonia is not very combustible, and ammonia flames will extinguish by minute turbulence in the air. Ammonia flames will only propagate upward but cannot burn downward. Furthermore, the required spark ignition energy of stoichiometric ammonia/air flames is extremely high, 680 mJ compared to 0.4 mJ for n-hexane or 0.3mJ for n-heptane. One of the main disadvantages of hydrogen as an automotive fuel is its extreme sensitivity to spark ignition during handling, the required ignition energy being only 0.02 mJ. In terms of fire hazard, hydronitrogen fuels compare very favorably with other alternative fuels.

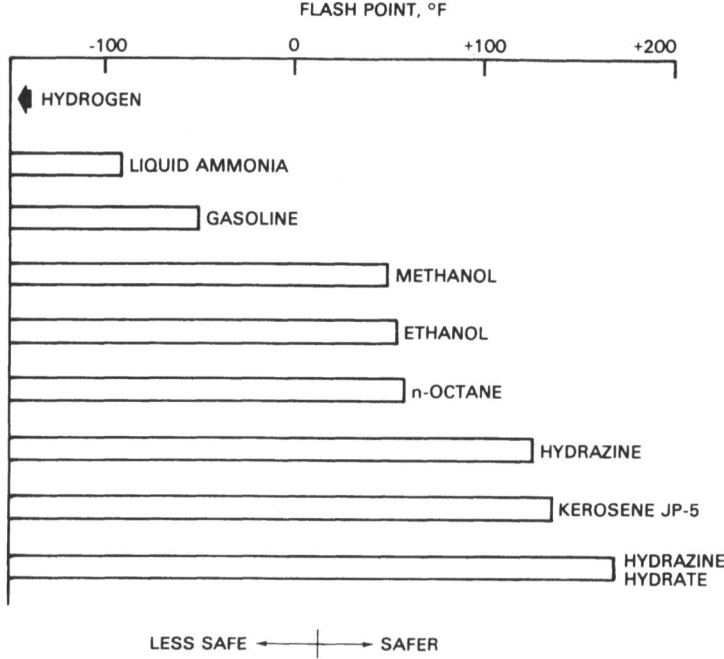

Fig. 6. Flash point of hydrocarbon and nonconventional fuels.

Spills of hydronitrogen fuels are easily decontaminated and fires are easily extinguished with water because the fuels are completely miscible with water. Once diluted below 40% in water, hydrazine will not burn at all.

Hydronitrogen fuels are not compatible with a number of metals and materials which are commonly used in today's gasoline-driven automobiles. In particular, corrosion problems must be expected with brass, copper, zinc, magnesium and aluminum alloys. This incompatibility has somewhat delayed demonstration tests with hydronitrogen fuels because of the extra effort required in converting or protecting conventional fuel pumps and carburetors. However, hydronitrogen fuels are compatible with 300- or 400-series stainless steels and a large number of polymers which cannot be used with gasoline.

TABLE 2

Comparison of Fire Hazard Properties in Air

	Limit of Flammability % by vol.		Stoichiometric Composition % by vol.	Flash Point °F	Autoignition Temperature In Glass °F
	lower	upper			
Gasoline	1.2	7.1	$\sim$ 1.5	-50	878
Kerosene JP-4	1.3	8	$\sim$ 0.9	134*	464
Methanol	6.7	36	12.25	$\sim$ 60	725
Ethanol	3.3	19	6.53	$\sim$ 57	689
Ammonia	15	28	21.83	>-80	1562**
Hydrazine	4.7	100	17.32	126	518***
TF-1				$\sim$ 100	Unknown
Hydrogen	4.0	75	29.53		752

* JP-5
** In Steel: 1204°F
*** In Steel: 315°F

Data from reference 11.

ENGINE OPERATION

Engine operation considerations discussed here are primarily directed toward internal combustion engines, even though hydronitrogen fuels also offer the unique potential of being used in other types of engine drives, e.g., electric motors with fuel cells. Topics to be discussed include the type of engine in which hydronitrogen fuels can be burned, the mode of introducing and igniting the fuel, the startup procedures, the possible compression ratios, specific fuel consumption and potential emission problems. Unfortunately, the evaluation of hydronitrogen fuels is still in its infant stage and answers to all these questions are not yet available.

Prior to beginning experimental engine evaluation of hydronitrogen fuels, the theoretical combustion temperature and pressure for adiabatic and isochoric combustion of fuel-air mixtures have been calculated at RRC using the NASA/Lewis chemical equilibrium computer program. As shown in Fig. 7, the maximum flame temperature for all fuels occurs at the stoichiometric equivalence ratio. The theoretical flame temperature of TF-1 is lower than that of the other fuels. However, temperature alone must not be the sole measure of energy content. The higher heat capacity and heat capacity ratio of steam in the combustion products makes the work which can be performed by expanding the gas higher than one would expect from the flame temperature alone. The same effect has been observed when injecting water into hydrogen-fueled internal combustion engines. A high flame temperature is not desirable because it will increase engine wear and formation of NO_x.

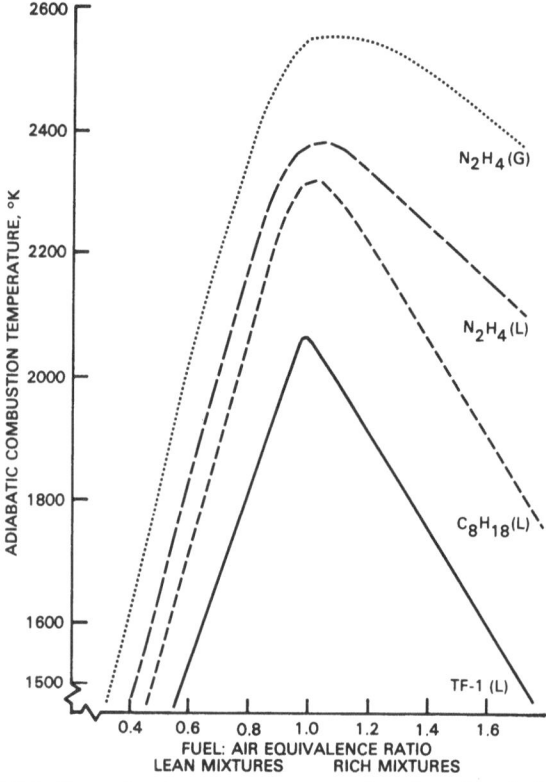

Fig. 7. Adiabatic combustion temperature as a function of equivalence ratio.

A similar set of calculations was performed for peak pressure at constant volume combustion, Fig. 8. The initial pressure should have been chosen higher than 1 atmosphere, depending on the compression ratio. However, repeating this calculation for higher initial pressure would only shift the curves as a group and the relative position is expected to remain the same. Again, peak pressure of TF-1 is slightly below that of octane or hydrazine. Using these data, one could calculate mean effective pressure and theoretical thermal efficiency for the Otto engine combustion cycle.

A large number of engine tests have been conducted with ammonia and some conclusions about TF-1 can be drawn by similarity. However, TF-1 and similar hydrazine mixtures promise to solve certain operating problems discovered during tests with ammonia alone. One advantage of ammonia which became apparent during these early tests was that it can be operated at very high compression ratios (Table 3).

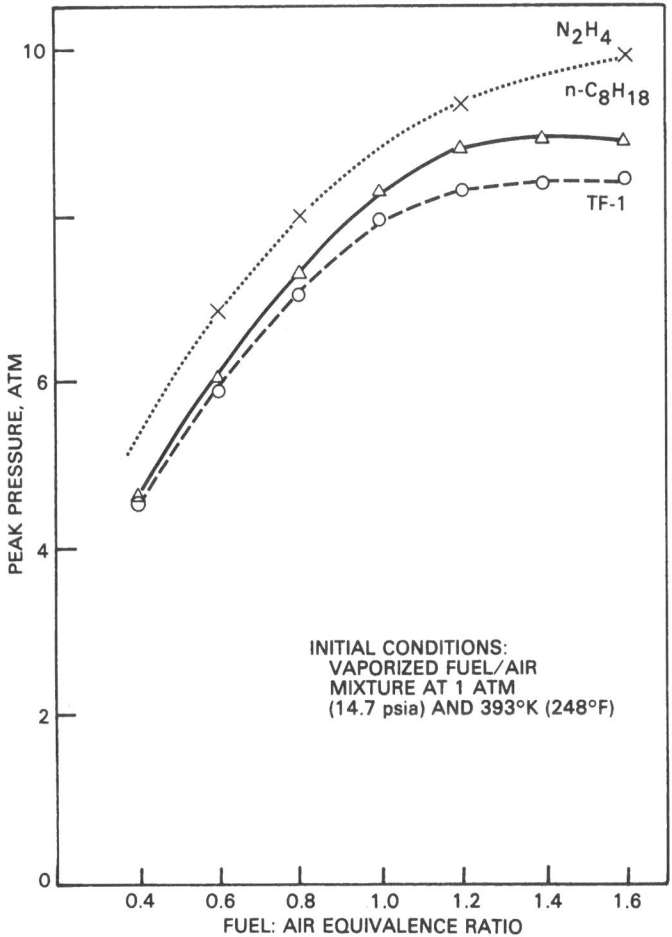

Fig. 8. Peak pressure for adiabatic constant volume combustion.

However, ammonia was difficult to ignite, and stronger ignition coils or dual spark plugs were required. Furthermore, it was very difficult to achieve complete combustion because the flame velocity of ammonia is low (1), and the quenching diameter is large. Hydrazine has a higher flame velocity and a smaller quenching diameter and is expected to result in more complete combustion. Only trace amounts of ammonia could be detected in the exhaust of TF-1 engine tests conducted at RRC. These could have originated during the fuel transition and may have adhered to the exhaust line with other condensation products.

TABLE 3

Compression Ratio for Ammonia Fuel

Author	Range Investigated	Recommended C.R.
Starkman et. al. (1)	8 – 10	
Gray et. al. (5)	8 – 16	
Cornelius et. al. (2)	9.4 – 18	12 (supercharged)
Graves et. al. (12)	10.2	
Sawyer (3)	10.0	
Pearsall and Garabedian (13)	12 – 30	12 – 16

Until a more systematic fuel evaluation program using CFR single cylinder research engines can be conducted, RRC has performed a preliminary hydronitrogen fuel demonstration in a single-cylinder four-cycle internal combustion engine driving an electric generator. The electric generator was instrumented as a dynamometer, and specific fuel consumption was measured for different electric load conditions. The engine was instrumented to measure inlet pressure, cylinder pressure, cylinder temperature, and fuel consumption. Hydrazine was injected directly into the throat of the carburetor after starting the engine on gasoline. The transition from gasoline to hydronitrogen fuel was smooth and no carburetion problems were encountered. So far, four binary hydrazine/water mixtures with 0, 10, 20, 30 and 40% water, and the ternary fuel TF-1 have been evaluated. Specific fuel consumption of binary mixtures expressed as pound of fuel consumed per horsepower-hour, passed through a minimum between 20 and 30% water (Fig. 9). The lowest specific fuel consumption of all hydronitrogen fuels was obtained with TF-1, although the SFC was more than twice as high as that of gasoline (Fig. 10).

These tests were only of a preliminary nature and additional tests in multicylinder engines are planned. Exhaust samples will be analyzed in those tests for a variety of air fuel ratios and loads.

A unique feature of hydrazine-based hydronitrogen fuels is the self-starting capability of these fuels in a monopropellant mode. Decomposition of hydrazine to

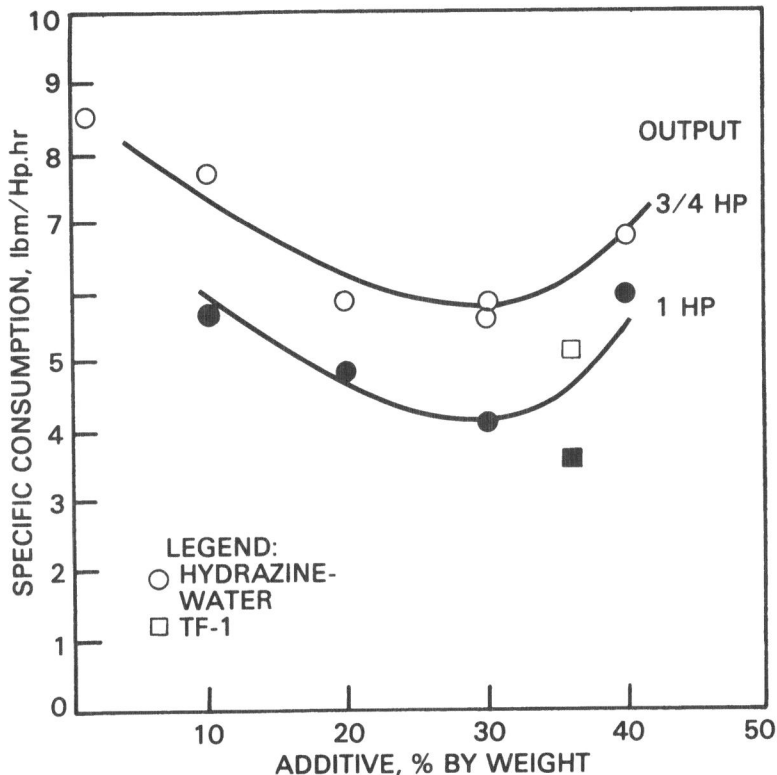

Fig. 9. Specific fuel consumption of hydrazine mixtures.

ammonia, hydrogen, and nitrogen according to

$$N_2H_4 \rightarrow \frac{4}{3}(1-X)\,NH_3 + \frac{2X+1}{3}\,N_2 + 2\,XH_2 + (1499 - X \cdot 823)\ BTU/lb$$

in which
X - ammonia dissociation fraction
X for TF-1 is expected to be in the range of 0.4 to 0.6

takes place on the surface of certain catalysts, providing hot gas at high pressures without the need of air or oxygen for combustion. As illustrated in Fig. 11, a starter fuel tank would contain a small amount ($\sim$ 2 ounces) of TF-1. An initial amount would be fed into the gas generator by opening the starter valve and depressing the piston which could be coupled with the accelerator pedal. Because it is undesirable to store the fuel under pressure, a bootstrap system could be used. In this fashion, a small fraction of the exhaust gas of the gas generator is branched off to exert pressure on a differential area piston. The piston area on the fuel side is smaller than the piston area on the hot gas side, thus augmenting the feed gas pressure to

References p. 339.

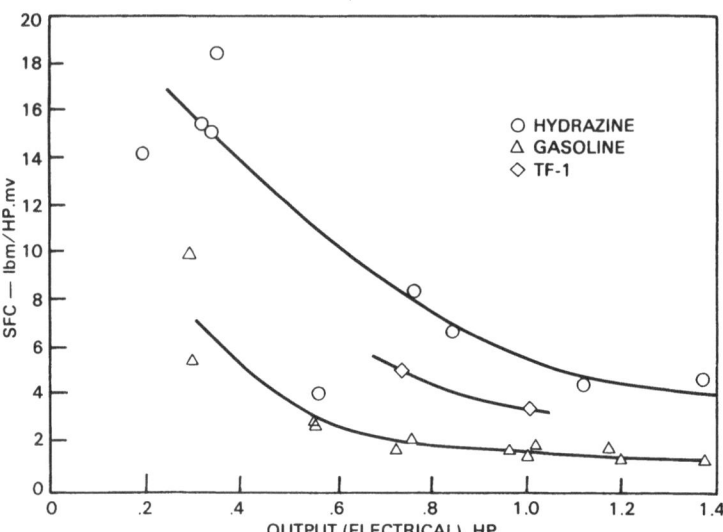

Fig. 10. Measured fuel economy as a function of output.

overcome the pressure drop in the reactor. The starter fuel reservoir would fill itself during engine operation and remain filled after the engine is turned off. When the engine is started, the starter valve is opened and the three-way valve at the inlet valve of at least two cylinders (only one shown) is set to position "S." Obviously some provision has to be made to overcome the dead center point of the cylinder which is first pressurized with the starting gas. The starting of diesel engines with solid propellant gas generator cartridges is routinely carried out in cold climate areas, using the same principle. Because it would be undesirable to vent the hot exhaust of the starting cylinders, it could be mixed with air and fed to the remaining cylinders which are about to commence operation in the air-breathing spark-ignited mode. In these cylinders, the hot gas would aid in the evaporation of fuel and its hydrogen content would make the initial mixture more flammable. After the air-breathing cylinders are operating, the three-way valve would be switched to the "O" position, to allow the starting cylinders to join the rest of the cylinders in an air-breathing mode.

The potential weight saving by eliminating the starter motor and part of the battery weight is evident from Table 4, showing component weight fractions of some typical 1975 engines. Eliminating the starter and cutting the weight of the battery and generator in half because of the reduced load requirements ("Plan A") would save 9% of the weight of a typical four-cylinder engine. The weight of the gas generator and controls replacing the starter would be less than two pounds. If the battery and generator were replaced by a hydrazine-air fuel cell ("Plan B"), even more, namely approximately 12% of the L-4 engine weight could be saved depending on the weight of the fuel cell. Light-weight hydrazine-air fuel cells have been developed by the U. S.

Army. The mileage of an automobile is expected to improve if the engine does not have to drive a generator all the time, regardless if it is charging the battery or not. However, some or all of the weight savings described above may be lost due to the greater fuel weight if hydronitrogen fuels are used to replace gasoline.

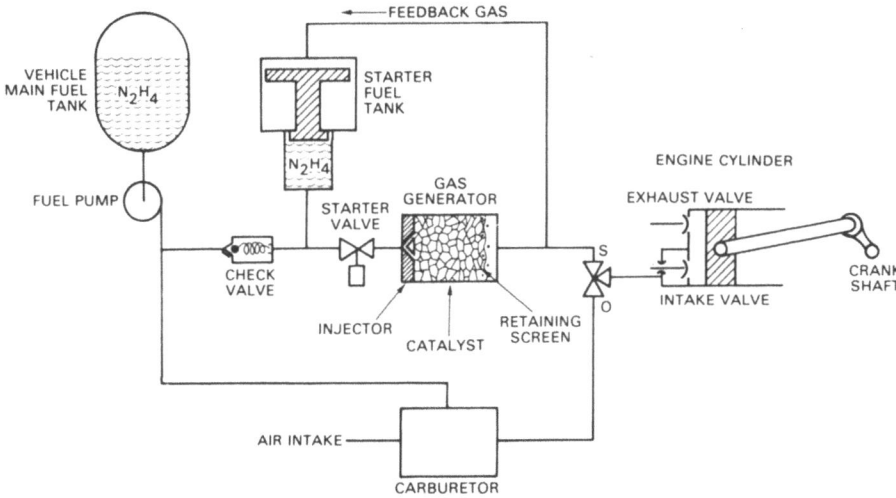

Fig. 11. Schematic of monopropellant self-starting internal combustion engine.

TABLE 4

Weight Breakdown of Current Automobile Engines

Basic Engine		Weight — Pounds				Weight Saving — %	
		Engine plus Transmission*	Battery	Starter	Alternator	Plan A	Plan B
Vega 140	1-bbl L-4	336	31.0	15.7	10.0	9.2	14.4
Buick 231	2-bbl V-6	471	33.2	18.8	10.0	7.6	11.6
Oldsmobile 350	4-bbl V-8	657	29.0	19.4	10.2	5.4	8.2

*Includes all items on or in engine and transmission when installed in car; such as oil, coolant, starter and alternator. Does not include battery or radiator.

References p. 339.

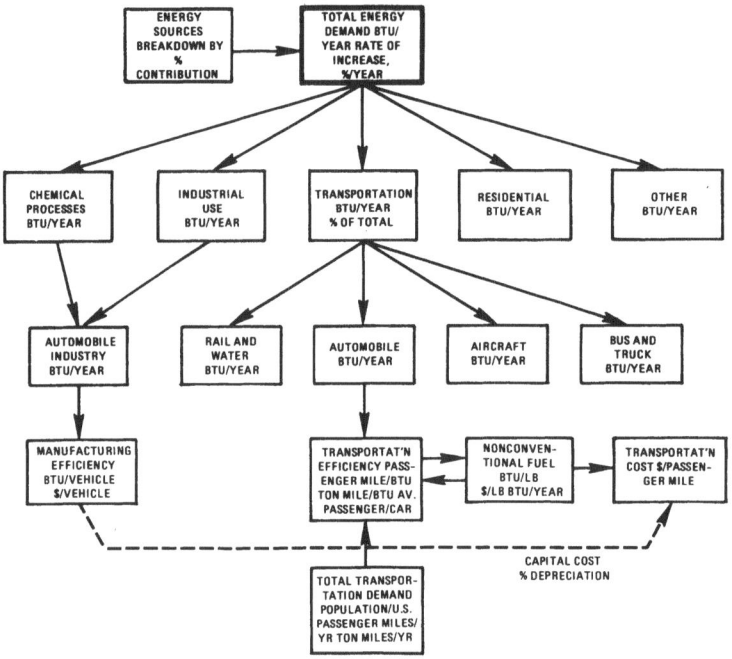

Fig. 12. Simplified economic model — transportation energy.

CONCLUDING COMMENTS

Analytical studies and preliminary experimental tests have identified hydronitrogen fuels as viable candidates for the post-fossil fuel era for automotive propulsion. Additional testing is required to clearly answer questions with regard to specific fuel consumption and thermal efficiency in an internal combustion engine. The implementation of a hydronitrogen fuel transportation economy depends on the realization of novel synthesis methods which would allow economically competitive production of fuels from air and water (with electrical energy provided by another source). In particular, it depends on the degree to which the hydronitrogen system is competitive with other synthetic fuel candidates.

It is proposed to develop an economic model (Fig. 12) which will enable us to predict the crossover point of hydronitrogens versus hydrocarbons as a function of time. Input data into this model would be the predicted escalation in prices of hydrocarbon fuels from various sources, the source mix, and the predicted rate of consumption. Important parameters would include the energy mix, the energy cost for the production of synthetic fuels, and the possible benefits to be derived from utilization of waste heat or dual purpose plants. The ultimate evaluation should be based on dollars/mile driven. The price of a vehicle and its lifespan will also enter into this evaluation. Primary emphasis would be on passenger automobiles, but other modes of transportation will also have to be included. A common fuel for all vehicles would result in simplification of the distribution system over today's systems with three types of gasoline, and several types of diesel and jet fuel to be shipped, stored, and dispensed separately.

ACKNOWLEDGEMENT

Part of the fuel evaluation work conducted herein was sponsored by General Motors Corporation under Contract R-68600. The experimental testing of hydrazine as automotive fuel at Rocket Research Corporation was conducted by D. R. Poole (1969) and F. X. McKevitt (1974).

REFERENCES

1. E. S. Starkman, et al., "Ammonia as a Spark Ignition Engine Fuel: Theory and Application," SAE Paper No. 660155, Society of Automotive Engineers.

2. W. Cornelius, L. Huellmantel and H. Mitchell, "Ammonia as an Engine Fuel," SAE Paper No. 650052, SAE Transactions, 74, 300-15, 1966.

3. R. F. Sawyer, et al., "Oxides of Nitrogen in the Combustion Products of an Ammonia-Fueled Reciprocating Engine," SAE Paper No. 680401, Society of Automotive Engineers.

4. R. Sutton and E. S. Starkman, "Oxides of Nitrogen in the Engine Exhaust with Ammonia Fuel," California University, Berkeley, Report TS-66-4, TR-7, Contract DA 04-200-AMC-791-X, June 1966, AD 640444, N67-21718, p. 27.

5. J. T. Gray, et al., "Ammonia Fuel — Engine Compatibility and Combustion," SAE Paper No. 660156, 1966.

6. R. D. Reed, "RPRV's — The First and Future Flights," Astronautics and Aeronautics 26-42, April 1974.

7. M. K. Hubbert, "Energy Resources for Power Production, in: Environmental Aspects of Nuclear Power Stations," International Atomic Energy Agency, Vienna, 1970, pp. 13-43.

8. Synthetic Fuels Panel: "Hydrogen and Other Synthetic Fuels," AEC TID 26136 N73-33738, September 1972.

9. D. V. Sokolskii, et al., "Reduction of Nitrogen to Hydrazine with Zinc in Presence of Vanadium and Magnesium Compounds," Russ. J. Gen. Chem., 42, 1415-7, 1972.

10. E. S. Starkman and G. S. Samuelsen, "Flame-Propagation Rates in Ammonia-Air Combustion at High Pressure," 11th Symp. Comb., A67-33844, 1967, pp. 1037-45.

11. M. G. Zabetakis, "Flammability Characteristics of Combustible Gases and Vapors," U. S. Bureau of Mines Bulletin 627 (1965).

12. R. L. Graves, et al., "Ammonia as a Hydrogen Carrier and Its Application in a Vehicle," Hydrogen Economy Miami Conference, pgs. SB-15 to SB-23 (March 1974).

13. T. J. Pearsall and C. G. Garabedian, "Combustion of Anhydrous Ammonia in Diesel Engines," SAE Paper No. 670938 (1967).

DISCUSSION

M. C. Hardin *(Detroit Diesel Allison Division, General Motors Corp.)*

You made a point that the catalytic monopropellant atomization system can eliminate the starter system. Can the TF-1 mixtures be catalytically activated? Does that mixture have a significant amount of propellant characteristics as far as its safety in general usage?

Schmidt

Yes, TF-1 is still a very powerful monopropellant in spite of the fact that you have added ammonia and water to the hydrazine. The flame temperature is still high enough that you get plenty of evaporation. I don't recall the dew point of the products, but it's at least 400-500°F so you would have equivalent to 400-500°F steam to drive the starter.

S. Gratch *(Ford Motor Company)*

I have three questions. First, you pointed to the low flame temperature as an indication of less formation of oxides of nitrogen. I feel that is only appropriate when the oxides of nitrogen are formed from the air. Do you have any basis to think the same would be true from pure nitrogen? I don't think this would be the case. Secondly, your comparison of toxicity based on volatility seems to me to be only appropriate for toxicity by inhalation. My understanding is that hydrazine is toxic also by absorption into the skin, and in that case I don't understand the relevance of volatility. Thirdly, you mentioned the weight saving because of the elimination of the battery. But, isn't it a fact that the weight of fuel consumed would be at least double for the TF-1 fuel, and more than sufficient to offset any weight advantage?

Schmidt

It is true that the volatility basis applies only to inhalation. But as far as the skin hazard is concerned, I'm always comparing it to battery acid, and automobile mechanics are trained to handle battery acids. So it would be just about the same ball park.

On the question of nitric oxide formation, we don't have any emission data yet. But the flame temperature is really lower than any other fuel that's currently available.

Unidentified

The fact that you have an abundant source of nitrogen in the fuel means that you don't need high temperatures to have lots of nitric oxide.

Schmidt

In answer to your third question, you do not consume very much fuel to start the engine. We used only about 2 oz.

Gratch

But, you have to use the fuel to run the engine and you have to carry that with you.

IMPACTS OF SYNTHETIC LIQUID FUEL DEVELOPMENT FOR THE AUTOMOTIVE MARKET

E. M. DICKSON and E. E. HUGHES

Stanford Research Institute (SRI) Menlo Park, California

ABSTRACT

This study has investigated the environmental, social, economic, and institutional impacts of producing synthetic fuels derived from coal and oil shale. From discussions with energy industry sources, the study concluded that the blending of synthetic and natural crude oils is a more realistic option than the independent development of a separate synthetic fuels industry. The study, therefore, focused on the resource—to—synthetic crude portion of the process chain because differences from present practice would be most evident there.

Because these sources of crude oil are not yet economically competitive — even with today's high oil prices — a key aspect of the study has been an examination of the profitability of the synthetic fuels enterprise, the business risks deriving from unstable policies and economic conditions, and the process by which decisions to implement the industry will be made.

The study has concluded that the following factors are especially critical to synthetic liquid fuel development — whether destined for automotive or other use:

- Reclamation of mined lands
- Possible water shortages in the arid West
- Air pollution from conversion plants
- Extremely rapid community growth rates in rural areas
- Socio-economic instability
- Conflicts between traditional local and new industrial interests

- Loss of local autonomy

- Economic risk mitigation for fuel producers.

INTRODUCTION

The technical feasibility to produce synthetic liquid fuels for automotive use has been demonstrated and evaluated by many companies and research workers. However exploration of the societal and environmental impacts of deployment of these synthetic fuels technologies has scarcely begun. We present here, in abbreviated form, some of the findings of our technology impact assessment of synthetic liquid fuels from coal and oil shale. The study (1), sponsored by the Environmental Protection Agency, was performed by an interdisciplinary team of physical scientists, engineers, economists, an ecologist, a sociologist, and an attorney.

The study has been concerned solely with synthetic gasolines, distillates, and methanol from coal; and synthetic gasolines and distillates from oil shale. Furthermore, to preserve the usefulness of large investments already in place, and to maximize flexibility in fuels processing and distribution, we emphasized synthetic crude oils rather than synthetic final products.

Figs. 1 and 2 show, in simplified form, the synthetic fuels system (based upon syncrudes) and the existing natural petroleum system, respectively. Industry has placed emphasis upon producing a syncrude that can simply be blended with supplies of natural crude. This blending means that the consumer of the final product has no way of knowing (and little reason of caring) whether the final product originated with oil or coal. Thus, the main impacts to be studied are those that result from resource extraction and conversion.

Unfortunately some of the impact areas and considerations relevant to the development of a synthetic liquid fuels industry that we studied can not be covered in this paper. The most notable omission is our extensive analysis of the impact upon air quality.

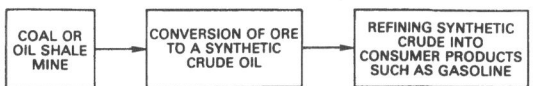

Fig. 1. Synthetic liquid fuels production system.

MAXIMUM CREDIBLE IMPLEMENTATION

Many speculations have been advanced in recent years concerning future levels of production of synthetic fuels from coal and oil shale. To set an upper limit on the possible impacts that would result from production of these fuels, we produced an implementation scenario depicting the maximum credible rate at which the synthetic

References p. 362.

liquid fuels industry could be expected to develop. This implementation scenario and its impacts are the main subjects of this paper. It is extremely important to recognize that this scenario is *not a prediction of what will occur* but is an attempt to elucidate the maximum possible impact situation.

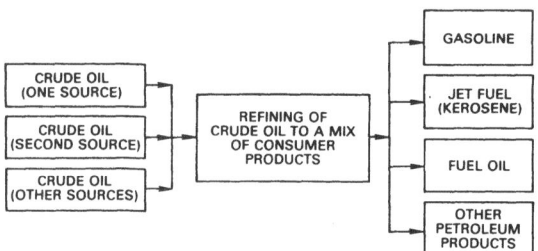

Fig. 2. Natural petroleum products production system.

The maximum credible implementation scenario is derived from a hypothesized growth schedule for a synthetic liquid fuels industry as presented in Table 1. The growth schedule indicates a slow start for synthetic liquid fuels with negligible production before 1985, followed by a rapid growth until the year 2000. The relatively slow start stems from the present situation in the oil industry regarding: (a) the increased activity to find and produce energy from conventional petroleum sources, and (b) the steady increase in cost estimates for synfuel plants. As a result, the oil industry can be expected to postpone construction of synthetic liquid fuel plants in favor of investment in more familiar resources.

The scenario projects accelerated growth for oil shale processing after 1980, and for the coal-based fuels after 1985. Such growth, of course, assumes that the first plants are successful, both technically and economically. This assumption is made solely to facilitate construction of a scenario that depicts the maximum rate at which an industry could be deployed subject only to physical and general economic constraints. Of course, other real world constraints, such as water availability, would lead to a lower actual rate of deployment.

The maximum credible implementation scenario reflects several judgements regarding the relative states of development of the three basic synthetic liquid fuel technologies (2,3). Oil shale technology is essentially ready for commercial deployment. Tests have been made on a scale large enough to confirm the feasibility of the technology and guide the design of a large plant. Future improvements in the technology (excluding the possibly significant case of *in-situ* technology) are not expected to be pronounced enough to render obsolete a plant begun today. Hence, our maximum credible scenario for oil shale shows two 50,000 B/D plants in 1980, and an addition of four 100,000 B/D plants by 1985. After 1990, oil shale production is shown leveling off as a reflection of anticipated water shortages.

TABLE 1

Hypothesized Growth Schedule of Synthetic Liquid
Fuels Industry

Fuel Description	Number of Plants				
	Year				
	1980	1985	1980	1995	2000
Syncrude from coal					
30,000 B/D* plant	0	3	7	7	0
100,000 B/D plant	0	0	3	13	40
Total production (10^6 B/D)	0	0.09	0.5	1.5	4.0
Syncrude from oil shale					
50,000 B/D plant	2	2	2	0	0
100,000 B/D plant	0	4	14	20	20
Total production (10^6 B/D)	0.1	0.5	1.5	2.0	2.0
Methanol from coal					
50,000 B/D plant	2	2	2	0	0
100,000 B/D plant	0	5	19	50	80
Total production** (10^6 B/D oil equivalent)	0.05	0.3	1.0	2.5	4.0

* *B/D = Barrels per Day*
** *To a close approximation, the energy content of a barrel of methanol is half that of a barrel
of oil.*

The commercial production of methanol and syncrude from coal are restrained
relative to oil shale to reflect the anticipated benefits of further research,
development, and demonstration work on processes of making syncrude from coal,
and the market uncertainties concerning introduction of methanol for large-scale use
as a fuel. The status of the technology for production of methanol from coal is similar
to that of syncrude from shale — basically ready for first generation commercial
production. The more advanced development of methanol compared with coal
syncrude production derives from the similarities of producing methane and methanol
from coal, and the greater attention that SNG technology has received in the last
decade compared with coal liquefaction technology.

In both the oil shale and the methanol cases, the actual realization of the schedules
of Table 1 required that present uncertainties be resolved soon in a way that
encourages development of synthetic fuels. Several recent events make it questionable
whether the maximum credible production levels for 1985 can still occur: (a) the
announcement by the Colony Development Operation that it will not start the
construction originally planned for spring 1975 on its 50,000 B/D oil shale plant at
Parachute Creek in Colorado, (b) the lack of enthusiasm for oil shale displayed in the
"Project Independence Blueprint" published by the Federal Energy Administration
(FEA), and (c) commercial scale uses of methanol as a fuel will have to be apparent

soon to justify the deployment of the 300,000 B/D (oil equivalent) production level by 1985. The most likely candidate uses of methanol emerging before 1985 are fuel for electric utilities (especially as fuel for turbine or combined cycle generators) and automotive fuel for fleet vehicles.

SCENARIOS AND SCALING FACTORS

The projected fuel production schedules shown in Table 1 have been used to derive the scenarios in Tables 2, 3 and 4 for coal syncrude, methanol from coal, and oil shale syncrude respectively. The scaling factors shown in the tables are used to account for the quantities of capital, labor, steel, and land required for the construction and operational phases of each of the building blocks used in these scenarios. Table 5 provides similar information on the surface coal mines need to support the syncrude and methanol production from coal.

TABLE 2

Syncrude from Coal: Maximum Credible Implementation Scenario

Data and Assumptions	Scenario for Year				
	1980	1985	1990	1995	2000
Production Schedule					
Cumulative capacity (million B/D)	0	0.09	0.5	1.5	4.0
Number of Plants					
Small (30,000 B/D)	0	3	7	7	0
Large (100,000 B/D)	0	0	3	13	40

Inputs and Outputs		Scaling Factors for a 100,000-B/D Plant (in units specified)	Year				
Items	Units		1980	1985	1990	1995	2000
			Cumulative Amount				
Construction							
Capital	10^9 1973 $	0.67	0	0.60	3.4	10	27
Labor	10^3 man-years	7.3	0	6.6	37	110	290
Steel	10^3 tons	110	0	100	560	1700	4400
Land	10^3 acres	1	0	0.9	5.1	15	40
Production			Annual Amount				
Operating costs	10^6 1973 $/year	130	0	140	780	2300	6200
Labor force	10^3 people	1.4	0	1.3	7.0	21	56
Coal	10^6 tons/year	18	0	16	90	270	720
Water	10^3 acre-ft/year	29	0	26	145	435	1160
Electric power	MW	140	0	130	700	2100	5600

FINANCING DEVELOPMENT OF THE INDUSTRY

As can be seen in Tables 2, 3 and 4, synthetic liquid fuels plants are expensive. To determine whether the petroleum industry could finance synthetic fuel endeavors from internal sources of funds, we performed a cash flow analysis for two possible futures of the petroleum industry — conventional fuels only, and conventional plus synthetic fuels. The quantity of conventional fuels assumed is that of the Domestic

TABLE 3

Methanol from Coal: Maximum Credible Implementation Scenario

Data and Assumptions	Scenario for Year				
	1980	1985	1990	1995	2000
Production Schedule					
Cumulative capacity					
(million B/D oil equivalent)*	0.05	0.3	1.0	2.5	4.0
Number of Plants					
Small (50,000 B/D)	2	2	2†	0	0
Large (100,000 B/D)	0	5	19	50	80

Inputs and Outputs		Scaling Factors for a 100,000 B/D Plant* (in units specified)	Year				
Items	Units		1980	1985	1990	1995	2000
			Cumulative Amount				
Construction							
Capital	10^9 1973 $	0.59	0.59	3.5	11.8	29.5	47.2
Labor	10^3 man-years	7.5	7.5	4.5	150	375	575
Steel	10^3 tons	100	100	600	2000	5000	8000
Land	10^3 acres	1	1	6	20	50	80
			Annual Amount				
Production							
Operating costs	10^6 1973 $/year	70	70	420	1400	3500	5600
Labor force	10^3 people	0.9	0.9	6.4	18	45	72
Coal (Western)	10^6 tons/year	13	13	78	260	650	1040
Water	10^3 acre-ft/year	15	15	90	300	750	1200
Electric power	MW	100	100	600	2000	5000	5000

* The energy of a barrel of methanol is half that of a barrel of oil.

† Arrow indicates that small plants are enlarged and enter large plant classification.

Oil and Gas subscenario of the Historical Growth Scenario of the Energy Policy Project of the Ford Foundation (4), and the quantity of synthetic fuel is that of the maximum credible scenario above. Several assumptions were made:

- The historical after-tax return on investment continues into the future and applies to both conventional and synthetic fuel aspects of the business. This implicitly assumes that governmental policies or market conditions have made synthetic fuels profitable.

- Depreciation credits approximate those of recent years as a percentage of assets.

- The historical debt/equity ratios remain valid.

- Historical payout of profits is maintained.

References p. 362.

TABLE 4

Syncrude from Oil Shale: Maximum Credible Implementation Scenario

Data and Assumptions	Scenario for Year				
	1980	1985	1990	1995	2000
Production Schedule Cumulative capacity (million B/D)	0.1	0.5	1.5	2.0	2.0
Number of Plants					
Small (50,000 B/D)	2	2	2*	0	0
Large (100,000 B/D)	0	4	14	20	20

Inputs and Outputs		Scaling Factors for a 100,000 B/D Plant (in units specified)	Year				
Items	Units		1980	1985	1990	1995	2000
			Cumulative Amount				
Construction							
Capital	10^9 1973 $	0.75	0.75	3.8	11.3	15.0	15.0
Labor	10^3 man-years	5.4	5.4	27	81	108	108
Steel	10^3 tons	90	90	450	1350	1800	1800
Land	10^3 acres	0.6	0.6	3.0	9.0	12	12
			Annual Amount				
Production							
Operating costs	10^6 1973 $/year	80	80	400	1200	1600	1600
Labor force	10^3 people	1.7	1.7	8.5	25.5	34.0	34.0
Shale	10^6 tons/year	54	54	270	810	1080	1080
Water	10^3 acre-ft/year	16	16	80	240	320	320
Electric power	MW	170	170	850	2250	3400	3400
Land	10^3 acres/year	0.15	0.15	0.75	2.25	3.0	3.0

* *Arrow indicates that small plants are enlarged and enter large plant classification.*

The calculations follow the algorithm of Hass et al. (5), and have been carried out for three annual rates of inflation: 0, 5, and 8 percent. The base year for the value of the dollar is 1973. Figs. 3 and 4 show the results for the 8 percent calculations.

For 0 percent annual inflation, the industry would be able to self-finance the synthetic fuel endeavors, but at 5 and 8 percent annual inflation rates, considerable demand for external capital would arise. The underlying reason for this difference is the inability of depreciation credits to fully recover the capital needed to replace facilities in an inflationary economy. This arises because depreciation credits are based on initial costs in historical dollars, while depreciation credits have only current dollar purchasing power.

RESOURCE DEPLETION

Normally, proposed synthetic fuels ventures are subjected to an economic analysis to determine whether they should be pursued. Recently, the term "net energy" analysis has been coined to describe the examination of the energy return on energy

TABLE 5

Surface Coal Mines Needed for Syncrude Plus Methanol Production

Data and Assumptions	Scenario for Year				
	1980	1985	1990	1995	2000
Production Schedule					
Cumulative capacity (million tons/year)	13	94	350	920	1760
Number of mines (5 million tons/year)	3	19	70	184	352

Inputs and Outputs		Scaling Factors for a 5 Million Ton/Year Mine (in units specified)	Year				
Items	Units		1980	1985	1990	1995	2000
Construction			Cumulative Amount				
Capital	10^9 1973$	0.03	0.09	0.57	2.1	5.5	10.6
Labor	10^3 man-years	0.25	0.75	4.75	17.5	46.0	88.0
Steel	10^3 tons	3	9	57	210	552	1060
Land*	Acres	10	30	190	700	1840	3520
Production			Annual Amount				
Operating costs	10^6 1973 $/year	12	36	228	840	2210	4220
Labor force	10^3 people	0.1	0.3	1.9	7	18	35
Water	10^3 acre-ft/year	0.15	0.45	2.85	10.5	27.6	52.8
Electric power	MW	10	30	190	700	1840	3520
Land*	10^3 acres/year	0.25	0.75	4.75	17.5	46	88

* Land for buildings, storage and handling facilities, parking, etc. This is not land for mining.

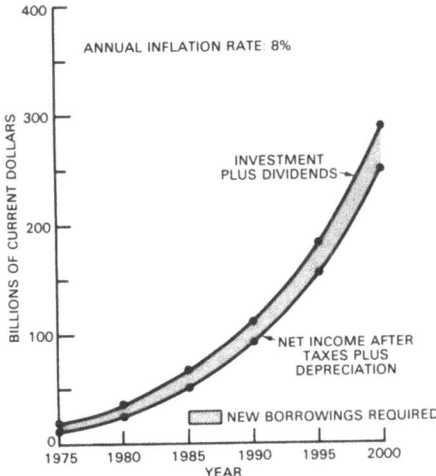

Fig. 3. Projected cash flow for domestic oil and gas industry - no synthetic liquid fuels - at an eight percent annual rate of inflation.

References p. 362.

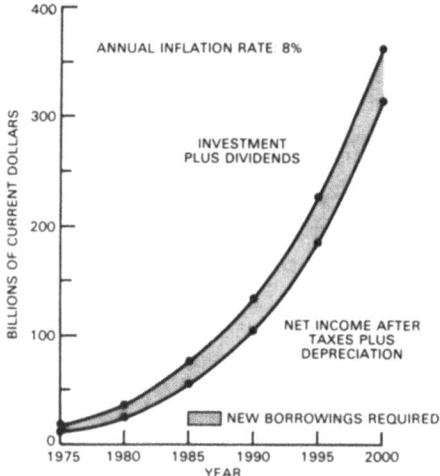

Fig. 4. Projected cash flow for domestic oil and gas industry - conventional activities plus synthetic liquid fuels - at an eight percent annual rate of inflation.

invested in an energy technology. Fig. 5 shows one form of this analysis schematically. In the figure, E_{res} refers to the energy content of the resource, E_{fuel} to the energy content of fuels used, E_{mat} to the energy used to produce materials used to build the plant or consumed in production, and E_{prod} is the energy content of the final fuel product. It is the intent to account for all indirect as well as direct energy inputs.

Net energy analyses are especially useful in comparing alternative approaches to using the same resource, but are less useful in comparing similar fuels produced by

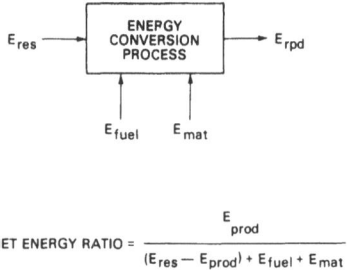

$$\text{NET ENERGY RATIO} = \frac{E_{prod}}{(E_{res} - E_{prod}) + E_{fuel} + E_{mat}}$$

Fig. 5. Flow diagram for definition of net energy ratio.

different resources. Figs. 6 through 8 show some of the calculations performed at SRI by Dr. Robert Steele for the resource-to-fuel chains relevant to this paper. In the figures, rectangles describe an activity, circles refer to the energy content of products, ovals give direct energy inputs, and triangles give the sum of direct plus indirect energy inputs. It can be seen, for example, that for coal liquefaction, the energy input into purchased chemicals and catalysts is comparable to the direct energy input of purchased electricity (both, however, are overshadowed by the coal input). Fig. 6 is one of several modules shown in Fig. 8.

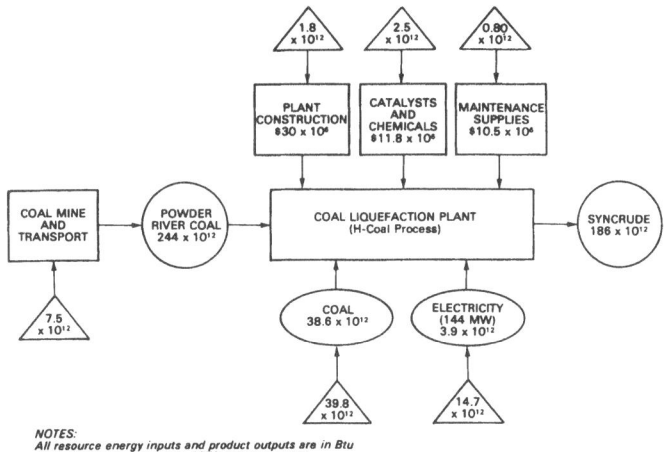

NOTES:
All resource energy inputs and product outputs are in Btu
All dollar figures are in late 1973 dollars per year

Fig. 6. Annual energy inputs for constructing and operating a 100,000 barrel per day H-Coal Process coal liquefaction plant.

Our net energy ratio results are presented in Table 6, where it can be seen that the synthetic crude oil option is a more conservative use of coal resources than methanol. A similar calculation for synthetic liquid products derived from in-situ coal gasification would make an interesting comparison. The net energy analysis is useful in helping set public policy towards synthetic fuels, but speaks only indirectly to the issue of resource depletion.

TABLE 6

Net Energy Ratio

Option	Syncrude	Product
Oil Shale	2.3	1.7
Coal Liquefaction		
Illinois Coal	1.8	1.4
Wyoming Coal	1.5	1.1
Methanol	- -	0.59
Four Corners Coal		

By far, the majority of the commercially significant oil shale reserves (25 to 30 gal/ton of shale) are found in the Piceance Basin in Western Colorado. Unlike oil shale, coal is widely distributed in the U.S. Table 7 shows a tabulation of recently

References p. 362.

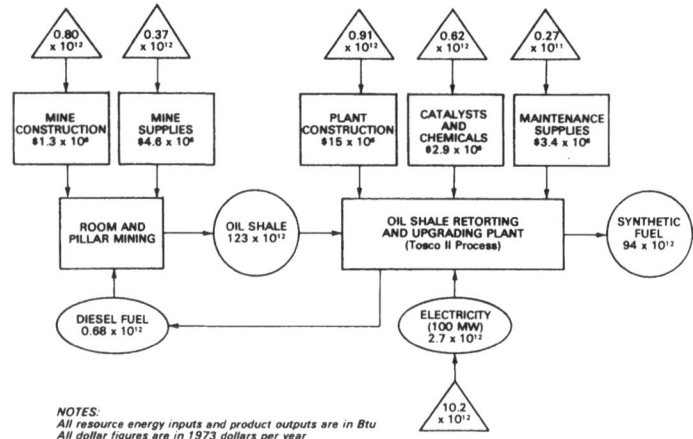

Fig. 7. Annual energy inputs for constructing and operating a 50,000 barrel per day oil shale mining, retorting and upgrading complex.

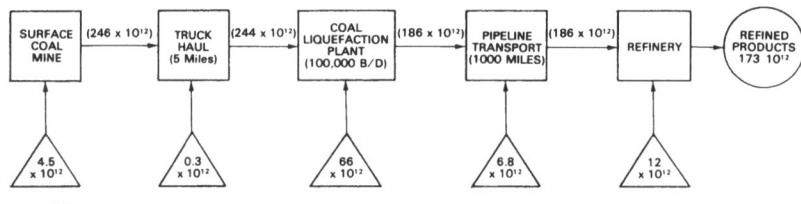

Fig. 8. Annual energy inputs for converting western surface-mined coal to refinery products in the midwest.

published strippable coal reserves, and the number of coal liquefaction plants that these reserves could sustain. Since synthetic fuels will require low cost feedstocks to be economically competitive (at least initially) with conventional petroleum fuels, strippable coal has been emphasized in the study. Clearly, strippable reserves would be able to sustain this study's maximum credible production scenario for several plant lifetimes. However, when other coal demands are also taken into account, there is a good chance that strippable reserves will be nearing depletion early in the 21st century.* This suggests the need to develop both *in-situ* recovery techniques, and improved methods of underground mining (especially since present methods cannot efficiently mine the very thick, deep seams of coal found in the West).

However, it is important to note the distinction between resources and reserves. Reserves are the fraction of resources that are economically recoverable with state-of-the-art technology at any given time. Hence, both changes in the market price of a mineral, and the technology available can alter estimates of reserves, while resource estimates can be changed only with new discoveries.

TABLE 7

States and Regions with Strippable Coal Reserves
Sufficient to Support a Large Synthetic
Fuels Industry *

	Strippable Reserves 10^9 Tons	Number of 100,000 B/D Plants Sustainable for 20 Years at 20 x 10^6 Tons/Year
Montana	43	110
Wyoming	24	60
North Dakota	16	40
Illinois/Western Kentucky	16	40
West Virginia/ Eastern Kentucky	8.7	22

* *Source: "Demonstrated Reserve Base," U.S. Bureau of Mines (1974).*

SOCIAL AND ECONOMIC IMPACTS

Quite clearly, if the impacts of producing synthetic liquid fuels are mainly associated with extracting the resource and converting it, then the impacts will be felt especially acutely in the regions where those activities are located. Fig. 9 shows the major coal regions of the United States. The areas within the circles have been chosen for the bulk of our analysis.

Fig. 9. Major coal regions of the United States.

Fig. 10 shows the specific coal counties expected to be especially impacted, and the oil shale counties of Colorado as well. These are the regions where: considerable surface mining, as depicted in Fig. 11, will occur; water requirements for fuel conversion activities will be large; and the influx of people will be large. The fuel

References p. 362.

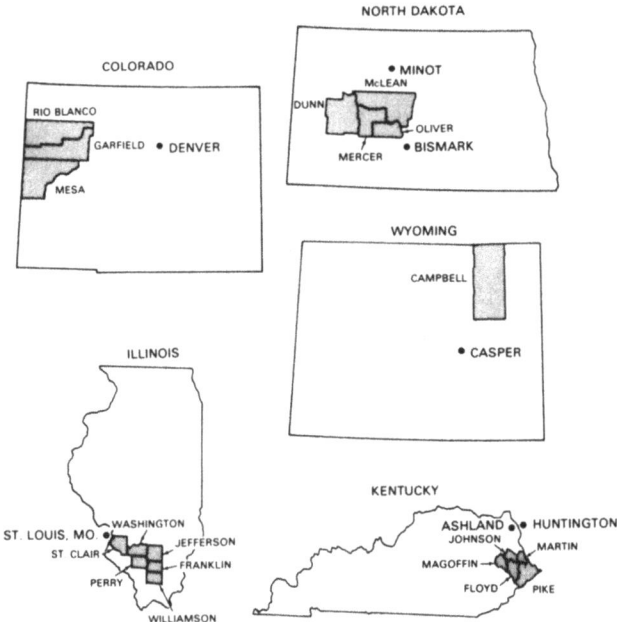

Fig. 10. Counties used for economic impact discussions.

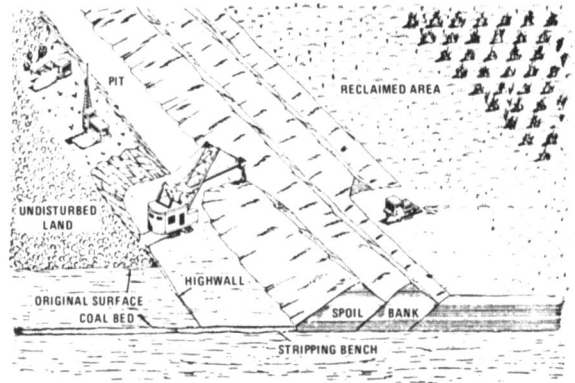

Fig. 11. Area strip mining with concurrent reclamation.

production schedule for the maximum credible scenario has been allocated in Table 8 to the various states shown in Table 7 and Fig. 9 to have suitable coal reserves. For simplicity, the entire oil shale schedule has been assigned to the Colorado counties shown in Fig. 10, since these contain most of the suitable oil shale deposits.

There would be both directly and indirectly related populations associated with each resource extraction and fuel conversion activity. This population attraction is shown in Fig. 12; it is handled numerically by the use of a population multiplier which is applied to each primary job created in the extraction or conversion activites.

We have concluded that a multiplier of 6.5 is about correct for these industries and these regions.

TABLE 8

Hypothesized Locations of Synthetic Liquid
Fuel Plants for Various Time Frames

Units for table entries are as follows:

Coal syncrude plants:	S = 30,000 B/D
	L = 100,000 B/D
Methanol plants:	S = 50,000 B/D (methanol)
	L = 100,000 B/D (methanol)
Surface mine:	5 million tons/year
Underground mine:	1 million tons/year
Water	10^3 acre feet/year

	Cumulative Quantities				
	Year				
State	1980	1985	1990	1995	2000
Wyoming					
Coal syncrude	0	2S	3S, 2L	3S, 5L	13L
Methanol	0	0	2L	8L	13L
Surface mines	0	2	14	42	81
Water	0	15	103	268	520
North Dakota					
Methanol	1S	1S, 2L	1S, 5L	13L	21L
Surface mines	2	9	20	47	76
Water	8	38	83	195	315
Illinois					
Coal syncrude	0	1S	1S, 1L	1S, 3L	7L
Methanol	0	1L	4L	9L	14L
Surface mines	0	1	3	8	14
Underground mines	0	9	40	93	161
Water	0	23	93	218	385
Kentucky					
Coal syncrude	0	0	1S	1S, 1L	4L
Methanol	1S	1S, 1L	1S, 3L	7L	10L
Surface mines	1	1	3	7	13
Underground mines	0	10	23	52	87
Water	8	23	60	138	250

For each type of plant, the activity must be separated into construction and operation phases because, in general, the work forces are composed of different people possessing different skills. Fig. 13 shows profiles of the total (direct plus indirect) population associated with the various types and sizes of synthetic liquid fuel production activities. These profiles are useful to construct population growth scenarios for the impacted regions.

References p. 362.

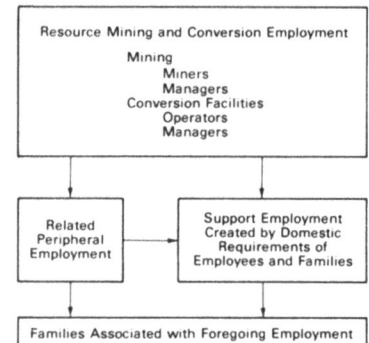

Fig. 12. Schematic of the population multiplier concept.

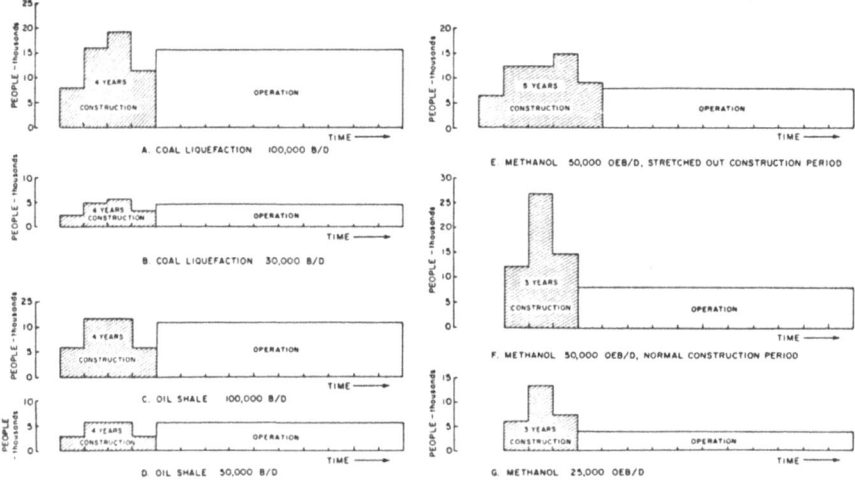

* OEB/D = Oil Equivalent Barrels per Day

Fig. 13. Total population associated with individual plant construction and operation building blocks. All building blocks include the mines that supply the plants. The actual labor force is multiplied by 6.5 to account for induced secondary employment and families. The data for these building blocks come from the scaling factors derived for the Maximum Credible Implementation Scenario.

Fig. 14 depicts the population growth that would be induced by the portion of the maximum credible scenario assigned to Campbell County Wyoming, the county that contains most of the famed Powder River Basin coal reserve. The population growth would be about 9 percent per year from the synthetic liquid fuels industry alone. If other, quite likely, uses of coal were also considered, the growth rate would be much higher. It should be noted that the mines themselves account for a very small portion of this growth which is dominated by the conversion facilities.

Fig. 15 shows the growth rate for Garfield and Rio Blanco counties in Colorado stemming from the oil shale development of the maximum credible scenario. The

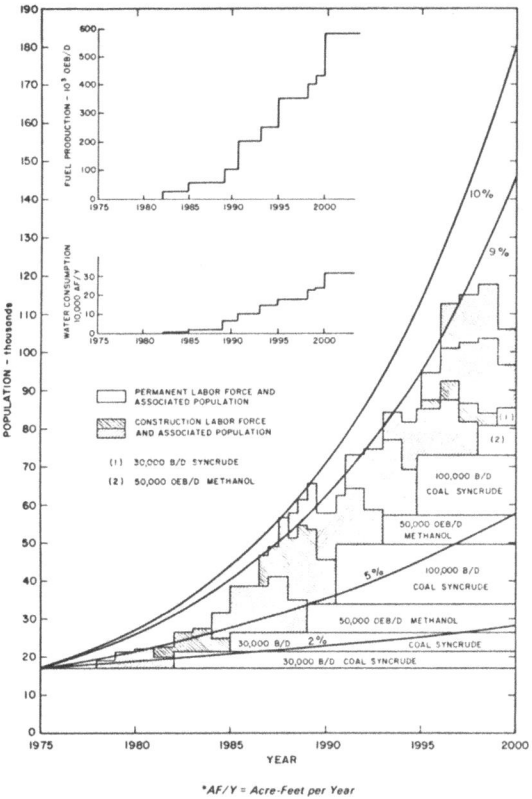

Fig. 14. Effects of the Maximum Credible Implementation Scenario upon population in Campbell County, Wyoming. Assumes that one quarter of all the Scenario's development in Wyoming occurs in Campbell County. This assumption is expected to be on the low side.

growth rate is about 17 percent per year. However, unlike Campbell county where other coal related activities could be expected to add further to the growth rate, these projections are probably a good representation of the maximum impact situation because there is no other use of oil shale.

In both Figs. 14 and 15 the inset in the upper left of the figure shows the fuel production rate and water requirements. The former can be compared with the fuel production in the maximum credible implementation scenario to judge the relative contribution of the area considered. The water requirements are very significant and, as will be discussed later, likely to be the cause of much intensified debate about the effects of using this much water.

Historically, high growth rates have occurred many times in many locations of the United States. From a great deal of accumulated experience, planners generally regard an annual growth rate of about 5 percent as the upper limit of manageability in the sense that a reasonable quality of life can be maintained in the community. Higher

References p. 362.

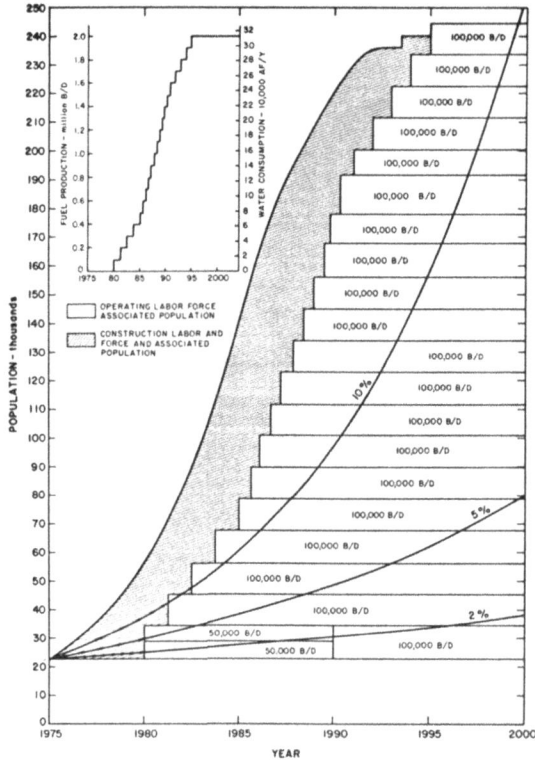

Fig. 15. Maximum Credible Implementation Scenario for oil shale development in Garfield and Rio Blanco Counties, Colorado. The resulting annual population growth rate is about 17 percent.

growth rates are usually viewed as the causes of many severe local fiscal and social problems.

A high growth rate usually results in fiscal problems for local authorities because tax revenues lag by several years the demand for public services such as roads, sewers, schools, police and fire protection. The problem stems from the influx of population that begins with the construction while the facilities do not pay full taxes until they are completed and in production. In an environment of prolonged growth, tax lag can become a chronic problem leading to a low quality of life for residents, and creating further social ills. Rock Springs, Wyoming is an example of a community where the recent addition of coal mines and a power plant, added to an already booming minerals extraction industry, led to severe social malaise. In Rock Springs, crime, divorce, suicide, alcoholism, and mental health problems are reported to be very high for a rural community of this size. Moreover, the rapid growth has created severe housing shortages, as well as sanitation and public health problems.

Because the impacts of the maximum credible scenario are so intensely felt at the local and regional level, several lower population growth scenarios have been prepared. Figs. 16 and 17 show profiles for 10 and 5 percent population growth-limited scenarios in the oil shale counties of Colorado. From the insets in these figures, it can be seen that protection of a reasonable quality of life for the people living in the region naturally results in diminished fuel production, compared to the maximum credible implementation scenario. In particular, the 10 percent growth constrained scenario gives about (1.5) million barrels per day in 2000 compared to the maximum credible implementation scenario's 2.0 million; it also gives about 200,000 barrels per day in 1985 – considerably less than the latest target in Washington of 350,000 barrels per day in 1985. The 5 percent constrained growth rate gives only 100,000 barrels per day in 1985, and 400,000 in the year 2000.

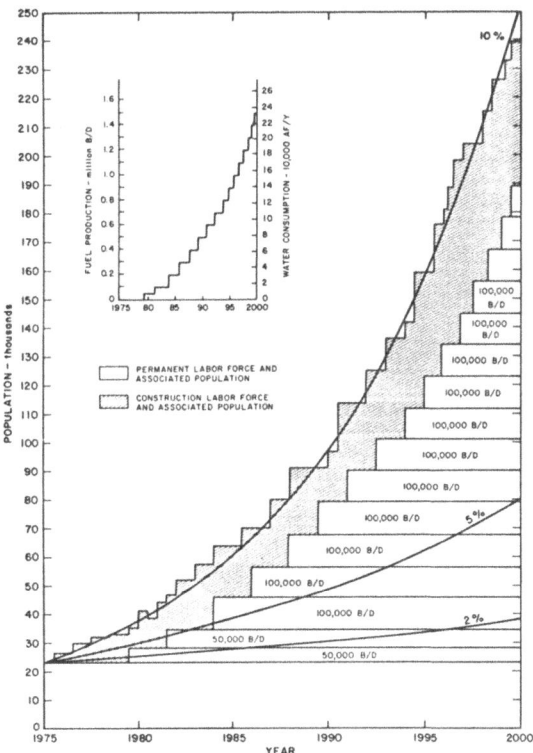

Fig. 16. Ten percent constrained population growth scenario for oil shale development in Garfield and Rio Blanco Counties, Colorado.

WATER

As Table 2 shows, a coal liquefaction plant uses a great deal of water *consumptively* – about 30,000 acre feet per year for a 100,000 barrel per day plant. *References p. 362.*

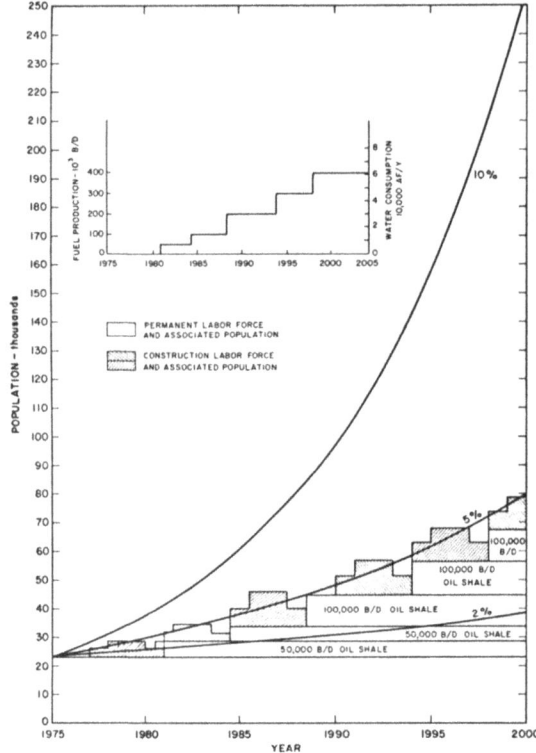

Fig. 17. Five percent constrained population growth scenario for oil shale development in Garfield and Rio Blanco Counties, Colorado.

About 80 percent of the water is consumed in evaporative cooling processes and over half the remainder is used as a chemical feedstock. Water-conserving designs could probably cut overall water consumption by a factor of two.

It is well known that the coal and oil shale resource-rich regions of the West are water-poor. This has become a central topic in discussions concerning resource development in the West. The demands for water have stirred controversy which often overshadows that stimulated by the use of the strip mining technique of resource extraction.

Because the West is so arid, water law there is based upon the concept of appropriative rights rather than the riparian rights of the East. Thus, it is sometimes the case that the owner of land through which water flows has no right to use that water. The concept of "first in time" being "first is right" is operative.

Over the years, a vast array of claims to water rights have accumulated, often with the result that if all rights were exercised and enforced, the streams would be more than 100 percent consumed. This absurd and impossible situation has been avoided in practice only because many claims to rights have not been exercised, and besides many water works that effect interbasin water transfers have been constructed.

At present, the major users of water in the regions under consideration are ranchers and farmers, and the dams of the Bureau of Reclamation have been constructed specifically to provide water for agriculture. Such water may not be legally available for energy projects. Sensitivity analyses have shown that the cost of synthetic liquid fuels is little affected by the cost of water. Thus, the synfuels industry could afford to pay enormous sums for water with only a miniscule effect on the cost of their products. The possibility that energy industries might be able to buy up water rights to the detriment of the agriculture that has been the basis for the regional lifestyle has become a matter of serious concern to present residents and some of their state governments.

There probably is enough water physically present in the West to supply a significant synthetic fuels business, but it is often neither in the correct place, nor legally available. The whole future of synthetic fuels development in the West is closely tied to the issue of water availability. The issue is clouded because much of the West is in the public domain, controlled by the Federal government, and Federal reserved water rights take priority. Thus, Federal policy towards synthetic fuels becomes a matter of paramount importance. Moreover, the status of Indian water rights is in dispute and, according to some interpretations, their rights predate nearly all others. Consequently, Indian tribes may hold the key to the synthetic fuels industry.

Most likely the resolution of the water issue will not come in a single, sweeping, clarifying blow, but in an area-by-area, problem-by-problem basis over a long period of time.

CONCLUDING DISCUSSION

We believe that all the various options should be compared by public and private decision makers with respect to the following criteria:

- Resource intensiveness (fossil resource, water, labor, capital, land)
- Geographic concentration of development (affecting impacts upon human and other populations)
- Net energy ratio
- Alternative uses of resources involved
- Evolutionary aspects of the systems.

We believe that the most important "critical factors" that will determine the future of the synthetic liquid fuels industry are the following:

- Decisions to rely on conventional fuels or produce synthetic fuels

References p. 362.

- "Boom" population growth rates in production regions
- Mined lands reclamation
- Water availability and conflicts over use
- Financing the synthetic fuels industry
- Air pollution in production regions

In general, it seems that no single synthetic liquid fuels option is best in all respects and all options would have large environmental, social, and institutional impacts — especially in resource development and fuel conversion regions.

REFERENCES

1. E. M. Dickson, et al., "Impacts of Synthetic Liquid Fuel Development for The Automotive Market," report in preparation for the Environmental Protection Agency, release expected in early 1976.
2. E. E. Hughes, et al., "Maximum Credible Implementation Scenario for Synthetic Liquid Fuels from Coal and Oil Shale," 10th Intersociety Energy Conversion Engineering Conference, Newark, Delaware, August 17-22, 1975, pp. 658-666.
3. E. M. Dickson, "Decision-Making for Synthetic Fuels," 10th Intersociety Energy Conversion Engineering Conference, Newark, Delaware, August 17-22, 1975, pp. 651-657.
4. Energy Policy Project of the Ford Foundation, "A Time to Choose: America's Energy Future" (Ballinger Publishishing Co., Cambridge, Massachusetts), 1974.
5. J. E. Hass et al., "Financing the Energy Industry" (Ballinger Publishing Co., Cambridge, Massachusetts), 1974.

DISCUSSION

A. R. Sapre *(General Motors Research Laboratories)*

What kind of demand did you assume for petroleum crude and natural gas to calculate your cash flows for the oil industry?

Dickson

We assumed those that were in the high growth scenario in the Ford Foundation Energy Policy Project report.

Sapre

Did your 8% inflation rate apply to investment required and also to crude oil and natural gas prices?

Dickson

It applied to every dollar that was transacted in any given year.

Sapre

How did you come up with 8% inflation?

Dickson

We ran at the 0%, 5%, and 8%. It was scary enough at 8% that we really didn't want to try anything else.

Sapre

Did you try to do some type of input-output analysis for the entire system, including the synthetic fuels industry, to come up with constraints that would be put on the synthetic fuels industry because of competition for limited capital, labor, and materials, etc.?

Dickson

The graphs we showed you were an attempt to answer that question. However, we could not develop a big enough input-output matrix for the whole economy. That would be impossible. We tried to compare our numbers with other projections for capital demands in the nation. We concluded that the fraction which the oil industry would need would remain more or less the same as historical.

Sapre

You included energy required to manufacture plant and equipment, etc. into your energy ratio. Did you start with iron ore and then make all the equipment from there?

Dickson

Yes.

Sapre

You said the conversion processes are water intensive. Yesterday, Dr. Gary mentioned that large amounts of water are required to obtain shale oil. Are you talking about additional water requirements, or are you talking about the water requirements for mining and shale disposal?

Dickson

The answer to the first part is yes. More water still. And, yes, it does include mining and shale disposal.

J. P. Longwell *(MIT and Exxon)*

If you look at the possibility of desalinating some of the local brackish water, sometimes that's not too heavy a financial load on these projects.

Dickson

One of the issues I didn't mention on the water slide is that there is very little applicable law to ground water. The ground water resources in the oil shale region, for example, are unknown. There's some test borings and some ideas of how much water there is, but it's not very definitive. There's supposedly a lot of ground water in the Madison formation in the Powder River Basin and it's supposed to be saline and not usable for anything else. However, some of the recent wells that were drilled, turned out to be sweeter water than was being used presently in Gillette, Wyoming. So it's not clear that that's really going to work out.

J. B. Pangborn *(Institute of Gas Technology)*

Something that does not come out explicitly in your study is the development of the SNG industry, at least it doesn't appear explicitly. A lot of people think this is a more firm commercialization for the future than coal liquefaction or making methanol from coal. How would one account for the technical and the social, and the economic effects of this occurring in addition to what you have already described?

Dickson

It turns out that it's very fortuitous in some of these cases that, if you look at the capital requirements, the materials requirements, the geographical locations, and the population profiles for building the plants, they're almost identical to those for a commercial-size SNG plant. So the modules that we built up for impacts on this study for synthetic liquid fuels can also be interpreted for the SNG industry.

Pangborn

Can we have both?

Dickson

We could have both if we could cope with the social problems and if we could cope with the fact that the strippable reserves would be depleted very much faster than is usually considered.

E. E. Ecklund *(U.S. Energy Research & Development Administration)*

This is a comment. Gentlemen, if we didn't know before why we need to conserve energy, and why we need to make more fuel-efficient vehicles, we sure ought to know now.

R. A. Husted *(U.S. Department of Transportation)*

Would you comment on the possibility and cost implications of bringing water in from someplace where it's in large supply?

Dickson

We certainly agree that the cost of water is almost of no importance. Some cost work and some sensitivity analyses have been done on the price of synthetic fuels as opposed to the price of water input. The water costs pennies per barrel of fuel. You can pay four or five or ten times as much as agriculture can pay for water and still not inconvenience or hardly even let the consumer of synthetic fuel know the difference. The real costs that are important in your question are the social costs of interbasin water transfers. The economic costs could be absorbed into the project and wouldn't be the thing that makes or breaks the project.

We've also looked at transporting the coal outside the regions where it was mined and we found that there are too many multiplicities in places you mine coal and take it to. We also looked at whether or not a coal slurry pipeline would save water, and we looked at a dual pipeline where we sent the dirty water back and used it again and again. Well, you really don't use it again and again, you lose about 25% every time you go around because the coal stays wet at the delivery end. We found that you would be paying hundreds of dollars per acre foot, in effect, for that water which is much more than agriculture is paying. Thus, you could go out and buy agricultural water rights, if you could get them.

R. W. Hurn   *(U.S. Energy Research and Development Administration)*

The report of the synthetic fuels commercialization group clearly defines oil shale as the preferable option. This was reportedly prepared by SRI. What is your comment?

Dickson

SRI is a beast with many heads. That study had to do, I think, with the financial situation. The Exxon and IGT reports certainly showed that if you wanted to prioritize these fuels in terms of the cost of producing them, oil shale is better. If you want to look at our net energy analysis, that's also in agreement. However, that panel didn't take much cognizance of environmental or social effects, or even water availability in a realistic way.

S. G. Liddle *(General Motors Research Laboratories)*

In view of the population problem and the water problem, wouldn't it be better to start building these types of plants in the eastern United States?

Dickson

For oil shale, you can only do it out west. For coal, our analysis indicates that the first place to go would be southern Illinois. You can restore the mined lands better, and actually get them back into farming. You have adequate water resources. You don't strain any of the water resources. The population base of the region is adequate to provide labor and to absorb influxes without stressing the system nearly so much as some of the western possibilities. However, you can't build up a very large industry in southern Illinois because there isn't sufficient coal there to sustain it for a long time period.

E. E. Spitler *(Chevron Research Company)*

You mentioned the automotive demand would be roughly five million barrels per day. Is that automotive in the broadest sense or only automobiles?

Dickson

That's automotive in the sense of autos, trucks and buses.

SESSION IV — SUMMARY

**The summary was omitted
due to insufficient time.**

PARTICIPANTS

Adt, Jr., R. R.
University of Miami
Coral Gables, Florida

Agarwal, P. D.
Research Laboratories, GMC
Warren, Michigan

Agnew, W. G.
Research Laboratories, GMC
Warren, Michigan

Aiman, W. R.
Research Laboratories, GMC
Warren, Michigan

Albers, W. A.
Research Laboratories, GMC
Warren, Michigan

Amann, C. A.
Research Laboratories, GMC
Warren, Michigan

Anderson, C. J.
Lawrence Livermore Laboratory
Livermore, California

Antonius, K. T.
Research Laboratories, GMC
Warren, Michigan

* * *

Beaubien, S. J.
Shell Oil Company
Houston, Texas

Beckham, J. L.
Cadillac Motor Car Division, GMC
Detroit, Michigan

Beckman, E. W.
Chrysler Corporation
Detroit, Michigan

Belding, J. A.
Energy Research and Development
Administration
Washington, D. C.

Bennethum, J. E.
Research Laboratories, GMC
Warren, Michigan

Benson, J. D.
Research Laboratories, GMC
Warren, Michigan

Bernhardt, W. E.
Volkswagenwerk AG
Wolfsburg, Germany

Bidwell, J. B.
Research Laboratories, GMC
Warren, Michigan

Blazaitis, J. F.
Cadillac Motor Car Division, GMC
Detroit, Michigan

Bleil, C. E.
Research Laboratories, GMC
Warren, Michigan

Bolt, J. A.
University of Michigan
Ann Arbor, Michigan

Brandes, J. G.
Detroit Diesel Allison
Division, GMC
Detroit, Michigan

Brinkman, N. D.
  Research Laboratories, GMC
  Warren, Michigan

Burgett, R. W.
  AC Spark Plug Division, GMC
  Flint, Michigan

Butler, F. G.
  Oldsmobile Division, GMC
  Lansing, Michigan

Butterworth, A. V.
  Research Laboratories, GMC
  Warren, Michigan

Buzan, L. R.
  Research Laboratories, GMC
  Warren, Michigan

* * *

Campau, R. M.
  Ford Motor Company
  Allen Park, Michigan

Cantwell, E. N.
  E. I. du Pont, de Nemours and
    Company, Inc.
  Wilmington, Delaware

Caplan, J. D.
  Research Laboratories, GMC
  Warren, Michigan

Carroll, T. J.
  Corporate Product Planning
    Group, GMC
  Detroit, Michigan

Caton, J. A.
  Research Laboratories, GMC
  Warren, Michigan

Chenea, P. F.
  Research Laboratories, GMC
  Warren, Michigan

Chloupek, F. J.
  Atlantic Richfield Company
  Harvey, Illinois

Clark, E. L.
  Energy Research and Development
    Administration
  Washington, D. C.

Collman, J. S.
  Research Laboratories, GMC
  Warren, Michigan

Colucci, J. M.
  Research Laboratories, GMC
  Warren, Michigan

Cornelius, W.
  Research Laboratories, GMC
  Warren, Michigan

* * *

Daniel, W. A.
  Research Laboratories, GMC
  Warren, Michigan

Davison, J. W.
  Phillips Petroleum Company
  Bartlesville, Oklahoma

Debbink, J. D.
  Corporate Product Planning
    Group, GMC
  Detroit, Michigan

de Boer, P. C. T.
  Cornell University
  Ithaca, New York

Dickson, E. M.
  Stanford Research Institute
  Menlo Park, California

Dryer, F. L.
  Princeton University
  Princeton, New Jersey

* * *

Ecklund, E. E.
  Energy Research and Development
    Administration
  Ann Arbor, Michigan

Elliott, M. A.
Energy Consultant
Houston, Texas

Escher, W. J. D.
Escher Technology Associates
St. Johns, Michigan

Evans, A. R.
Research Laboratories, GMC
Warren, Michigan

* * *

Fleming, J. D.
Research Laboratories, GMC
Warren, Michigan

Fones, T. H.
Caterpillar Tractor Company
Peoria, Illinois

Frend, M. A.
Shell Oil Company
Southfield, Michigan

Friday, J. R.
Continental Oil Company
Ponca City, Oklahoma

Furey, R. L.
Research Laboratories, GMC
Warren, Michigan

* * *

Gallopoulos, N. E.
Research Laboratories, GMC
Warren, Michigan

Gardels, K. D.
Research Laboratories, GMC
Warren, Michigan

Gary, J. H.
Colorado School of Mines
Golden, Colorado

Gast, R. A.
Research Laboratories, GMC
Warren, Michigan

Gilbert, A. W.
Chevrolet Motor Division, GMC
Warren, Michigan

Gratch, S.
Ford Motor Company
Dearborn, Michigan

Green, F. L.
Manufacturing Staff, GMC
Warren, Michigan

Grube, W. L.
Research Laboratories, GMC
Warren, Michigan
* * *

Hall, C. A.
Ethyl Corporation
Ferndale, Michigan

Hammond, D. C.
Research Laboratories, GMC
Warren, Michigan

Hanson, D. T.
Texas A & M University
College Station, Texas

Hardin, M. C.
Detroit Diesel Allison
Division, GMC
Indianapolis, Indiana

Harding, K. J.
Amoco Chemicals Corporation
Chicago, Illinois

Harrington, D. L.
Research Laboratories, GMC
Warren, Michigan

Harrington, J. A.
Ford Motor Company
Dearborn, Michigan

Hartley, D. L.
Sandia Laboratories
Livermore, California

Hartman, J. L.
Research Laboratories, GMC
Warren, Michigan

Heffner, F. E.
Research Laboratories, GMC
Warren, Michigan

Heidacker, W. C.
Pontiac Motor Division, GMC
Pontiac, Michigan

Hemphill, R. F.
Federal Energy Administration
Washington, D. C.

Henein, N. A.
Wayne State University
Detroit, Michigan

Heywood, J. B.
Massachusetts Institute of
Technology
Cambridge, Massachusetts

Hietbrink, E. H.
Research Laboratories, GMC
Warren, Michigan

Hilden, D. L.
Research Laboratories, GMC
Warren, Michigan

Hill, J. C.
Research Laboratories, GMC
Warren, Michigan

Hinton, M. G.
The Aerospace Corporation
Los Angeles, California

Hittler, D. L.
American Motors Corporation
Detroit, Michigan

Hodgson, J. W.
University of Tennessee
Knoxville, Tennessee

Hollyer, R. N.
Research Laboratories, GMC
Warren, Michigan

Holzwarth, J. C.
Research Laboratories
Warren, Michigan

Hubbard, J. E.
GMC Truck & Coach Division
Pontiac, Michigan

Huellmantel, L. W.
Research Laboratories, GMC
Warren, Michigan

Huffman, H. C.
Union Oil Company of
California
Brea, California

Hughes, E. E.
Stanford Research Institute
Menlo Park, California

Hunstad, N. A.
Research Laboratories, GMC
Warren, Michigan

Hurn, R. W.
Energy Research and Development
Administration
Bartlesville, Oklahoma

Hurter, D.
Arthur D. Little, Inc.
Cambridge, Massachussetts

Husted, R. A.
Department of Transportation
Washington, D. C.

* * *

Jamerson, F. E.
Research Laboratories, GMC
Warren, Michigan

James, R. E.
  Corporate Product Planning
    Group, GMC
  Detroit, Michigan

Johnson, E. M.
  Texaco Inc.
  Beacon, New York

Johnson, R. T.
  University of Missouri
  Rolla, Missouri

Joseph, B. W.
  Research Laboratories, GMC
  Warren, Michigan

* * *

Kant, F. H.
  Exxon Research and Engineering
    Company
  Linden, New Jersey

Krieger, R. B.
  Research Laboratories, GMC
  Warren, Michigan

Kwolek, S. J.
  Gulf Research and Development
    Company
  Pittsburgh, Pennsylvania

* * *

Landis, J. R.
  Chevrolet Motor Division, GMC
  Warren, Michigan

Larson, J. G.
  Research Laboratories, GMC
  Warren, Michigan

Lauriente, M.
  Department of Transportation
  Washington, D. C.

Lee, R. E.
  Corporate Product Planning
    Group, GMC
  Detroit, Michigan

Lessard, R. D.
  United Technologies Research Center
  East Hartford, Connecticut

Liddle, S. G.
  Research Laboratories, GMC
  Warren, Michigan

Lipkea, W. H.
  Research Laboratories, GMC
  Warren, Michigan

Longwell, J. P.
  Massachusetts Institute of
    Technology
  Cambridge, Massachusetts

Lyn, W. T.
  Cummins Engine Company, Inc.
  Columbus, Indiana

* * *

MacDonald, J. S.
  Research Laboratories, GMC
  Warren, Michigan

MacDonald, W. E.
  Marathon Oil Company
  Findlay, Ohio

Marshall, E. F.
  Sun Oil Company
  Marcus Hook, Pennsylvania

Martens, S. W.
  Environmental Activities
    Staff, GMC
  Warren, Michigan

Mason, H. F.
  Chevron Research Company
  Richmond, California

Mason, W. T.
  Research Laboratories, GMC
  Warren, Michigan

Mattavi, J. N.
  Research Laboratories, GMC
  Warren, Michigan

Matthews, C. C.
    Research Laboratories, GMC
    Warren, Michigan

Matula, R. A.
    Drexel University
    Philadelphia, Pennsylvania

McCuen, N. H.
    Corporate Product Planning
        Group, GMC
    Warren, Michigan

McDonald, R. J.
    Research Laboratories, GMC
    Warren, Michigan

McLean, W. J.
    Cornell University
    Ithaca, New York

McReynolds, L. A.
    Phillips Petroleum Company
    Bartlesville, Oklahoma

Meisel, S. L.
    Mobile Research and Development
        Corporation
    New York, New York

Mellor, A. M.
    Purdue University
    West Lafayette, Indiana

Mick, S. H.
    Engineering Staff, GMC
    Warren, Michigan

Miller, F. R.
    Texas A & M University
    College Station, Texas

Moreau, R. A.
    Research Laboratories, GMC
    Warren, Michigan

Morel, T. A.
    Research Laboratories, GMC
    Warren, Michigan

Mowers, R. G.
    Energy Management Section, GMC
    Detroit, Michigan

Muench, N. L.
    Research Laboratories, GMC
    Warren, Michigan

Myers, P. S.
    University of Wisconsin
    Madison, Wisconsin

* * *

Nagel, B. E.
    Research Laboratories, GMC
    Warren, Michigan

Naylor, M. E.
    Transportation Systems
        Division, GMC
    Warren, Michigan

Niepoth, G. W.
    Engineering Staff, GMC
    Warren, Michigan

* * *

Pangborn, J. B.
    Institute of Gas Technology
    Chicago, Illinois

Parks, F. B.
    Research Laboratories, GMC
    Warren, Michigan

Pefley, R. K.
    Santa Clara University
    Santa Clara, California

Penner, S. S.
    University of California at
        San Diego
    La Jolla, California

Perry, Jr., R. H.
    Mobil Research and Development
        Corporation
    Paulsboro, New Jersey

Peters, B. D.
  Research Laboratories, GMC
  Warren, Michigan

Plassman, E.
  TUV Rheinland
  Germany

Polen, G. R.
  Buick Motor Division, GMC
  Flint, Michigan

* * *

Quader, A. A.
  Research Laboratories, GMC
  Warren, Michigan

Quick, R. M.
  GM of Canada, Limited
  Oshawa, Ontario, Canada

* * *

Reiland, W. H.
  Sun Oil Company
  Marcus Hook, Pennsylvania

Remus, R. R.
  General Motors Overseas Operations
  Detroit, Michigan

Roberts, M. A.
  Ford Motor Company
  Dearborn, Michigan

Rothery, R. W.
  Research Laboratories, GMC
  Warren, Michigan

Russell, J. L.
  General Atomics Company
  San Diego, California

* * *

Sapre, A. R.
  Research Laboratories, GMC
  Warren, Michigan

Schilke, N. A.
  Research Laboratories, GMC
  Warren, Michigan

Schmidt, E. W.
  Rocket Research Corporation
  Redmond, Washington

Schwing, R. C.
  Research Laboratories, GMC
  Warren, Michigan

Scott, W. M.
  Ricardo Consulting Engineers
  Sussex, England

Shannon, H. F.
  Exxon Research and Engineering
    Company
  Linden, New Jersey

Shapis, W.
  Richard P. Mueller and
    Associates, Inc.
  Baltimore, Maryland

Sheridan, D. C.
  Research Laboratories, GMC
  Warren, Michigan

Siegla, D. C.
  Research Laboratories, GMC
  Warren, Michigan

Skellenger, G. D.
  Research Laboratories, GMC
  Warren, Michigan

Smith, J. R.
  Research Laboratories, GMC
  Warren, Michigan

Smith, P. G.
  Research Laboratories, GMC
  Warren, Michigan

Snyder, R. W.
  The Standard Oil Company (Ohio)
  Cleveland, Ohio

Sorenson, S. C.
University of Illinois
Urbana, Illinois

Sovran, G.
Research Laboratories, GMC
Warren, Michigan

Spitler, E. E.
Chevron Research Company
Richmond, California

Spreitzer, W. M.
Research Laboratories, GMC
Warren, Michigan

Starkman, E. S.
Environmental Activities
Staff, GMC
Warren, Michigan

Stebar, R. F.
Research Laboratories, GMC
Warren, Michigan

Steiner, J. C.
Research Laboratories, GMC
Warren, Michigan

Stettler, R. J.
Engineering Staff, GMC
Warren, Michigan

Stivender, D. L.
Research Laboratories, GMC
Warren, Michigan

* * *

Teague, D. M.
Chrysler Corporation
Detroit, Michigan

Thomson, R. F.
Research Laboratories, GMC
Warren, Michigan

Tison, R. R.
Institute of Gas Technology
Chicago, Illinois

Tracy, J. C.
Research Laboratories, GMC
Warren, Michigan

Tudor, W. E.
Pontiac Motor Division, GMC
Pontiac, Michigan

Tuesday, C. S.
Research Laboratories, GMC
Warren, Michigan

Turunen, J. R.
General Motors Overseas Operations
Detroit, Michigan

Turunen, W. A.
Research Laboratories, GMC
Warren, Michigan

* * *

Vickers, P. T.
Research Laboratories, GMC
Warren, Michigan

Voecks, G. E.
Jet Propulsion Laboratory
Pasadena, California

* * *

Wadsworth, F. T.
Cities Service Oil Company
Cranbury, New Jersey

Wallace, W. A.
International Harvester
Company
Melrose Park, Illinois

Wanttaja, G. E.
Research Laboratories, GMC
Warren, Michigan

Wentworth, J. T.
Research Laboratories, GMC
Warren, Michigan

Williams, H. A.
Electro-Motive Division, GMC
La Grange, Illinois

Williams, F. A.
University of California at
San Diego
La Jolla, California

Winger, J. G.
Chase Manhattan Bank
New York, New York

Withrow, L. L.
Rochester, Michigan

SUBJECT INDEX